Introduction to Aeronautics: A Design Perspective

Second Edition

Steven A. Brandt
U.S. Air Force Academy
Colorado Springs, Colorado

Randall J. Stiles
Colorado College
Colorado Springs, Colorado

John J. Bertin
U.S. Air Force Academy
Colorado Springs, Colorado

Ray Whitford
Royal Military College of Science
and the Cranfield Institute of Technology
United Kingdom

EDUCATION SERIES
Joseph A. Schetz
Series Editor-in-Chief
Virginia Polytechnic Institute and State University
Blacksburg, Virginia

Published by
American Institute of Aeronautics and Astronautics, Inc.
1801 Alexander Bell Drive, Reston, VA 20191-4344

American Institute of Aeronautics and Astronautics, Inc., Reston, Virginia

5

Library of Congress Cataloging-in-Publication Data

Introduction to aeronautics : a design perspective / Steven A. Brandt . . . [et al.]. – 2nd ed.
    p.   cm.   –   (Education series)
  Includes bibliographical references and index.
  ISBN 1-56347-701-7 (alk. paper)
  1. Airplanes–Design and construction.   2. Aeronautics.   I. Brandt, Steven A., 1953-   II. Title.
III. Series: AIAA education series.
  TL671.2.I64   2004
  629.134′1–dc22   2004005844

Foreword to the Second Edition

This second edition of *Introduction to Aeronautics: A Design Perspective* by Steven A. Brandt, Randall J. Stiles, John J. Bertin, and Ray Whitford is an updated, comprehensive treatment of a very important subject in the field of aerospace education. The first edition has proven to be a valuable part of the AIAA Education Series, and we are delighted to welcome this new edition to the series.

The AIAA Education Series aims to cover a very broad range of topics in the general aerospace field, including basic theory, applications, and design. A complete list of titles published in the series can be found at http://www.aiaa.org. The philosophy of the series is to develop textbooks that can be used in a college or university setting, instructional materials for intensive continuing education and professional development courses, and also books that can serve as the basis for independent self-study for working professionals in the aerospace field. We are constantly striving to expand and upgrade the scope and content of the series, and so suggestions for new topics and authors are always welcome.

Joseph A. Schetz
Editor-in-Chief
AIAA Education Series

Foreword to the First Edition

This book is the culmination of an effort to infuse design into the introductory course in aeronautics taught to all cadets at the United States Air Force Academy. Design is an inherently motivational activity, because it permits the student to apply creativity and developing technical knowledge to the task of meeting performance, cost, and other specifications for a device, in this case an airplane, for human needs. The design process is iterative and does not result in only one right answer; thus it develops the student's ability to frame and resolve ill-defined problems as well as fosters intellectual curiosity. Design problems are also ideal for solution by student teams that encourage the development of interpersonal skills so critical to success in a variety of professional endeavors. Traditionally, design was viewed as a process that could only be undertaken once one had obtained a certain level of technical knowledge imparted in a series of courses in physics, mathematics, and engineering science. By following the traditional approach, the student was deprived of the motivational experience that is design until late in the postsecondary curriculum. More recently, educators have come to realize that even with the limited background available to them, beginning students can greatly benefit from the challenge of simple design projects with necessary knowledge and tools introduced as required to keep the process going. It was this realization that spurred revision of the Academy's introductory aeronautics course and the development of this accompanying text. This introductory course now includes simple whole aircraft design projects and projects that explore modifications to existing designs to improve performance. These projects motivate and facilitate students' learning by challenging them to apply theory to practical situations.

It has long been the goal of this department to develop and publish from time to time books that contribute to the education and practice associated with the broad field of aeronautical engineering. These publications, starting with *Aircraft Engine Design* by Jack D. Mattingly, William H. Heiser, and Daniel H. Daley and continuing with *Hypersonic Aerothermodynamics* by John J. Bertin and *Hypersonic Airbreathing Propulsion* by William H. Heiser and David T. Pratt, are all in the AIAA Education Series. I sincerely hope that this new volume in the series will continue our tradition by serving as an up-to-date and valuable contribution to the fascinating discipline we call aeronautical engineering.

Michael L. Smith
Professor and Head
Department of Aeronautics
United States Air Force Academy

Table of Contents

Preface

This textbook and the accompanying AeroDYNAMIC software were created for use as resources for teaching basic design methods in conjunction with an introductory course in aeronautics. That these two topics can coexist and reinforce each other in an introductory-level course has been proven by 11 years of experience with a course taught to the entire student body of the United States Air Force Academy (USAFA). The book was written for use in teaching students in their first three years of post-secondary education, so understanding of the derivations and analysis methods is facilitated by prior knowledge of calculus, classical physics, and engineering mechanics. However, the material is presented at a level that should be understandable by advanced high school students.

The text begins with an entire chapter devoted to the motivation, methods, and history of engineering design in general and aircraft design in particular. The design methods described are commonly used throughout industry, but most of the examples given are specific to aircraft design. The reasoning behind placing the design chapter first is identical to the reasoning that motivates teaching design in conjuction with basic aeronautical engineering concepts. Typically, engineering students spend their first two or three years learning basic math, science, and engineering analysis methods without having much opportunity to see where all this knowledge fits in the work that engineers do. Most engineering students have chosen engineering as their major because they want to learn how to design. Experience at USAFA has shown that learning about design first motivates students to learn the theory and analysis methods that they need in order to design. Chapter 1 does not describe any analysis methods in detail, but simply describes where analysis fits in the design process. Subsequent chapters introduce analysis methods and describe why they are necessary tools of the aircraft designer. In other words, Chapter 1 establishes a framework or outline of the aircraft design process, then Chapters 2–8 fill in the analysis methods needed in each design step.

Chapter 2 introduces analysis methods, which apply to static fluids, and describes how these methods can be used in the construction of pressure measurement devices and a model of the atmosphere that engineers use as a standard reference. Chapter 3 teaches analysis of moving fluids and shows how these methods are used to predict the lift and drag exerted by fluids on bodies moving through them. Chapter 4 expands the lift and drag prediction methods to include three-dimensional effects of finite wings and whole aircraft, and gives an example of how these methods are used to predict the aerodynamic characteristics of a modern jet fighter aircraft. In Chapter 5 the reader learns how knowledge of an aircraft's aerodynamics, engine thrust capabilities, and the laws of physics are used to predict its performance capabilities, or in other words the ability of the aircraft to meet a customer's needs. Chapter 5 also discusses constraint analysis, a method used to choose aircraft characteristics such that the aircraft can meet customer needs.

Chapter 6 introduces stability and control concepts, and further describes static longitudinal stability analysis as a simple example of how the static and dynamic stability of an aircraft in all six degrees of freedom is analyzed. Chapter 7 discusses the fundamentals of structural design and analysis and presents a simple structural sizing example. Chapter 8 deals with sizing, determining how large an aircraft needs to be in order to fly a particular mission with a specified payload. This discussion includes an introduction to mission analysis methods that are used to determine the amount of fuel required by an aircraft to fly a mission. Chapter 8 also discusses a simple method for predicting aircraft purchase and operating costs.

Once students have mastered the basic aeronautical engineering methods presented in Chapters 1–8, they are adequately equipped to tackle simple aircraft design problems. Chapter 9 explains how all of the methods are used in the conceptual aircraft design process. The AeroDYNAMIC software that complements this text provides user-friendly aircraft geometry modeling and design analysis tools that greatly facilitate the working of such problems. Chapter 10 is offered to support students' first design experiences by presenting case studies of the development of three well-known aircraft designs and a discussion of future aircraft design possibilities. These discussions focus on the impact of technology, politics, and the economy on each aircraft's design process.

The material in this textbook and the AeroDYNAMIC software is sufficient to support a university-level course of three to six semester hours. At USAFA, Chapters 1–6 are used in a three-semester-hour course taught to the entire student body. The material in Chapters 7–10 is taught at USAFA in later courses taken only by aeronautical engineering majors.

Steven A. Brandt
May 2004

Acknowledgments

This book is the product of the efforts of many individuals who have taught aeronautical engineering at the United States Air Force Academy over the past 30 years. In particular the materials, which have been developed and used to teach an introductory aeronautics course to the entire student body, have formed the central structure of this book. The list of all those individuals who have contributed to this textbook and to the supporting summaries, examples, software, slideshows, and movies now has grown so large that it is impossible to name everyone. This textbook truly is a product of the entire U.S. Air Force Academy Department of Aeronautics. This book is dedicated to the whole department, with sincere thanks from the primary authors.

Nomenclature

A = stream tube cross-sectional area

$\mathcal{R}$ = aspect ratio, b^2/S

a = speed of sound in air, $\sqrt{\lambda R T}$; linear acceleration, dV/dt

b = wing span

C_D = finite wing or whole aircraft (three-dimensional) drag coefficient, D/qS

C_{D_i} = finite wing or whole aircraft (three-dimensional) induced (due to lift) drag coefficient

C_{D_0} = whole aircraft (three-dimensional) zero-lift drag coefficient

C_{f_e} = equivalent skin-friction coefficient

C_L = finite wing or whole aircraft (three-dimensional) lift coefficient, L/qS

C_M = finite wing or whole aircraft pitching moment coefficient, M/qSc

c = wing or airfoil chord; specific fuel consumption used for reciprocating engines

$\bar{c}$ = wing mean aerodynamic chord

c_d = airfoil (two-dimensional) drag coefficient, d/qS

c_l = airfoil (two-dimensional) lift coefficient, l/qS

c_m = airfoil pitching moment coefficient, m/qSc

c_t = thrust-specific fuel consumption used primarily for turbine engines

c_{root} = wing root chord

c_{tip} = wing tip chord

D = wing or aircraft total drag

D_i = wing or aircraft induced drag or drag due to lift

D_0 = wing or aircraft zero-lift drag

d = airfoil drag

E_{WD} = wave drag efficiency parameter, compares actual to ideal wave drag

e = wing span efficiency factor

e_O = aircraft Oswald's efficiency factor

F = force

f = airspeed compressibility correction factor

h = altitude

i = incidence angle

k, k_1 = induced drag term coefficient, $1/(\pi e \mathcal{R})$

k_2 = camber influence coefficient in induced drag term

L = finite wing or whole aircraft (three-dimensional) lift

l = airfoil lift; overall length, especially of the fuselage or the entire aircraft

$\mathcal{L}$ = whole aircraft rolling moment

M = wing or whole aircraft pitching moment; Mach number, V/a

m = airfoil pitching moment

N = normal force, yawing moment

n = load factor

P = static pressure; power, TV

P_{avail} = power available

P_{req} = power required

P_S = specific excess power, $V(T - D)/W$

q = dynamic pressure, $\frac{1}{2}\rho V^2$

R = range; ideal gas law proportionality constant ($P = \rho R T$)

Re = Reynolds number, $\rho V x/\mu$

r = turn radius

S = reference planform area, usually the area of the wing planform

S_c = canard planform area

S_t = horizontal tail planform area

s = distance along a path

T = thrust; temperature

T_{avail} = thrust available

T_{req} = thrust required, D

t = time

u = x-axis component of velocity

V = velocity (magnitude)

V_H = horizontal tail volume ratio, $S_t l_t/Sc$

$\mathbf{V}_\infty$ = freestream velocity vector

v = y-axis component of velocity

w = z-axis component of velocity

X = force component along the x axis

x = axis of coordinate system, which is frequently aligned with the aircraft's longitudinal axis

Y = force component along the y axis

y = axis of coordinate system, which is frequently aligned positive out the aircraft's right wing

Z = force component along the z axis

z = axis of coordinate system, which is frequently aligned positive down

Greek

α = angle of attack; thrust lapse, T/T_{SL}

α_a = absolute angle of attack, $\alpha - \alpha_{L=0}$

$\alpha_{l=0}$ = airfoil zero-lift angle of attack

$\alpha_{L=0}$ = wing zero-lift angle of attack

β = sideslip angle; weight fraction, W/W_{TO}

Γ = dihedral angle

γ = flight-path angle; ratio of specific heats, 1.4 for air

δ = control surface deflection angle

ε = downwash angle

η = mechanical efficiency

Λ = wing sweep angle

λ = wing taper ratio

μ = air viscosity; rolling friction coefficient

ρ = air density

τ = shear stress
ϕ = bank angle
ω = angular velocity, rate of rotation

Math

∞ = largest possible number, used as a subscript to denote a great
     distance away
$\parallel$ = parallel to
$\perp$ = perpendicular to

1
Design Thinking

"A scientist discovers that which exists.
An engineer creates that which never was."

Theodore von Kármán

1.1 Introduction

Imagine the thrill Clarence L. "Kelly" Johnson (Fig. 1.1), founder and Chief Engineer of the Lockheed "Skunk Works," felt the first time the YF-12 (forerunner of the SR-71) flew faster than 2000 mph, or the satisfaction Harry Hillaker, Chief Designer of the F-16, felt when his creation became the most numerous jet fighter in the U.S. Air Force and in nine other air forces around the world. Aeronautical engineers live for the opportunity to design an aircraft and then, after years of work, to see it fly. The purpose of this textbook is to teach you how aircraft designers create the aeronautical marvels we take for granted. In the process, you will learn the basic concepts of aeronautical science and develop a working knowledge of the equations that govern aircraft behavior. You will come to understand how an aircraft is shaped and optimized by countless design decisions so that it will be able to perform a specific mission. You will also learn why an aircraft designed for a particular mission looks and performs the way it does. The basic concepts, theories and analysis methods you will learn will provide a solid basis for more advanced studies in aeronautical or aerospace engineering.

Throughout this text, the emphasis will be on design. The governing equations and analysis methods will be discussed in terms of their place in the design process. The analysis will be specific to aeronautics, but the design thinking and methods you will learn through design problems and examples are applicable to any type of creative problem-solving situation.

1.2 Design Method

The purpose of all engineering is to build something. The task can be large or small: the Sears Tower or a doorstop. The product can be hardware, software, or even something that has no physical existence at all, such as an organizational plan. Many engineers work solely with the actual building or production of a product, but most are involved in planning and design.

A product must be planned so that when it is built the construction process is efficient and makes the best use of resources, and the product meets the requirements set forth for it. The process of planning the physical characteristics and construction methods of a product is called *design*. In the process of creating a new product, design is the most important phase. The carpenter's adage, "Measure

Fig. 1.1 Kelly Johnson and the YF-12 (courtesy of National Air and Space Museum, NASM A-2065, Smithsonian Institution).

twice, cut once," applies doubly to design because the dimensions and materials of the components that will comprise a product must first be carefully planned and then measured, fabricated, and assembled in order to make a successful product.

The files of consumer protection organizations are filled with examples of poorly designed and/or poorly constructed products. It is the primary business of engineers to plan products that will meet their requirements and to ensure the products are constructed as planned. In most cases, the engineer who designs the product is not the one who builds it, so that communication of the design to those who will build it is another important step in the design process.

1.2.1 Design Process

So how does one begin to design a product? Because design is really just creative problem solving, the classic problem-solving method often called the scientific method is a great place to start. The steps in the scientific method are commonly[1] understood to be 1) define the problem, 2) collect data, 3) create a hypothesis, 4) test the hypothesis, and 5) if the hypothesis fails the tests, return to step 2.

For the design process, the steps must be modified somewhat. In defining a design problem, one must specify what function the product is to have, what constraints limit possible design choices, and what performance the product must achieve. The data collected for a design will include information on specific requirements and limits of the problem, characteristics of similar products, available technology, and the analysis methods available for evaluating the design. In place

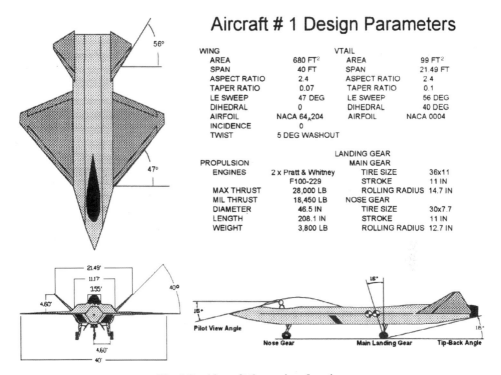

Aircraft # 1 Design Parameters

WING		VTAIL	
AREA	680 FT²	AREA	99 FT²
SPAN	40 FT	SPAN	21.49 FT
ASPECT RATIO	2.4	ASPECT RATIO	2.4
TAPER RATIO	0.07	TAPER RATIO	0.1
LE SWEEP	47 DEG	LE SWEEP	56 DEG
DIHEDRAL	0	DIHEDRAL	40 DEG
AIRFOIL	NACA 64₄204	AIRFOIL	NACA 0004
INCIDENCE	0		
TWIST	5 DEG WASHOUT		

PROPULSION		LANDING GEAR	
		MAIN GEAR	
ENGINES	2 x Pratt & Whitney	TIRE SIZE	36x11
	F100-229	STROKE	11 IN
MAX THRUST	28,000 LB	ROLLING RADIUS	14.7 IN
MIL THRUST	18,450 LB	NOSE GEAR	
DIAMETER	46.5 IN	TIRE SIZE	30x7.7
LENGTH	208.1 IN	STROKE	11 IN
WEIGHT	3,800 LB	ROLLING RADIUS	12.7 IN

Fig. 1.2   Aircraft three-view drawing.

of the hypothesis, the designer creates a design or a design concept. This creative process is also called *synthesis*. The design must be described, recorded, and in most cases communicated to others. Fig. 1.2 is an example of a typical aircraft three-view drawing and data block, a common way to record and communicate important design details.

The test of the design concept will be analysis or experiments to determine if it is manufacturable, economical, and in compliance with the design requirements. Establishing criteria for cost, manufacturability, and design performance therefore must be part of step 1, defining the problem, and also step 4, testing the success of the design. These criteria for design success are called *measures of merit*. Finally, step 5 must be modified to reflect the fact that a typical design process will involve many iterations through the cycle from collecting data through analysis, with each analysis producing more data to help further refine the design. The function of step 5 in the design process is to make a decision about whether the design will work, what parts of the design are deficient and must be changed, or about which of several designs or design ideas is the best and should be further developed. In some cases, step 5 can ask whether the design requirements and measures of merit are reasonable, and if not, recommend changes. With these modifcations from the scientific method, the design process steps can be stated as follows: 1) define the problem, 2) collect data, 3) create or synthesize one or more design concepts,

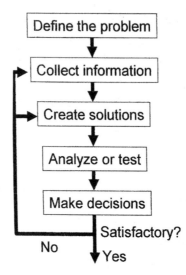

Fig. 1.3   The design method.

4) perform analysis, and 5) make decisions. If changes are needed, return to step 1 or 2 (see Fig. 1.3).

1.2.2 Design Cycle

Note that the decisions in step 5 will certainly send the designer back to step 2 for many cycles until the design is completely defined, refined, and built. This design cycle has been described by Leland Nicolai[2] of the Lockheed Advanced Aeronautics Company (also known as the Skunk Works) using a diagram similar to Fig. 1.4. Nicolai groups the steps of the design process into three actions: synthesis, analysis, and decision making. He points out the sharp differences between the relatively unstructured, often intuitive creative thinking required for synthesis and the highly structured methodical thinking required for analysis and decision making. Designers and design teams must be masters of both types of thought processes to be successful.

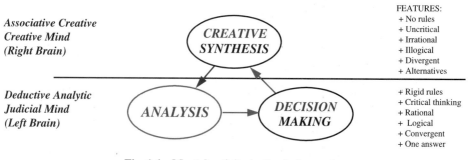

Fig. 1.4   Mental activity in the design cycle.

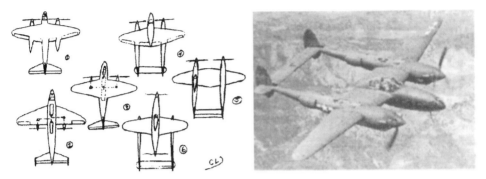

Fig. 1.5   Configuration sketches used in the early conceptual design phase of the Lockheed P-38 (courtesy Lockheed Martin).

1.2.3   Design Phases

The design process is usually described as having three phases. The initial phase is called *conceptual design*. The goal of conceptual design is to select a workable concept and optimize it as much as possible. Figure 1.5 illustrates sketches made by Kelly Johnson in 1937 of nine possible configuration concepts for a twin-engined fighter, which eventually became the Lockheed P-38 Lightning. Notice that the sketches, though crude, clearly communicate the fundamental configuration choices that Johnson's design team considered. A photo of a production P-38 is included in Fig. 1.5 for comparison.

Once a concept is selected, further iterations through the design cycle are needed to select the materials and work out the dimensions, structures, and functions of the design. Computer simulations are performed, and physical models of the design are built and tested. This phase is usually called *preliminary design*. Once the preliminary design phase is complete, detail design begins. In the *detail design* phase the product is prepared for production. The design is described in complete detail, and the process by which it will be manufactured is also designed. A detailed set of drawings, a materials list, and a detailed cost estimate are prepared. In later iterations through the design cycle, a prototype is often built and tested. The results of these tests are fed back into further design cycles to improve the performance and manufacturability of the design. Even after the design is in production, the design cycle continues. Information that is learned from continued testing of prototypes and initial operation of the early production models is typically used to further refine and improve later versions of the product. In some cases a design will continue to be improved for many years and even decades after the first versions have been produced and gone into service.

1.2.4   Optimization

The goal of refining a design is not just to meet the requirements, but to make the best possible or optimum design. The concept of optimizing a design to its full potential is crucial to creating a superior product. In many cases, the results of steps 4 and 5 will force the designer back to step 1 because it is determined that one of the original design requirements is unreasonable or that two of the requirements conflict with each other. When this happens, the measures of merit must be adjusted to be reasonable but with performance that still satisfies the customer. The measures of merit may also change because the customer's requirements change. This is particularly common for a design that continues to evolve for many years after initial production.

The measures of merit for a design are primarily established by the customer. In fact, the real answer to *Where does one begin to design a product?* is *With the customer*. Indeed, the entire design/build process begins and ends with the customer. Many designs that have failed have done so because the customer was forgotten somewhere along the way, and as a consequence, the product failed to meet the customer's needs. Whenever possible, the customer must be involved and giving feedback in every step of the design and construction process. In some cases part of step 1, defining the problem, will include defining who is the customer. For instance, in designing an academic course in aeronautics, is the customer the student or the organization that will eventually employ that student?

1.3 Design Example

To see how this process works, consider a very simple design problem, let us say the design of a pencil. A pencil is a well-known product that has been around for many centuries, yet new ones are designed every year.

1.3.1 Problem Definition

First, define the problem, and list all of the requirements that this pencil must meet. The customer can be a student, a school, or even an advertising agency. Each customer will have a different idea of what measures of merit they will want for the pencil. Because the pencil must be held to be used, the size and shape of the human hand (and the variability of those dimensions from person to person) will constrain the size and shape of the pencil. The need to operate for long periods without the availability of a pencil sharpener might be important. Appearance (conventional, unusual, or attractive) might be important. The performance required of the pencil might include the ability to produce a certain fineness or quality of line for a set distance. User comfort might be important, as might eraser capacity. Of course, cost must be kept to an absolute minimum. However, depending on the customer, the cost that is most important might either be the cost of a single unit or the cost of using pencils over some period of time. Other criteria can include flammability, heat tolerance, toxicity, chemical stability, disposability, and similar environmental and safety concerns. Many designers use a selection matrix such as Table 1.1 to help them tabulate and rank order the design requirements and measures of merit. This matrix is also used to record and evaluate the results of the analysis in step 4 and allows a direct comparison between performance of several designs. It then forms the primary basis for the decisions in step 5. A few of the possible design concepts for a pencil are listed on the selection matrix to illustrate how it is used.

Although an actual selection matrix for this design problem would have many more columns, Table 1.1 does illustrate how such a matrix would be used. Clearly the weight given to each of the selection criteria on the matrix will have a profound effect on the final decision. These weight values are established based on the customer's needs and priorities.

Table 1.1 Pencil design concept selection matrix

Concept	Semester cost (sc), $2/sc (×0.30)	Line width (w), 5 mm/w (×0.25)	Eraser volume (v), v/1 cm³ (×0.20)	Sharpener? 1 = does not need (×0.15)	Unit cost (uc), $0.25/uc (×0.10)	Total (large = good)	Choice
Importance factor	(×0.30)	(×0.25)	(×0.20)	(×0.15)	(×0.10)	Total weight = 1.0	—
			Scores: raw (weighted)				
Fixed Wooden	1 (0.3)	0.5 (0.125)	0.25 (0.05)	0	1 (0.1)	(0.575)	—
Fixed Plastic	0.4 (0.12)	0.6 (0.15)	0.25 (0.05)	1 (0.15)	0.50 (0.05)	(0.52)	—
Disposable Mechanical	0.4 (0.12)	0.8 (0.20)	0.25 (0.05)	1 (0.15)	0.25 (0.025)	(0.545)	—
Reusable Mechanical	0.5 (0.15)	1 (0.25)	1 (0.20)	1 (0.15)	0.0625 (0.00625)	(0.75625)	—
Mechanical (large eraser)							×

Measures of merit

1.3.2 Data Collection

In step 2 you must gather data on the problem to help establish what values the measures of merit should have and what constraints the operating environment may enforce. Details such as the dimensions and shape of the average human hand, the degree to which these dimensions vary in the population, the physiology of writer's cramp, the size and shape of other pencils on the market, and the amount an average writer must erase are all important. Always include cost and maintainability as constraints on the design. In this case the customer might have a good idea what they are willing to pay for a pencil, but in other situations a market survey might be needed. It is often very helpful to look at how other designers have solved similar design problems, and so a look at history and the current competition might be in order.

The data you have available to you for the design process actually include everything you have ever learned and every experience you have ever had. You never know what combination of background and experience will produce a great design idea or a new way to apply an existing technology. The data you need for the design project might also include some skills or knowledge that you have not yet learned. A good designer finds out what expertise is needed to tackle a particular problem and acquires those skills, either through research and training or by hiring an assistant with the necessary experience.

1.3.3 Synthesis

Step 3 is the creative step of creative problem solving. It is at this point in the process where new ideas emerge and new products are created. Once you have immersed yourself in the facts and constraints of the problem, give yourself some time to come up with creative solutions. Many designers find it helpful at this point to exercise, or relax, or even "sleep on it." Some designers keep a notebook by their bedside for use in writing down ideas that come to them in the middle of the night. It seems that by allowing the conscious mind to rest or be distracted with a simple activity, the creative part of your brain is better able to make the connections between all of the data you have assembled. Draw on the whole range of your experiences. Some of the best design solutions come from completely unexpected combinations of ideas that no one else has ever thought of. For instance, a military aviator working on a spacecraft project came up with the idea that aerial refueling of horizontal-takeoff and -landing spacecraft would greatly reduce the weight needed for the vehicle's landing gear. Because the vehicle could take off and land at very light weight and then take on the majority of its propellants once airborne, the landing gear could be built to take much lower loads. This saved weight. Analysis showed that the weight saved on landing gear would be very significant. This would allow the proposed design to be much smaller and less expensive while still carrying the same payload as a horizontal-takeoff spacecraft concept that would take off with all of its propellants onboard. The engineer who thought of this idea combined his experience with aerial refueling with his understanding of the spacecraft landing-gear design problem.[3]

The success of the creative process can be greatly enhanced by tackling it as a team. In this way, the combined experiences and insight of the entire team can be brought to bear on the problem. An unworkable idea by one member of the team might spark another member to think of an idea which ultimately proves to be the most successful. This phase of the process is often called brainstorming or *ideation*. It is very important that you not attempt to evaluate the ideas as they are suggested. Save the evaluation process for step 4. During brainstorming, let the ideas flow, without criticism. It might be the idea you know will not work that gives rise to the best solution of all.

In most cases the design concept must be recorded and communicated. For the pencil, a drawing should be made with all components, dimensions, and angles clearly labeled. Materials, manufacturing, and marketing choices can be simply listed on the selection matrix, or they might need to be described in more detail. Especially for group work, clear communication of design ideas is essential to the success of the creative process.

1.3.4   Analysis and Decisions

Step 4 requires you to determine first how you will evaluate your design to see how well it meets the measures of merit. Use the list of analysis methods you made in step 2 to plan a sequence of tests you will perform. Be sure to involve the customer in this planning, and make sure you are testing every characteristic that the customer thinks is important.

Next, perform the tests and/or analyses, record the results in the selection matrix, and communicate them as appropriate. Based on the results of these evaluations, you can identify weaknesses of a design and select which of several design ideas is best at meeting the measures of merit. Because it is unlikely that the perfect solution will be found on the first time through the cycle, your decisions will probably send you back to collecting more data, or even redefining the problem to start more iterations through the cycle. Repeat this process as necessary until you have selected a pencil with all of the features and characteristics to meet the customer's needs.

1.3.5   Phases of Design

At this point, the design process for the pencil is far from complete. A design idea has been selected as likely to produce an acceptable solution, and so the design process is now probably done with the conceptual design phase and is ready to proceed with preliminary design. The pencil's design must continue to be defined, refined, and communicated through the preliminary and detail design phases before the product can begin to be manufactured.

1.3.6   Questions

At this point, we must ask "What have we just done?" We considered a simple problem, applied quite a bit of thought and effort, and arrived at a simple, sensible answer. Did we really need all of that analysis to come to that conclusion? Because there are so many variables and options in the design of something as simple as a pencil, the analysis was probably time well spent. And what about more complex design problems, like the design of a whole aircraft? Though the number of variables and decisions in designing an aircraft seems totally overwhelming, the coherent, systematic approach of the design method provides structure and direction to the process. The method assembles data, inspires creativity, identifies and focuses attention on decisions, and improves or eliminates unacceptable concepts. With each iteration of the design cycle, it moves toward a more and more completely defined and optimized design solution.

1.3.7   Ill-Defined Problems

One of the particular strengths of the design method is that it provides a way to tackle ill-defined problems. Ill-defined problems are those for which only partial information is available and more than one solution can be acceptable. Under that definition, nearly all design problems are ill defined. This lack of information that characterizes most design problems has led Billy Koen, an engineering professor at the University of Texas at Austin, to define the engineering (design) method[4] as follows: "The engineering method is a strategy for causing the best change in a poorly understood or uncertain situation within the available resources."

The design method leads you to resolve ill-defined problems by assembling and reviewing the information you know, creating and analyzing possible solutions to obtain the information you do not know, making decisions on the basis of your preliminary information, and repeating the cycle until you have selected and completely defined an acceptable (preferably the best) solution. An additional consequence of the cyclic nature of the process and the time required to perform it is that as more information is obtained about a design the time invested in advancing definition of a design to the point where it is possible to make a very detailed estimate of its characteristics also makes it very costly in time and effort to make large changes to that design. In other words, if you get enough detail about a design to determine that it will not work, you cannot afford the time and effort it would take to go back and redesign it from scratch. This dilemma is depicted in Fig. 1.6, which shows the design cycle as an upward spiral. The increasing height of the spiral indicates the increasing understanding of the problem, and the decreasing width of the spiral symbolizes the reduced ability to make changes in the design.

An example of the narrowing of available choices with progression in the development of a design is found in the history of the first attempt by the United States to build a supersonic transport (SST). In the 1960s a very aggressive SST design was

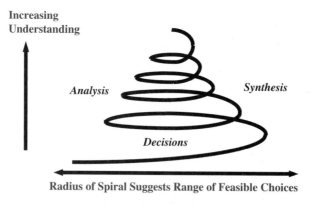

Radius of Spiral Suggests Range of Feasible Choices

Fig. 1.6   Design spiral.

being developed in the United States to compete with the Anglo–French Concorde. The American SST was initially planned to have variable sweep wings rather than a fixed-wing planform like the Concorde and the Soviet Tu-144. As design development progressed, analysis made it clear that the extreme weight penalty associated with the structure and pivoting mechanism required for the variable-sweep wing would make it impossible for the American SST to be profitable. Unfortunately, the cost of redesigning the aircraft with a fixed wing like the Concorde was also too expensive, and the entire program was canceled. Concern over the environmental impact of SST operations also contributed to the cancellation decision, but the high price tag for starting over sealed the program's fate.

1.4   Design and Aeronautics

What does all of the preceding text have to do with aeronautics? Aeronautics is the science of flight. Every aircraft that ever flew was designed, though the design process for some was much longer and more elaborate than for others. Like any product, an aircraft design is initiated because customers have needs. The customers might be airlines, corporations, private pilots, government agencies, or military services.

A customer's needs are normally specified as one or more design mission(s) that the aircraft must be able to fly and a set of constraints or performance requirements that the aircraft must meet. Figure 1.7 illustrates a typical design mission for a multi-role tactical fighter aircraft. Table 1.2 lists additional design constraints or requirements that might be specified for the aircraft. Note that the design mission and the constraint table specify performance that the customer requires from the aircraft. An aircraft design that fails to achieve these required performance levels will probably not be purchased by the customer.

The requirements are formally communicated to industry in a request for proposals (RFP). Receiving an RFP officially initiates the design and development process within a company. The company might have gotten a head start on this process though by being involved in previous feasibility studies and/or similar studies of its own. Not doing so places the company at a great disadvantage.

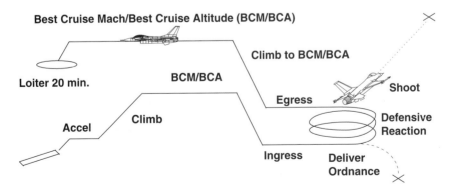

Fig. 1.7   Typical design mission for a multirole jet fighter.

Table 1.2   Typical minimum design requirements/constraints for a
multirole jet fighter

Item	Requirement
Combat mission radius	400 miles
Weapons payload	Two AIM-120
	Four, 2000-lb MK-84
	600 rounds 20-mm ammunition
Takeoff distance	2000 ft
Landing distance	2000 ft
Max Mach number	$M = 1.8$ at optimum altitude at W_{man}[a]
Instantaneous turn rate	18 deg/s at $M = 0.9$, 20,000 ft MSL[b] at W_{man}
Specific excess power	800 ft/s at $M = 0.9$, 5000 ft MSL at W_{man}
Sustained load factor	4 g at $M = 1.2$, 20,000 ft MSL at W_{man}
	9 g at $M = 0.9$, 5000 ft MSL at W_{man}

[a] The maneuver weight W_{man} is the aircraft weight with 50% internal fuel, two AIM-120 AMRAAM (Advanced Medium Range Air-to-Air) missiles, and full-cannon ammunition, but no air-to-ground weapons.
[b] The abbreviation MSL signifies altitude above mean sea level, the average elevation of the Earth's oceans.

1.4.1 Aircraft Design Phases: Conceptual Design

As depicted in Fig. 1.8, the process proceeds through conceptual, preliminary, and detail design phases. For an aircraft, conceptual design involves mostly paper and computer studies, with heavy emphasis on optimization and finding the best possible aircraft concept. The result of aircraft conceptual design is an aircraft configuration, described by drawings and models, with analysis results that predict what the aircraft will be able to do.

Conceptual design can be described as consisting of 12 tasks or activities, all of which are described in this textbook. The process begins with 1) customer focus, which involves exactly defining the design problem and determining in detail what the customer needs. Once the designers have a clear understanding of what the design must be able to do, they engage in 2) design synthesis to create innovative ideas that can be developed into superior products. Next, 3) geometry modeling allows them to represent the shape and physical characteristics of the design so it can be analyzed. Section 4.7 of this book includes a simple example of geometry modeling.

The first step in analyzing a conceptual aircraft design is 4) aerodynamic analysis, as described in Chapters 3 and 4 of this book. The results of aerodynamic analysis and 5) propulsion modeling (Sec. 5.3) are then used in 6) constraint analysis (Sec. 5.15). Constraint analysis ensures that the aircraft meets all of the performance requirements or constraints specified by the customer. These results are used in 7) mission analysis (Sec. 8.5), which determines how much fuel the aircraft concept must carry and how much thrust its engines must deliver to allow it to fly all of the required missions carrying the specified payloads. Then 8) weight prediction (Sec. 8.3) estimates the weight per unit area of the various components of the aircraft. The results of all of the preceding steps are used in 9) sizing, which predicts how large a particular design concept would have to be to meet all the

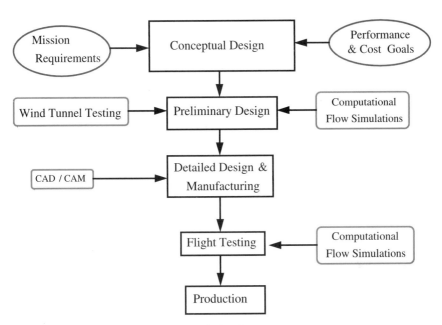

Fig. 1.8 Aircraft development process.

design requirements. Once a size is determined, 10) cost analysis (Sec. 8.11) gives feedback on whether the concept is likely to meet cost constraints.

These analyses can identify unreasonable or conflicting requirements that must be resolved to the customer's satisfaction before the design can proceed. Typically, some requirements are easier to meet than others. The most difficult constraints to meet have the strongest influence on the configuration and characteristics of the aircraft. These requirements are called *design drivers*.

Activities 1–10 are repeated many times to find the best configuration that allows the concept to achieve the customer's requirements for the least cost. This process is called 11) optimization. Once a final, optimized configuration is chosen, 12) performance analysis (Chapter 5) predicts how well it will meet all of the customer's needs.

The conceptual design phase also usually includes layout of an initial structural concept (Chapter 7). The designers also predict the concept's weight and center of gravity (Sec. 8.7) to ensure that it can be built light enough and that it will balance properly. A heavy emphasis is also placed at this early stage on designing the aircraft for producability and maintainability. Representatives of the customer(s) are kept closely involved throughout this and subsequent design phases, to ensure the evolving design continues to meet the customers' needs.

1.4.2 Aircraft Design Phases: Preliminary Design

In the preliminary design phase, the selected aircraft concept is developed and refined. Wind-tunnel, computer flow simulation, and in many cases subscale free-flight models are built and tested to ensure the aircraft will be efficient and controllable in the air. Flight and avionics simulators are used to prove and refine the aircraft's handling qualities, instrument displays, and pilot comfort and visibility. The aircraft structural concept is greatly refined, materials for the major components are specified, and the structure is analyzed to ensure all components have adequate strength without excess weight. More refined weight and cost predictions are made, and weight and cost budgets for the aircraft are maintained. The budgets are used to enforce goals for aircraft weight and cost. In many cases, if a component's predicted weight or cost increases as its design becomes more refined savings must be found somewhere else in the design. These savings can be achieved by the use of advanced technology for some components or possibly by eliminating a component or feature. As the data from these studies become available, they are fed back through the design cycle and used to improve and correct design characteristics. All of the analyses performed in the conceptual phase are repeated for the more refined design to ensure all constraints and requirements are still being met. Detailed physical and CAD/CAM drawings and three-dimensional databases are created. These databases are used to construct maintenance simulators, many of which now use virtual reality, to ensure all required maintenance functions can be performed safely and efficiently. Maintenance and system simulations are also used to design and test the built-in-test (BIT) software and computer hardware that will automate much of the troubleshooting of aircraft malfunctions. At the completion of the preliminary design phase, the aircraft's configuration is quite well-defined and is not likely to change significantly.

1.4.3   Aircraft Design Phases: Detail Design

In the detail design phase, the aircraft is prepared for production. Every one of the thousands or millions of components, every fastener, every connector, every knob, and every linkage must be designed and/or specified in exact detail. Most modern aircraft companies use CAD/CAM systems to keep track of all of the parts, ensure they fit and move properly, specify how and of what materials they will be made, and update databases on weight and cost. Flying prototypes are built and test flown. Structural, avionics, and systems test prototypes are also built and tested. Wind-tunnel and computer analysis continues and is gradually reducing the need for prototypes. The main goal of this testing is to ensure everything works as planned, so that production of the aircraft can proceed. The CAD/CAM database is used to plan and control all phases of the manufacturing process, including assembly line design and control of automated milling and machining equipment.

1.4.4   Production

Generally, flight testing of the prototype(s) and factory tooling for production proceed in parallel. Early production models can also be used in the flight-test program. Flight test primarily determines that the aircraft is safe to fly and that it meets or exceeds the performance required by the customer. After flight test, early production models go through service trials to verify that the aircraft is suitable for the use the customer plans. If flight tests or service trials identify problems with the aircraft, changes are made on the assembly line, and retrofit modifications are made to aircraft already built. These changes naturally involve many more iterations through the design cycle. This process continues throughout the service life of the aircraft, with changes being required as new problems are found or as the customer's needs change. New versions of the aircraft can be produced and/or existing aircraft can be modified. These changes are often grouped into a major rebuild or midlife improvement.

1.4.5 Disposal

With wise use of upgrades and modifications, the service life of a modern aircraft typically extends beyond 20 years. But sooner or later every aircraft reaches the end of its useful life. Metal fatigues and cracks, joints and moving parts wear out, composites delaminate, and even the upholstery wears thin. Frequently the threat, the technology, and/or the mission change so dramatically that the aircraft is no longer useful. Whether obsolete or just worn out, the aircraft must be removed from service.

When an aircraft is retired from service, it can be stored as a reserve or a source of spare parts, but eventually the owner must dispose of it. Recycling of aircraft aluminum is routine, but some of the new composite materials and components such as lithium batteries require special procedures when disposing of them. The cost of (or in some cases profit from) disposing of an aircraft is a small but significant part of its life-cycle costs. The aircraft designer must give due consideration to this issue, especially if some materials or components used in the aircraft will pose environmental hazards.

1.5 Brief History of Aircraft Design

As an additional aid in establishing the context for your study of aircraft design, consider briefly its history. This section will acquaint you with the most significant personalities, aircraft, and events in the history of aviation and aircraft design. As you read, note how aircraft designs were equally influenced by customer needs and available technology. Also note the profound influence aircraft designs have had on the recent history of the world.

1.5.1 Early Years

Man's early thinking about flying was undoubtedly inspired by watching and seeking to imitate birds. The first men to fly were probably Chinese, flying in large manned kites perhaps as much as two millennia ago. Kites are aircraft that generate lift from pressure changes as air flows past them, but that are tethered to the ground and rely on the wind to provide the necessary airflow. They were probably developed originally as toys, but some inspired ancient engineer recognized their military potential. Manned kites were developed to meet military needs for elevated platforms from which to observe their enemies. Kites might also have been used to drop soldiers or spies behind enemy lines.

In Europe, small kites were developed in the 13th and 14th centuries, but practical, man-carrying kites were not used militarily until the early 1900s. Most Europeans who thought about manned flight imagined flapping-wing vehicles called ornithopters with human muscle providing the motive power. In the late 15th century, the famous artist Leonardo da Vinci made many sketches of such vehicles and the mechanisms needed to translate human arm and leg motions into the flapping of wings. Unfortunately, all of these dreamers and experimenters did not have sufficient knowledge of aerodynamics to do analysis to determine if a man could generate the power those vehicles would require to sustain flight. They also failed to understand the requirements for stability and control. Though many man-powered ornithopters were built, none were successful.

The first Europeans to fly were Francois Pilatere de Rozier and the Marquis d'Arlandes, who became airborne in a hot-air balloon built by Etienne and Joseph Mongolfier in 1783 (Fig. 1.9). In the same year, Professor Jacques Charles and Marie-Noel Robert were the first to fly in a hydrogen-filled balloon. Balloons of both types quickly became attractions at fairs and other public displays. Like kites, they were also adopted for military observation duties. When the advent of long-range artillery established the requirement for lofty vantage points from which observers could evaluate and correct the artillery's aim, hydrogen-filled balloons were most commonly chosen. Balloons served in many 19th-century wars, including the Franco–Prussian War and the U.S. Civil War. They were still used extensively as observation platforms in World War I and as obstacles to low-flying aircraft in World War II.

Balloons were acceptable as fair attractions or tethered observation platforms, but potential customers who would use lighter-than-air vehicles for transportation required the ability to move in any desired direction, even against the wind, at a reasonable speed. A vehicle with these capabilities was achieved by adding to a balloon or group of balloons a steering mechanism and a steam engine with a propeller. Such a vehicle is called an airship or dirigible. A French engineer, Henri Giffard, was the first to fly such a vehicle in 1852. Development of internal combustion engines soon provided airships with lighter, more powerful propulsion. Airships were used for passenger transportation and also as bombers in World War I. The German airship *Graf Zeppelin* flew around the world in 1929. It and the *Hindenburg* (Fig. 1.9) made regular passenger flights between Europe and the United States until 1937, when the *Hindenburg* was destroyed by a hydrogen explosion.

Fig. 1.9   Mongolfiers' first human-carrying free balloon and the *Hindenburg* (courtesy National Air and Space Museum, SI 94-12454 and SI 96-15169, Smithsonian Institution).

The airplane, a winged heavier-than-air vehicle with a propulsion system based on engine-driven propellers rather than flapping wings, was first conceived in its modern form by Sir George Cayley. Cayley was a remarkable scientist and engineer, one of the first to tackle the problem of manned flight scientifically. He was the first to propose *separate mechanisms for generating lift and thrust*, rather than the combined lift and thrust produced by flapping wings. In the early 19th century, he built a device for testing the lift and drag of wing shapes by attaching them to the end of a pivoting arm, which moved then rapidly through the air. He also worked out the *basic requirements for aircraft stability* and built several successful child-carrying gliders with such original inventions as spoked wheels like those found today on bicycles. Cayley's testing and analysis correctly suggested to him that steam engines of the day did not have high enough thrust-to-weight ratios to enable human flight. He spent much of his effort developing alternative airplane propulsion systems, but in the end was unsuccessful. However, his work provided the solid foundation from which the Wright brothers succeeded in making the first sustained human-carrying flight of a practical heavier-than-air vehicle less than 50 years after his death.

1.5.2 Wright Brothers

Sir George Cayley's work inspired much experimentation with human flight in Europe and the United States. Small airplane models powered by twisted rubber bands attached to propellers became very popular and helped establish an intuition if not a mathematical basis for aircraft stability requirements. Several steam-powered airplanes were built, and some made short hops but not sustained flights. At this point, only two basic elements were lacking: an adequate propulsion system and an understanding of aircraft control requirements.

Beginning in 1891, a German engineer named Otto Lilienthal began building and personally flying in his own gliders. His intent was to develop a powered ornithopter, but his glider flights were more useful in establishing an *understanding of requirements for aircraft control*. Lilienthal achieved control of his gliders by shifting his weight. Unfortunately, in 1896 this method of control proved inadequate when a wind gust upset one of his gliders. He was unable to regain control and was fatally injured in the ensuing crash.

A French-born American, Octave Chanute, included Lilienthal's work in a history of aviation experimentation published in 1894. Chanute tested several human-carrying gliders beginning in 1896, but more importantly he became the foremost promoter of human flight in the United States at the turn of the century. When Wilbur Wright contacted him in 1901, Chanute enthusiastically provided information, assistance, advice, and encouragement. He also was well acquainted with Samuel Pierpoint Langley, the Wright brothers' closest competition for making the first human-carrying powered flight.

Langley had a head start on the Wright brothers, and as Secretary of the Smithsonian Institution, had won a $50,000 contract from the U.S. government (prompted by the Spanish–American War) to develop a flying machine. He employed Charles Manley and Stephen M. Balzer as mechanics. Their most important achievement was to build the most advanced gasoline engine of its day, with a power-to-weight ratio that was well in excess of what was needed to allow

Fig. 1.10   Langley aerodrome on its houseboat launcher (courtesy National Air and Space Museum, SI 2003-35050, Smithsonian Institution).

human flight. Unfortunately, Langley failed to consider the need for control of the aircraft. His airplanes (or "aerodromes" as he called them) lacked landing gear because they were launched by catapult from atop a houseboat in the Potomac River (Fig. 1.10). Langley successfully flew several small aerodromes powered by steam engines, but the structure of his full-size human-carrying model failed on launch, and the machine plunged into the river. It was probably just as well because had the aerodrome been successful at commencing sustained flight its lack of any control system could have resulted in more serious consequences for the occupant, Charles Manley.

Langley's final attempt was made just days before the Wright brothers' first powered flight. Wilbur and Orville Wright had approached the problem of human flight very scientifically and had originally used tables of air pressure data published by Lilienthal. Their initial glider experiments convinced them Lilienthal's data were erroneous, and so they built their own wind tunnel and tested a number of airfoil shapes. They debated the cause of Lilienthal's fatal accident and finally decided on the need for moveable control surfaces, which allowed them much greater control than just shifting weight. So convinced were they of the paramount need for control, they deliberately built their airplanes to be unstable. As a result, they had to make constant control adjustments to keep their planes on a desired path. Extensive gliding experience gave the Wrights an understanding of *airmanship*, the science and art of flying, which was unequaled anywhere else in the world at the time. This understanding or feel for flying allowed the Wrights to build an airplane that not only could carry an adult human but also one that the pilot could maneuver in any desired direction. This capability was a distinct advantage

Fig. 1.11   Wright Flyer, 1903 (courtesy Library of Congress).

of the Wright machines over Langley's or any other airplanes up to that time. Unable to find a willing engine manufacturer, the Wrights built their own gasoline engine. Its power-to-weight ratio was not nearly as good as Manley's engine, but it was adequate. By repeated cycles of testing, analysis, and refinement of their designs, plus the flying training and insight they received from their glider flights, the Wright brothers finally achieved the first sustained flight of a heavier-than-air aircraft carrying an adult human on 17 December 1903 (Fig. 1.11).

In the years that followed, many engineers and inventors in addition to the Wrights began building successful manned airplanes. Most notable of these were Glenn Curtiss in the United States and Henri Farman, Gabriel Voison, Louis Blériot, and Alberto Santos-Dumont in Europe. These other designers chose to make their aircraft stable. As a result, they were easier to fly though less maneuverable than the Wright "flyers," and they became more popular. The Wrights sold relatively few of their unstable, difficult-to-control airplanes. They became embroiled with Curtiss in patent litigations over their control systems, a dispute that was finally settled by a merger of the Wright and Curtiss aeroplane companies. At the same time, aeronautical science was progressing rapidly in Europe, so that practical military observation aircraft were available early in World War I.

1.5.3   World War I

The requirement for fast, far-ranging observation, fighting, and bombing aircraft to contest the skies and support the land forces in World War I led to tremendous expansion and acceleration of aviation science and industry. The United States lagged far behind developments in Europe. In 1915 the National Advisory Committee for Aeronautics (NACA) was formed to help U.S. aviation science catch up. The most advanced aerodynamic studies were performed in Germany and led to the development of advanced airfoil shapes. Whereas the wings of most planes since the Wright brothers had relatively thin cross sections (airfoils), the new German sections were thicker (Fig. 1.12) and had better lift-to-drag ratios and higher values of maximum lift coefficient C_{Lmax}. The thicker sections also had structural

Fig. 1.12   Comparison of the relatively thin airfoil on Sopwith Camel and other Allied aircraft during World War I with the thicker airfoil used by the German Fokker DVII (courtesy Shell Companies Foundation via National Air and Space Museum, SI 83-8559 and SI 90-6720, Smithsonian Institution).

advantages. These aerodynamic refinements gave German aircraft superior performance, even though their engine technology lagged behind the Allies. So impressive was the Fokker DVII, a German fighter which appeared late in the war, that it was made a prize of war by the Versailles Treaty. The Germans were required to provide a number of Fokker DVIIs to the Allied nations for evaluation and duplication. Several Fokker DVIIs were flown by the U.S. Army Air Corps in the early 1920s.

1.5.4   Airlines and Long-Distance Flights

Spurred by the explosive advances during World War I in aerodynamics, propulsion, and structures and the huge increase in the number of trained flyers, aviation expanded rapidly in the postwar years. Airlines and airmail services proliferated. Barnstormers flew from town to town, popularizing aviation by giving rides and air shows. Prizes were offered by various newspapers, industrialists, and aviation organizations for long-distance flights. One of the best-known of these was the first nonstop flight from New York to Paris. Most of the aircraft competing for this prize were large because conventional technology required a large airplane to have the required range. An American airmail pilot named Charles Lindbergh contracted with the Ryan Airplane Company to build a relatively small, aerodynamically clean monoplane (having one wing as opposed to the biplanes used by the Wright Flyer and the Fokker DVII) with a large fuel tank and a new high-efficiency engine. The single wing, with its thick airfoil and lack of external bracing wires, gave the plane a relatively high lift-to-drag ratio (L/D). The large fuel tank allowed a high fuel weight fraction, and the new Wright Whirlwind engine had a relatively low specific fuel consumption (SFC) and high reliability. Careful analysis had shown that these design features and new technology would give the plane *The Spirit of St. Louis* the necessary range. Lindbergh made the flight solo in 33 hours and 39 min, 20–21 May 1927 and instantly became a world celebrity (Fig. 1.13).

Flights such as Lindbergh's further popularized aviation and proved the feasibility of world-wide passenger flights. Airlines grew and demanded aircraft with greater speed, range, and payload. Industry responded with improved aerodynamics, all-metal monoplane airliners, retractable landing gear, and improved engines. NACA performed numerous wind-tunnel studies and published extensive data on the aerodynamic characteristics of families of airfoil shapes and on methods for

Fig. 1.13 *The Spirit of St. Louis* (courtesy National Air and Space Museum, NASM A-47054, Smithsonian Institution).

reducing drag. One of the most outstanding aircraft developed during this period was the Douglas DC-3, the venerable Gooney Bird (Fig. 1.14). The DC-3 was a progressive development and enlargement of previous Douglas twin-engined all-metal airliners. Its two 1200-hp Pratt and Whitney Twin Cyclone Engines allowed it to carry up to 28 paying passengers for 2000 miles at 200 miles per hour. DC-3s were extremely popular with the airlines and were built in large numbers as military transports for World War II.

1.5.5 World War II

The return of global war in 1939 added fuel to an already frantic pace of aviation development. The newest fighter planes produced in Germany and the United Kingdom in 1939 were obsolete by 1941. The United States initially lagged behind in military aircraft technology, but experience with advanced airliners and military

Fig. 1.14 Douglas DC-3 (courtesy National Air and Space Museum, SI 95-8328, Smithsonian Institution).

Fig. 1.15   Two long-range fighters of World War II: Mitsubishi A6M2 and the North American P-51 Mustang (U.S. Air Force photographs).

airplane orders from Europe soon allowed American technology to catch up in many areas. German research in jet engines, rockets, and high-speed aerodynamics enabled production of jet- and rocket-powered fighters and bombers such as the Messerschmitt 262, which far outperformed Allied aircraft. The vast scale of Allied aircraft production, however, overwhelmed the German Air Force.

A technological achievement of particular interest during World War II was the development of a long-range fighter aircraft no bigger nor less maneuverable than its contemporaries. Two such aircraft are notable for the different approaches designers took toward meeting the requirement. The Japanese Mitsubishi A6M Zero-Sen (Fig. 1.15) achieved the necessary high fuel fraction without sacrificing performance by using lightweight structures and eliminating armor protection for the pilot. The North American P-51 Mustang sacrificed performance and handling qualities at heavy fuel weights instead. By the time Mustangs reached their distant combat areas, the extra fuel was burned off, and their performance and maneuverability equaled or exceeded that of their short-ranged opponents.

1.5.6 Jets

Advances in jet-powered aircraft during World War II continued postwar with development of several generations of jet fighters, bombers, and transports. Improved understanding of high-speed flight was made possible by supersonic wind tunnels and ambitious flight tests. The victors of World War II also took advantage of test data, aircraft prototypes, and scientists captured in Germany at the end of the war. By the early 1950s, the first generation of supersonic jet fighters was developed (Fig. 1.16). The Cold War prompted governments to continue to ask for aircraft that could fly higher, farther, and faster with little concern for the cost.

By the 1970s, jet fighter performance was limited primarily by the maximum acceleration and extremes of altitude pilots could endure. Eventually, economic concerns forced a more pragmatic approach to military aircraft procurement, so that any military aircraft built in the future will have to comply with strict cost constraints. In many cases cost has become more important than performance. A new consideration that has influenced the design of the latest generation of combat aircraft is low observables or "stealth" technology. Designing aircraft to reduce their radar, infrared, and visual signatures has significantly changed their appearance (Fig. 1.17).

Fig. 1.16 First-generation supersonic fighters: Convair F-106 Delta Dart and McDonnell–Douglas F-4E Phantom II (U.S. Air Force photographs).

Jet airliners, beginning with the De Haviland Comet and Boeing 707, have literally changed the world. Travel between far-distant capitals and trade centers has become commonplace. Supersonic airliners, the Anglo–French Concorde, and the Soviet Tu-144, have been developed and used extensively. A new-generation high-speed civil transport (HSCT) is being planned in the United States. Airline customers have become impatient with flight times between cities that would have seemed incredible 50 years ago. Airlines are also seeking new aircraft with improved technology that will lower their operating costs and allow them to remain competitive in a rapidly growing market for low-cost air fares. Fuel-efficient high-bypass-ratio turbofan engines, computer-designed airfoils, and advanced materials and construction methods have made this possible.

A major figure in aircraft design from World War II through the 1970s was Clarence L. "Kelly" Johnson. A brilliant and creative designer, Johnson first gained fame as the designer of the Lockheed P-38 Lightning twin-engined fighter, as

Fig. 1.17 Jet tanker/transport, fighter, and stealth bomber. KC-10 military tanker version of McDonnell–Douglas DC–10 commercial transport prepares to refuel a B-2 Stealth Bomber as an F-16A, a second-generation supersonic fighter, flies in formation (U.S. Air Force photograph).

shown in Fig. 1.4. He was chosen to head the Lockheed Special Projects Division, also known as the Skunk Works. The Skunk Works designed and built a prototype of the Lockheed P-80, America's first operational jet fighter, in only 143 days. Johnson went on to lead the Skunk Works in designing such diverse and innovative aircraft as the F-104 Starfighter Mach 2 fighter, C-130 Hercules turboprop transport, U-2 high-altitude spy plane, XFV-1 vertical-takeoff fighter, and Mach 3+ YF-12A (Fig. 1.1) fighter and SR-71 Blackbird spy plane. Johnson was twice awarded the Collier Trophy recognizing outstanding achievement in aeronautics. The first trophy was for the F-104 in 1958 and the second for the YF-12A in 1963. U.S. President Lyndon Johnson presented him the Medal of Freedom (highest U.S. civilian honor) in 1964.

1.6 Conclusion

Design is an essential part of aeronautics, and aeronautics permeates the aircraft design process. How natural then to learn the two disciplines together. As your appreciation grows for the complexities of the problems that must be tackled to design a successful aircraft, hopefully so will your facility with the fundamental methods used to solve those problems. The result should be a heightened understanding of both design and aeronautics and a solid preparation for further study.

1.7 Chapter Summary

At the end of the main body of material of each chapter, this book provides a brief summary of the major concepts and analysis methods covered in the chapter. Use these summaries to review the material and make sure you understand it. If you find you do not understand a concept, return to the appropriate section and reread it, taking special care to follow each step of any examples provided. You can also find additional explanations of difficult concepts in the "More Details" section that immediately follows this summary.

1.7.1 Aircraft Design

Aircraft are designed and built in response to a *need* from the customer (airlines, pilots, government agencies, military services).
Needs come from the customer as a result of the following: 1) changes in doctrine, tactics, training, demographics, politics, regulations [e.g., new Federal Aviation Administration (FAA) regulation allowing extended twin-engine over-water operations]; 2) shortcomings in existing systems (e.g., three- and four-engine airliners too big and too expensive to operate on many overwater routes); 3) need to introduce new operational capabilities [global positioning system (GPS)-guided weapons]; and 4) need to introduce new technologies (stealth, supercruise).
Needs are translated into *requirements* by the *user*:
  1) User ALWAYS defines the requirements (but designer can help).
  2) Design missions are used to help define requirements (Fig 1.7, p. 11).
  3) Requirements should be quantifiable.
  4) Compromise needed for conflicting requirements (fuel/range vs weapons/ payload).
  5) Trade studies are accomplished to help determine the "best" solution to the requirements (Joint Strike Fighter vs Block 60 F-16).

All designs will have "critical requirement(s)" or "design driver(s)." A design driver is the most restrictive and most important requirement, and so it will most heavily influence the design. Examples: 30-mm cannon on A-10, low-observable flying wing design of B-2A, high-altitude cruise/low-altitude penetration role of B-1B.

In recent years, cost has become one of the most important requirements. Aircraft prices have risen dramatically as a result of improved technologies, higher materials and labor cost, and smaller numbers of aircraft being produced.

Design process—Five-step process.

1) Define the problem, define the product's purpose, translate this purpose into performance parameters, and establish measures of merit for product cost, manufacturability, and performance.

2) Collect data on limits and requirements, existing designs, available technologies, and analysis methods.

3) Create or synthesize one or more design concepts, describe them in words, parameters diagrams, and drawings.

4) Select the types of analysis that must be performed to evaluate the design(s), perform analyses and record and communicate the results.

5) Make decisions: Will it work? Which idea is best? What must be changed? Are the measures of merit reasonable? If changes are needed, return to step 1 or 2 as appropriate and repeat the cycle.

Design cycle—Three-step, iterative cycle that provides an orderly method to integrate ideas and technology into aircraft and a structured method to solve ill-defined problems.

1) Synthesis
   a) Brainstorming
   b) Multiple conceptual designs to meet requirements
2) Analysis
   a) Testing concepts against the requirements
   b) Collecting information about designs
   c) Determining pros and cons of each design
   d) Flight testing is a critical part
3) Decision making
   a) *Should* be based on customer's assessment of the design
   b) Actual factors include cost, schedule, performance, other alternatives, contractor's past experience

The design process continues even after a design is operational to 1) incorporate new technology (avionics upgrades, new weapons), improve performance (new engines on KC-135 and B-52), and 3) reduce maintenance costs (B-2A radar absorbent material).

Design phases—User should be involved throughout all phases.

1) Conceptual—"ideas," multiple designs, mostly paper and computer studies
2) Preliminary—"models," wind-tunnel testing, computer optimization
3) Detail—"prototype," flight testing, fly-offs, manufacturing process defined

1.7.2 Aircraft Design History

Major players and their contributions to aeronautics are as follows:
1) George Cayley, the true inventor of the airplane (1773–1857) (England)
   a) First to propose separate mechanisms for lift and thrust (until then everyone used da Vinci's ornithopters drawings, which had the same mechanism for lift and thrust)
   b) Built gliders and worked out basic requirements for stability and control
   c) Understood steam engine did not provide enough thrust for its heavy weight and so researched alternative propulsion systems
2) Otto Lilienthal, the glider man (1848–1896) (Germany)
   a) First successful man-carrying glider in 1891; died from injuries after glider crash
   b) Achieved control of his gliders by shifting his weight
3) Samuel Langley, Secretary of the Smithsonian Institution (1834–1906) (United States)
   a) $50,000 from U.S. government to develop an airplane (aerodrome)
   b) With mechanics Manley and Balzer, developed advanced gasoline engine of the time
   c) Aircraft design neglected the need for control; second failure on 9 December 1903
   d) Wright brothers closest competition
4) Wright brothers, Wilbur and Orville (1867–1912 and 1871–1948) (United States)
   a) Built wind tunnel and tested airfoils; built and flew gliders
   b) Learned from Lilienthal's mistakes; used movable control surfaces to control aircraft
   c) All aircraft designs intentionally unstable and required constant control from the pilot
   d) Built their own gasoline engine for their aircraft
   e) First successful human-carrying, heavier-than-air sustained flight (17 December 1903)
5) Clarence L. "Kelly" Johnson, Lockheed Aviation Legend (1910–1990) (United States)
   a) Leading role in design of 40 famous aircraft: P-38, F-80, C-130, F-104, U-2, SR-71
   b) Set up the Skunk Works in 1943 and built the prototype P-80 jet aircraft in 143 days
   c) Won Collier Trophy twice: F-104 in 1958 and YF-12A Mach 3 aircraft in 1963
   d) President Johnson presented him the Medal of Freedom (highest civilian honor) in 1964

1.8 More Details

Each chapter of this textbook will include a section at the end titled "More Details." These sections will provide additional details to help understand difficult concepts, describe additional analysis methods, and give examples of how the methods presented in this textbook can be used to solve actual conceptual aircraft

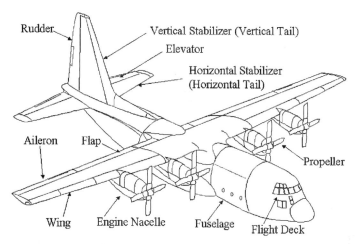

Fig. 1.18   C-130 components.

design problems. The additional details for this chapter are the nomenclature used to describe aircraft components. The additional analysis method for this chapter is called the *House of Quality*.

1.8.1   Aircraft Geometry and Nomenclature

The science of aeronautics has a vocabulary all its own. Many of these terms describe the geometry and components of aircraft. Figure 1.18 illustrates a Lockheed C-130 turboprop-powered tactical transport, with major components labeled.

Note that the tail of the aircraft has fixed stabilizing surfaces, the vertical and horizontal stabilizers. Hinged to these fixed surfaces are moveable control surfaces, the rudder and elevators. Likewise, the wing has moveable ailerons hinged along its rear or trailing edge, which move differentially (one moves trailing edge up when the other moves trailing edge down), and high-lift devices called flaps, which move together trailing edge down to increase the wing's lift.

The C-130's body or fuselage hangs from the underside of its wing. This makes it a high-wing aircraft. The pilot, copilot, and flight engineer sit in the flight deck area in the top front of the fuselage. Its four-engine nacelles house turboprop engines driving four-bladed propellers.

Figure 1.19 illustrates a Northrop F-5E Tiger II lightweight fighter. Although these aircraft are becoming obsolete, their configuration is typical of modern high-performance fighter aircraft. Instead of a fixed stabilizer and moveable elevator, the F-5E's designers combined these surfaces in an all-moving stabilator. Aircraft designers coined the word "stabilator" by combining the words "stabilizer" and "elevator."

Two afterburning turbojet engines buried in the rear fuselage power the F-5E. Air flows to the engines from inlets ahead of the wing through long ducts. After being compressed, mixed with fuel, burned, and expanded through a turbine, which powers the engine compressor, the hot gases shoot out the rear of the engine through carefully shaped nozzles.

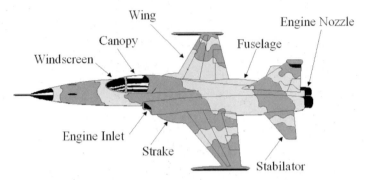

Fig. 1.19   Northrop F-5E Tiger II components.

A small high-lift device known as a strake sticks forward from the leading edge of the wing alongside the fuselage. This adds to lift by generating a vortex, something like a small horizontal tornado, which flows back over the top of the wing and reduces the pressure there. The pilot sits in a cockpit behind a clear plastic windscreen and under a clear canopy. The canopy is hinged at the rear and opens to allow the pilot to get in and out.

Figure 1.20 shows an unusual research aircraft, the X-31. It has a canard on its nose in place of the elevator of the C-130 and the stabilator of the F-5E. The moveable control surfaces hinged to the trailing edge of the wing serve both as elevators and as ailerons, so they are called "elevons."

The preceding paragraphs have provided what is by no means an exhaustive list of aeronautical geometry terms. As you study aeronautics and aircraft design, expect to encounter more unfamiliar terms. The engineers and scientists who first invented the science of aeronautics also invented words to describe what they were doing. In some cases, they borrowed terms from shipbuilding. However, in many more cases they invented new words to describe objects and ideas that no one had ever thought about before.

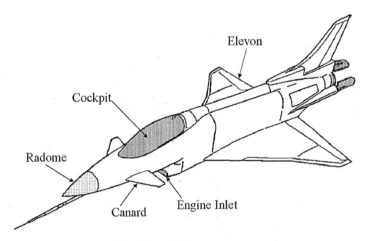

Fig. 1.20   X-31 components.

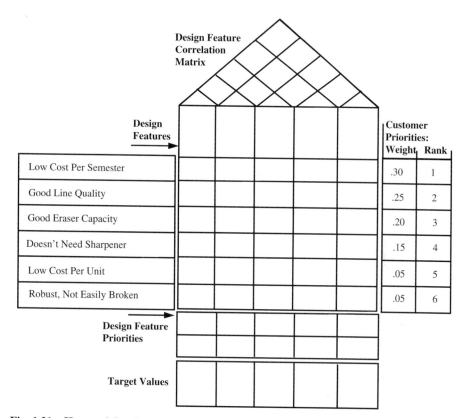

Fig. 1.21   House of Quality chart for a pencil with engineering student's needs and priorities.

1.8.2   House of Quality

Figure 1.21 shows a chart called the House of Quality, which is often used[5] as a worksheet for establishing the relative importance of the various design criteria or measures of merit displayed in a selection matrix. To understand how the House of Quality works, consider again the pencil design problem discussed in Sec. 1.3. Recall that the weights given to the various measures of merit in the selection matrix were based on customer needs and priorities. The House of Quality is used in discussions with a customer to record those needs and the priority the customer gives to each one. Then the chart is used in discussions with design and technology experts to determine what features or technologies a design should have in order to best meet the customer's needs.

The House of Quality chart also has a feature that helps to focus your creative thinking for certain problems. While the measures of merit are listed down the left side of the chart, parameters describing various characteristics of your design are listed across the top. The central grid of the chart is a correlation matrix that lets you identify how strong an influence each feature of your design has on your

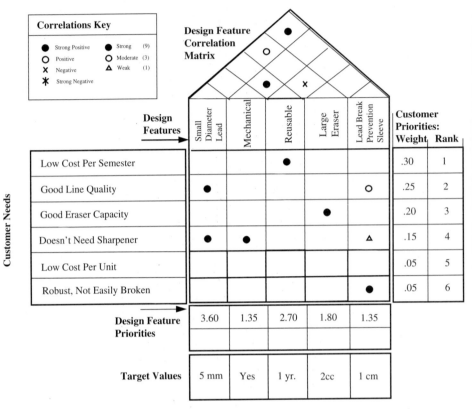

Fig. 1.22   Completed House of Quality for an engineering student's pencil.

design's ability to meet each measure of merit. The "roof" or "attic" area of the chart is another correlation matrix that allows you to identify which design features either enhance or counteract the effects of other features. There is space across the bottom of the chart to write in target values for parameters that describe the design features. Figure 1.22 shows the House of Quality chart for the pencil with the design features and correlation data for a college engineering student as the customer.

Symbols are used in the attic matrix to identify design features that enhance or interfere with the positive effects of other design features. No weights are assigned to those correlations. On the central matrix of Fig. 1.22, strong correlations are marked with a filled circle and given a value of nine. Moderate correlations are marked with an open circle and given a value of three, weak ones an open triangle and a value of one, and neutral or negative correlations a value of zero. These correlation values are multiplied by the weight given to each measure of merit and summed to yield the weight or importance of each design feature. The sums are written across the bottom of the chart, just above the design parameter target values. Figure 1.22 shows the design feature priority calculations on the House of Quality chart for the engineering student's pencil.

Example 1.1

Consider the House of Quality chart in Fig. 1.22. For the design feature "small diameter lead," strong correlations (multiplication factor 9) are indicated with the user needs "good line quality" and "doesn't need sharpener." The priorities given by the customer for these needs are 0.25 for "good line quality" and 0.15 for "doesn't need sharpener," and so the design feature priority is calculated as

$$\text{Design Feature Priority} = 0.25 \cdot 9 + 0.15 \cdot 9 = 3.6$$

Now consider the design feature "lead break prevention sleeve." It has a moderate correlation (multiplication factor 3) with "good line quality," a weak correlation (multiplication factor 1) with "doesn't need sharpener," and a strong correlation with "robust, not easily broken." The priorities given by the customer for these needs are 0.25 for "good line quality," 0.15 for "doesn't need sharpener," and 0.05 for "robust," and so the design feature priority is calculated as

$$\text{Design Feature Priority} = 0.25 \cdot 3 + 0.15 \cdot 1 + 0.05 \cdot 9 = 1.35$$

The calculations on Fig. 1.22 suggest that for an engineering student the most important pencil features are a small lead diameter and the capability of being refilled with lead and reused. The target value for lead diameter is 5 mm and for reusability is that it lasts one year in normal use. If a pencil concept does not come close to these values, the high weight given to those two features will probably eliminate that concept from consideration. Of course, a different customer would give different weights to the design requirements, and that would produce different weights for the design features. Consider how different the House of Quality would be if the customer was an advertising agency who needs pencils to give away to their clients, or to their clients' customers. See Ref. 5 for more details on the origin and use of the House of Quality.

The House of Quality chart is an extremely useful tool for organizing, prioritizing and tracking the inputs, efforts and decisions of diverse, multidisciplinary design teams. Because team members cannot be experts on every topic, the House of Quality charts serve to communicate what design parameters are important, and how close the evolving design's characteristics are to the target values. The House of Quality is not perfect, however. In particular, it lacks an explicit indication of the cost of each design priority and decision.

1.8.3 House of Quality in Aircraft Design

Now consider how the House of Quality chart might be used in an aircraft design problem. The chart is becoming a widely used tool for this process. Figure 1.23 is a House of Quality chart that might be made for a multirole jet fighter aircraft. The design features of Fig. 1.23 are not at the level of design decisions. Most of these design features are parameters whose values are obtained from engineering analysis and testing. They are frequently used as measures of merit for a given aircraft design, but they are an intermediate step between design decisions and customer needs.

The chapters that follow will describe the significance of these aeronautical measures of merit. In Chapter 2, the properties of air, the nature of the atmosphere, and the physics of fluids at rest are discussed. This description of the environment

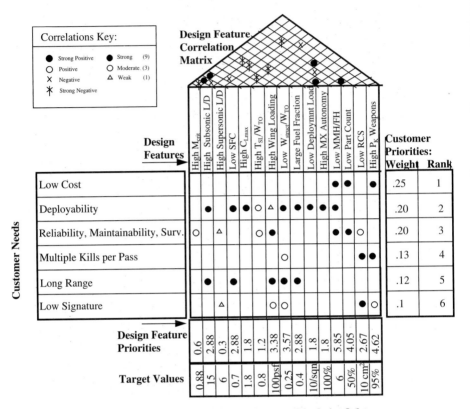

Fig. 1.23   House of Quality chart for a multirole jet fighter.

within which aircraft must operate provides the basis for analysis of aircraft performance capabilities, significant factors in the customer needs listed in Fig. 1.23. In Chapter 3 the physics of fluids in motion, especially fluids moving around and past solid bodies, are discussed. Pressure variations in fluid flowing around a body can create a net force on the body that varies with the body shape and orientation. This net aerodynamic force can be both beneficial and detrimental to the flight of an aircraft. At speeds below but near the speed of sound in air, the flow around a body is disrupted by shock waves. Aircraft that have high values of critical Mach number M_{crit} generally experience less flow disruption caused by shock waves, and the disruption is delayed to higher flight velocities.

For airplanes, flying vehicles that are heavier than air and that are supported in flight by pressure variations in fluid flowing over their wings, the component of the net aerodynamic force that counteracts the weight of the airplane and allows it to stay airborne is known as lift. Inevitably, there is also a component of the aerodynamic force that is parallel to and opposite to the direction of the aircraft velocity. This parallel component of the force is known as drag. It tends to slow the aircraft down. An airplane's engine(s) must produce thrust to overcome the drag and allow the airplane to sustain its speed. The ratio of the two components of the aerodynamic force, the lift and the drag, is a measure of how efficiently the airplane develops this force. In a sense, the L/D indicates the price one must pay (in thrust produced and fuel burned) to fly an aircraft of a particular weight. Chapter 4 discusses methods for predicting an airplane's L/D as well as the plane's M_{crit} and the effect of shock waves on its drag and L/D. Chapter 4 also describes the effect of an airplane wing's aspect ratio (a measure of how short and stubby or long and skinny the wing is) on the plane's maximum achievable lift-to-drag ratio $(L/D)_{max}$. Generally, high-aspect-ratio (long and skinny) wings give airplanes higher values of $(L/D)_{max}$. Unfortunately, high-aspect-ratio wings also weigh more, especially if they must be strong enough to allow the airplane to fly at very high speeds. One of the most important tasks of an airplane designer is selecting just the right wing aspect ratio so that $(L/D)_{max}$ is high enough but the portion of the airplane's takeoff weight devoted to structure W_{struct}/W_{TO} is low enough.

Chapter 5 discusses methods for predicting an airplane's performance once its aerodynamic and thrust characteristics are known. It also contains information on the various types of propulsion systems that an aircraft designer can choose. Just as L/D measures an airplane's aerodynamic efficiency, the engine's thrust-specific fuel consumption (TSFC, the ratio of fuel burn rate to thrust produced) measures the engine's propulsive efficiency. The net efficiency of the aircraft is influenced equally by L/D and by TSFC. Chapter 5 also makes it clear that for any aircraft configuration and atmospheric conditions there is a particular flight speed at which the airplane operates most efficiently. The airplane designer must shape the vehicle so that it operates most efficiently at the speeds and atmospheric conditions which the customer requires for the design mission. Chapter 5 also introduces methods for predicting C_{Lmax}, the aircraft's maximum lift coefficient. C_{Lmax} is a measure of the aircraft's ability to fly at slow speeds and therefore a measure of its ability to takeoff and land from short runways. A high value of C_{Lmax} indicates a wing that is able to generate a relatively large amount of lift for its size. This allows the aircraft to fly slower. Wing loading, the ratio of aircraft takeoff weight to wing area, indicates how much lift is required from each square foot of the wing. Airplanes with low wing loading generally are more maneuverable and have shorter takeoff and landing distances.

Chapter 6 deals with stability and control. No customer needs or design features in Fig. 1.23 deal with stability and control, yet good aircraft stability and handling qualities are essential. Standards for aircraft stability and control are generally set by Federal Aviation Regulations, Military Specifications, International Civil Aviation Organization Standards, etc. It is required that any aircraft built by an established manufacturer must comply with the appropriate standards. Chapter 6 discusses methods for predicting how large an aircraft's stabilizing surfaces and control surfaces must be to give it the necessary stability and also enough maneuverability to fly all maneuvers and missions required by the customer.

Chapter 7 discusses fundamental concepts of aircraft structures, and Chapter 8 deals with sizing, determining how large a particular aircraft configuration must be to fly the design mission. Generally, a larger aircraft can carry a larger payload (passengers, cargo, weapons, etc.) over a longer distance or range. Another way to increase an aircraft's range is to increase its fuel fraction, the ratio of fuel weight to total aircraft weight. This is difficult to do. Normally, either payload weight must be sacrificed or new structural designs, materials, or fabrication methods must save on structural weight.

The remaining design features in Fig. 1.23 deal with more advanced design considerations. The deployment load requirement is often expressed as the number of transport aircraft missions required to deploy and operate a squadron of aircraft at an austere (minimum local support facilities) operating base. MMH/FH is the maintenance man-hours per flying hour, a measure of the number of ground crewmen required to keep the airplane flying. RCS is the aircraft's radar cross section, expressed as the area of a good radar reflector, which would give the same strength of reflected radar energy. Aircraft with low radar cross sections can approach closer to search radar before they are detected. P_K is the probability of kill, the probability that each weapon will destroy its intended target. Precision guidance systems based on radar, optics, lasers, etc. have greatly increased the P_K of many bombs and missiles in recent years.

Knowledge of the relationship between airplane and propulsion system physical characteristics and their performance makes it possible to create a second-level or second-tier House of Quality chart. On this chart the design decisions about the aircraft's desired physical characteristics are filled in as design features, and the required performance parameters from Fig. 1.20 become the customer needs. Once this is done, the complete links between decisions and customer needs are charted. Figure 1.24 is an example of a second tier House of Quality for a multirole jet fighter.

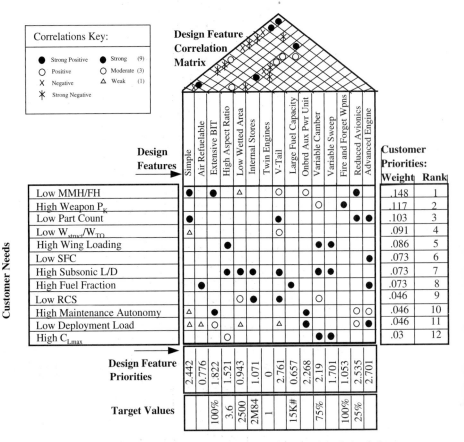

Fig. 1.24   Second-tier House of Quality chart for a multirole jet fighter.

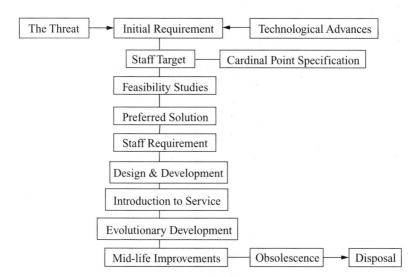

Fig. 1.25   Military aircraft procurement process.

1.8.4   Aircraft Procurement Process

The process by which design requirements are established and communicated varies depending on the customer, but all are quite similar. Consider as an example the procurement process for a military aircraft. Figure 1.25 is a block diagram depicting this process. As with any design process, procurement of a military aircraft begins with defining the problem or need. The military services are the customers for military aircraft. These organizations are in the business of creating, sustaining, and operating forces to meet the needs of their customer, the government of their nation. To do this, they need equipment and personnel that are capable of performing specific missions in environments, which often include hostile action by the forces of other nations. For this reason, the current and projected future capabilities and likely actions of expected hostile forces weigh heavily in decisions about what capabilities a new aircraft must have. The projected capabilities and actions of other nations and groups comprise what is called the threat. The current state of technology and new technology that is expected to be available in time to be used in a new design must also be considered, so that the design is not already obsolete before it is built.

Chapters 1–9 cover all of the basic concepts needed for a first experience with aircraft design. They also familiarize you with the four major specialties within the profession of aeronautical engineering: aerodynamics, flight mechanics, propulsion, and structures. Aerodynamics encompasses the topics in Chapters 2, 3, and 4. Flight mechanics includes performance, the topic of Chapter 5, as well as stability and control as discussed in Chapter 6. Propulsion is discussed briefly in Chapter 5, and Chapter 7 addresses some basic fundamentals of structural design and analysis. Chapter 8 describes the methods used to determine the size a particular aircraft configuration must be in order to meet a particular set of mission and performance requirements. Chapter 9 ties everything together in a summary and example of how all of the methods of the preceding chapters are used together in conceptual aircraft design. Chapter 10 presents examples of the design process for several different aircraft.

References

[1] Wilson, E. B. Jr., *An Introduction to Scientific Research*, McGraw–Hill, New York, 1952, pp. 21–28.

[2] Nicolai, L., "Designing the Engineer," *Aerospace America*, April 1992, pp. 72–78.

[3] Zubrin, R. M., and Clapp, M. B., "An Examination of the Feasibility of Winged SSTO Vehicles Utilizing Aerial Propellant Transfer," AIAA Paper 94-2923, June 1994.

[4] Koen, B. V., *Definition of the Engineering Method*, American Society for Engineering Education, Washington, DC, 1985.

[5] King, R., "Listening to the Voice of the Customer: Using the Quality Function Deployment System," *National Productivity Review*, Summer 1987, pp. 277–281.

Problems

Synthesis Problems

S-1.1 Motorcycle riders need a secure method of storing their helmets when their vehicle is parked, for example, in a university parking lot while the rider/student is in class. Brainstorm at least five possible ways to secure the helmet with the motorcycle while they are unattended. Remember, write down all ideas, no matter how outlandish they might seem at first.

S-1.2 Brainstorm at least five ways to eliminate the need for pencils, pens, and paper in college engineering classes.

S-1.3 A person making a journey by private aircraft generally has to use a taxi or public transportation to get from the landing airport to the final destination. Brainstorm at least five ways to combine in one vehicle the high-speed, long-distance capabilities of an airplane with the automobile's ability to deliver a passenger to a specific address in a crowded neighborhood.

S-1.4 The lift and propulsive forces of a bird are derived from a single mechanism—the bird's wings. Early attempts to fly were based on this same principle incorporated in machines called ornithopters. Who was the first to propose separate mechanisms for producing the lift and propulsive forces for flight?

S-1.5 What present-day aircraft combines the lift and propulsive forces in the same mechanism?

S-1.6 Why didn't the Wright brothers' aircraft achieve more worldwide use after its development?

S-1.7 Pick one of the following aircraft and describe what you think its design mission was like. Generally, aviation history books give good indications of why a famous aircraft was designed and what its performance capabilities were. Be careful. In some cases an aircraft became famous performing a mission that was completely different from the one it was designed for. Draw a diagram similar to Fig. 1.7 for your aircraft.

Wright Flyer	Lockheed P-38 Lightning
North American F-86 Sabre	Fokker D VII
Boeing B-17 Flying Fortress	North American F-100 Super Sabre
S.P.A.D. XIII	Douglas DC-3/C-47 Dakota
Convair F-106 Delta Dart	Sopwith Camel
Republic P-47 Thunderbolt	Boeing B-52 Stratofortress
Spirit of St. Louis	North American P-51 Mustang
Lockheed C-5 Galaxy	Boeing P-26 Peashooter
Messerschmitt 262 Schwalbe	Lockheed SR-71
Graf Zeppelin	Lockheed P-80 Shooting Star
McDonnell–Douglas F-15 Eagle	Republic F-105 Thunderchief

S-1.8 S-1.8 Pick one of the following types of aircraft, and write design specifications for it similar to Table 1.2. In general, the specifications should be written so that the aircraft is competitive with airplanes of the same era that had similar missions.

First flight across the English Channel	World War II fighter plane
Jet airliner	World War I fighter plane
World War II bomber	Sailplane
World War I bomber	Intercontinental jet bomber
Supersonic fighter	Private airplane
First nonstop flight across the Atlantic	Air–sea rescue amphibian

S-1.9 S-1.9 The House of Quality chart in Fig. PS-1.9 shows customer needs and priorities for an action motion picture. Fill in the design features section of the chart with what you feel are the five most important design features of a good action movie. Also fill in the two correlation matrices with what you believe are the appropriate symbols.

Analysis Problems

A-1.1 On the House of Quality chart you filled out in Problem S-1.9, calculate the Design Feature Priorities. What do these priorities tell you about designing a good action movie?

A-1.2 Pick two familiar action movies (e.g., the original *Star Wars* and *Raiders of the Lost Ark*), and rate your perceptions of the weight or emphasis given in each movie to the design features you identified in Problem S-1.9. Compare with the design feature priorities you calculated in Problem A-1.1. Can you identify ways in which the movies you chose could be improved? Try this exercise with one good movie and one you consider poor. Can you identify differences in the design feature priorities?

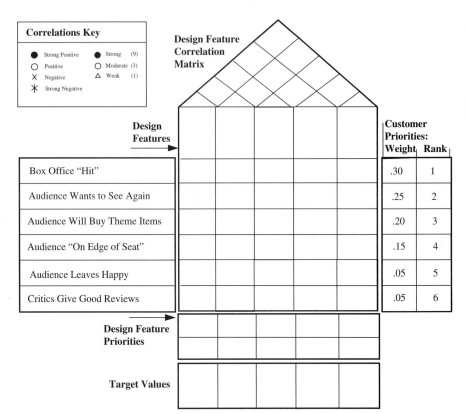

Fig. PS-1.9   House of Quality chart for action movies.

Note: This is another important use of House of Quality charts. Companies use these charts[5] to compare new designs with their previous products and with their competition. This information is useful to design engineers as they try to improve their designs. The comparisons can also be used by sales representatives to sell new products.

A-1.3 The House of Quality chart in Fig. PA-1.3 shows customer needs and priorities as well as design features and their correlations for a light general aviation aircraft. Calculate the design feature priorities.

A-1.4 Choose five aircraft with similar configurations but different sizes, the Boeing 737, 747, 757, 767, and 777 for instance, and make a graph plotting the maximum range of each aircraft vs its maximum takeoff gross weight. What trend do you notice?

A-1.5 For the aircraft you chose in Problem A-1.4, make a graph plotting each plane's maximum payload weight vs its maximum takeoff gross weight. Is the trend here similar to that in A-1.4?

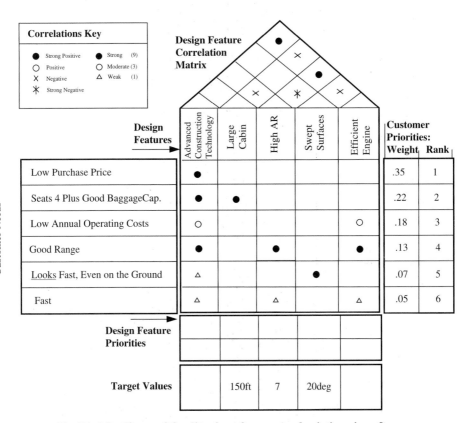

Fig. PA-1.3   House of Quality chart for a general aviation aircraft.

Note: Analysis of historical trends such as the relationship between range and size of a class of aircraft is an important part of the collecting data step of the aircraft design method. This information can be used to give the aircraft designer an initial estimate of how large a new aircraft design must be, based on the range and payload capabilities specified by the customer.

A-1.6   An airline requires a new airliner that can carry 300 passengers and has a range of 4000 n miles. Based on your analyses in Problems A-1.4 and A-1.5, estimate the maximum takeoff weight of the new aircraft.

2
Operating Environment

"It is our duty not to rest until we have attained to a perfect scientific conception of the problem of flight."

Otto Lilienthal

2.1 Design Motivation

2.1.1 Source of Aerodynamic Forces

Figure 2.1 depicts the forces acting on an aircraft in flight. The motion of the aircraft is determined by its mass and velocity and by the directions and magnitudes of these forces. Three of the four forces, lift, drag, and thrust, result from the interactions of the aircraft and its propulsion system with the air around it. Lift and drag are called *aerodynamic forces* because they result from the motion of the aircraft through the air. The aircraft designer shapes and optimizes the aircraft to control the production of these forces, maximizing the lift and thrust extracted from the air while minimizing the penalties paid in drag and fuel consumption. The characteristics of the air through which the aircraft is flying heavily influence the magnitudes of these forces. For this reason, the aircraft designer must understand the characteristics and interrelationships of the properties of air.

2.1.2 Standard for Comparison

The performance requirements listed in Table 1.2 specify altitudes and Mach numbers that must be achieved. To facilitate documentation of a particular aircraft's performance and to ensure fair comparisons between competing designs, a standard model for the Earth's atmosphere has been defined. This widely accepted model is called the standard atmosphere. By reference to it, any two engineers evaluating aircraft performance at a specified altitude will use identical atmospheric conditions. An understanding of how the standard atmosphere model is defined and how atmospheric conditions for an arbitrary altitude are determined is therefore an essential tool of the aircraft designer.

2.1.3 Basis for Altitude Measurement

Once a model for pressure variation in the atmosphere is established, a simple pressure-measuring device can be used to indicate to the pilot the altitude at which an aircraft is flying. Virtually all standard altimeters (altitude indicating instruments) are simply pressure gauges calibrated in units of altitude instead of pressure. Nearly all uses of indicated altitude in the operation and control of air traffic are based on this simple concept.

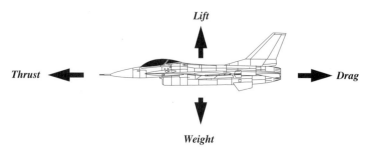

Fig. 2.1   Forces acting on an aircraft.

2.2   Characteristics of the Atmosphere

2.2.1   Language

A discussion of the atmosphere is greatly facilitated by first defining a few terms. Of particular interest to a study of the atmosphere are four fundamental quantities called *properties*, which describe the condition or state of the air. These four properties are density, pressure, temperature, and velocity.

Density is the amount of mass per unit volume. It is commonly designated by the symbol ρ and has units of kilograms per cubic meter (kg/m^3) or slugs per cubic foot (slug/ft^3.) A slug is the unit of mass in the English system of measurement. On Earth, a slug of matter weighs 32.2 lb because $W = mg$, where g is the acceleration of gravity. Slugs are units of mass, and pounds are units of force in the English system just as kilograms are units of mass and Newtons are units of force in the metric or SI system. If 1 m^3 of space contains 1 kg of air, then the average density of that air is 1 kg/m^3.

Air is composed of nitrogen, oxygen, carbon dioxide, and very small amounts of other gases. On a molecular level, the mass of air in a given space is the number of the various types of molecules in the volume multiplied by their molecular masses. These molecules are in constant motion, colliding with each other and bouncing off to collide again.

Temperature is a measure of the average kinetic energy of the air molecules as they move and collide with each other. If the average speed of the molecules is high, we sense this as a higher air temperature. Temperature is given the symbol T. The most commonly used units of temperature are degrees Celsius ($^\circ$C), degrees Fahrenheit ($^\circ$F), Kelvin (K), and degrees Rankine ($^\circ$R).

Pressure is the force exerted by a fluid (liquid or gas) per unit area. Air pressure in an inflated balloon pushes outward on the walls of the balloon and stretches them. If you attempt to squeeze the balloon, air pressure resists. Because the air exerts a force per unit area, you will have to push much harder to flatten a balloon with your whole hand than to just depress a portion of the wall of the balloon with your finger. The pressure (the force per unit area) multiplied by the area of the balloon surface you are depressing equals the force you must exert. Pressure is denoted by the symbol P and has units of Newtons per square meter or pounds per square foot.

On a molecular level, air pressure arises from countless collisions between individual air molecules and the molecules of the surface on which the pressure is exerted. The mass of the molecules, their speed when they collide with and bounce off the surface, and the rate at which these collisions occur determine the rate at which momentum is transferred from the air molecules to the surface. This rate of momentum transfer is what we call pressure.

Velocity is the net motion of the air. Whereas density, temperature, and pressure are scalars, velocity is a *vector*. It has both a magnitude and a direction. The velocity of an air mass is the average of all of the velocities of all of the individual molecules added vectorially. Wind is air in the atmosphere with a net velocity. The magnitude of the velocity is given the symbol V. Bold face will be used in this text to indicate a vector, so the symbol for velocity (direction and magnitude) is **V**.

Velocity is relative. That is to say, the velocity of any object, air, for instance, is measured relative to some frame of reference. For example, if you stand facing a 10-m/s breeze and hold up your hand, the air has the same velocity relative to your hand as if you were riding a bicycle at 10 m/s in still air (or for that matter if you rode a bicycle at 5-m/s directly into a 5-m/s breeze.) In each case, the frame of reference is fixed to your hand, and the air velocity is 10 m/s relative to it.

The four properties just mentioned can be used to describe the average conditions of any size volume of air. Because the values of these properties are not very uniform in the atmosphere, especially in the vicinity of a moving aircraft, it is useful to consider the limit of these values as the size of the volume being considered shrinks to zero. These limit values are called *point properties*. Point properties can have different values at every point in a volume. Henceforth in this text we will deal with density, temperature, pressure, and velocity as point properties.

The *equation of state* describes the relationship between the density, temperature, and pressure of a gas, or a mixture of gases such as air. Because temperature measures the average kinetic energy of the individual molecules and density measures the number and mass of molecules in a volume, the pressure (the rate of momentum transferred by the molecules) is proportional to the product of density and temperature. The proportionality constant is called the gas constant and is given the symbol R. The equation of state is written:

$$P = \rho R T \qquad (2.1)$$

Each gas or mixture of gases has a unique value of the gas constant. For air the value of R is given as

$$R = 287 \frac{J}{(kg)(K)} = 1716 \frac{(ft)(lb)}{(slug)(°R)}$$

Equation (2.1) does not always hold for all gases. Under conditions of very high density, the molecules interact with each other when they collide, and intermolecular forces are strong enough to cause a gas to fail to follow Eq. (2.1). Gases that obey Eq. (2.1) are called *perfect gases*, and the equation is often referred to as the *perfect gas law*. All gases discussed in this text will be assumed to be perfect gases.

Example 2.1

An aircraft's instruments measure an air temperature of 15°C and an air pressure of 100 kPa. What is the air density for these conditions?

Solution: Using the perfect gas law, Eq. (2.1):

$$\rho = \frac{P}{RT} = \frac{100,000 \text{ N/m}^2}{[287 \text{ J/(kgK)}](15°\text{C})} = \frac{100,000 \text{ N/m}^2}{[287 \text{ Nm/(kgK)}](288 \text{ K})} = 1.2 \text{ kg/m}^3$$

2.2.2 Hydrostatic Equation

Consider an infinitesimal mass of air in a static (all velocities are zero) atmosphere. As shown in Fig. 2.2, the height of the volume the air occupies is dh, and the horizontal area of the top and bottom faces of the volume are dA. The pressure on the lower surface is P, and to allow for variation in pressure in the air the pressure on the upper surface is $P + dP$. The mass of the air is the air density at that point multiplied by the volume. The weight of the air is the mass multiplied by the acceleration of gravity g. Because the mass of air is stationary, the net force on it must be zero. Summing the forces in the y direction:

$$\sum F_y = ma_y = 0$$

$$P \, dA - (P + dP) \, dA - \rho g (dh)(dA) = 0$$

$$dP = -\rho g \, dh \qquad (2.2)$$

Equation (2.2) is known as the *hydrostatic equation* because it describes a static fluid. The derivation for Eq. (2.2) is identical whether the fluid is air, another gas, or any liquid. The relationship in Eq. (2.2) between pressure and height in a fluid can be used as the basis for a number of useful tools for aeronautical engineers.

2.3   Pressure Measurement

2.3.1 Manometry

Consider a U-shaped tube filled with water as shown in Fig. 2.3. Each end of the tube is connected to a reservoir of air, with the two reservoirs at different pressures. If we break the fluid in the tube into two parts at point 1, it is clear that the pressure at point 1 must be the same as P_1, or the fluid in the lower portion

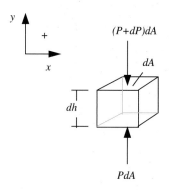

Fig. 2.2   Force balance on a static air mass.

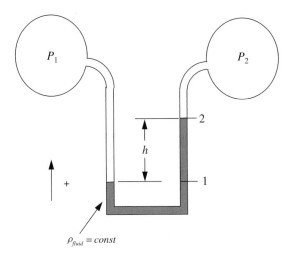

Fig. 2.3   Manometer.

of the tube would not be stationary. Next, apply Eq. (2.2) to the vertical column of fluid, which extends from point 1 to point 2. Unlike gases that obey Eq. (2.1), liquids such as the water in the tube do not change density significantly with small changes in pressure. We can therefore treat ρ as a constant and integrate Eq. (2.2) from point 1 to point 2.

$$dP = -\rho g\, dh$$

$$\int_1^2 dP = -\rho g \int_1^2 dh$$

$$P_2 - P_1 = -\rho g(h_2 - h_1) \tag{2.3}$$

The device depicted in Fig. 2.3 is called a *manometer*, and Eq. (2.3) is called the *manometry equation*. Manometers connected by tubing to openings called *pressure ports* on the surfaces of wind-tunnel models are used to measure pressure differences on the model. Until recently, it was very common in aeronautics laboratories to see banks of 50 or even 100 manometer tubes connected to pressure ports all over a model. Now, pressure transducers connected to automated data collection systems have replaced most manometers in many laboratories.

2.3.2   Barometers

A *barometer* is a special type of manometer in which one end (the higher end) of the tube is completely sealed off and filled with fluid. When the fluid in this end of the tube falls to an equilibrium level, the pressure P_2 is essentially zero. The height of the column of fluid then gives an absolute reading of the pressure P_1 rather than the difference between P_1 and P_2 given by a manometer. Barometers are usually filled with mercury and are often used to measure atmospheric pressure. These atmospheric pressures are often specified in inches of mercury, the height of the column of mercury in the barometer.

Example 2.2

A manometer filled with water is connected on one side to a pressure vessel that contains air at standard sea-level conditions. The other half of the manometer is open to ambient conditions in the room. If the height of the water in the manometer column that is connected to the sea-level conditions vessel is 12 cm lower than the water in the column that is open to the room, what is the air pressure in the room? (Note: The density of water is 1000 kg/m^3.)

Solution: Using Eq. (2.3), the pressure difference is given by

$$P_2 - P_1 = -\rho g(h_2 - h_1) = -(1000 \, \text{kg/m}^3)(9.8 \, \text{m/s}^2)(12 \, \text{cm})$$

$$= -(1000 \, \text{kg/m}^3)(9.8 \, \text{m/s}^2)(0.12 \, \text{m})$$

$$P_2 - P_1 = -1078(\text{kg m/s}^2)/\text{m}^2 = -1176 \, \text{N/m}^2$$

and so the pressure in the room is 1078 N/m^2 lower than standard sea-level pressure, so

$$P_{\text{room}} = P_{\text{SL}} - 1078(\text{kg m/s}^2)/\text{m}^2 = 101{,}325 \, \text{N/m}^2 - 1176 \, \text{N/m}^2$$

$$= 100{,}149 \, \text{N/m}^2 \cong 100 \, \text{kPa}$$

Example 2.3

A barometer filled with mercury is exposed to an atmospheric pressure of 2116.2 lb/ft^2. What is the height of the column of mercury in the barometer?

Solution: The sealed end of the barometer tube is a vacuum, and so the pressure there, let us call it P_1, is zero. The other end of the barometer is the zero reference height for the mercury $h_2 = 0$, and the air pressure at this end P_2 is 2116.2 lb/ft^2. Using Eq. (2.3), the height of the mercury column is given by

$$P_2 - P_1 = -\rho g(h_2 - h_1) = 2116.2 \, \text{lb/ft}^2 - 0 = -\rho g(0 - h_1)$$

$$h_1 = \frac{P_2}{\rho g} = \frac{2116.2 \, \text{lb/ft}^2}{(26.38 \, \text{slug/ft}^3)(32.174 \, \text{ft/s}^2)} = 2.493 \, \text{ft} = 29.92 \, \text{in.}$$

2.4 Standard Atmosphere

2.4.1 Definition

The tools are now in hand to define the standard atmosphere. The hydrostatic equation is used in defining the variation of pressure with altitude in the standard atmosphere. However, unlike the assumption for water density in the integration, which resulted in the manometry equation, air density in the atmosphere cannot be assumed to be constant. The atmosphere is assumed to be static so that velocity in the model is everywhere zero. The other three point properties, density, pressure, and temperature, must be determined everywhere in the atmosphere to define the model. With three unknowns, Eq. (2.2) is not sufficient. The equation of state can also be used, but that still leaves two equations and three unknowns. Therefore one of the three point properties must be specified everywhere in the model.

Observations of atmospheric conditions have been made for many decades, and averages of these conditions are used as a starting point for the standard atmosphere and to define a variation of temperature throughout the model. The starting point is established at *mean sea level* (MSL), an altitude representing the average elevation of the ocean's surface. Standard sea-level conditions are defined as follows:

$$T_0 \equiv 288.16 \text{ K} \equiv 518.69°\text{R}$$

$$P_0 \equiv 101,325 \text{ N/m}^2 \equiv 2116.2 \text{ lb/ft}^2$$

$$\rho_0 \equiv 1.225 \text{ kg/m}^3 \equiv 0.002377 \text{ slug/ft}^3$$

2.4.2 Temperature Model

An algebraic model is defined for the variation of temperature with altitude in the atmosphere. This model represents roughly an average of atmospheric temperatures measured at each altitude by countless weather balloon and sounding rocket flights. Figure 2.4 is a graph of this temperature model.

Notice that the temperature variation model is composed of segments in which the temperature is either constant or else increasing or decreasing linearly. The altitude bands where temperature is changing are called gradient layers. The gradient layer in the model that is closest to the Earth's surface represents the layer in the atmosphere known as the *troposphere*. The rate of decrease in temperature with altitude is called the temperature lapse rate. The altitude bands where the temperature is constant are called isothermal layers. The isothermal layer in the model just above the troposphere represents the lower part of the *stratosphere*. Nearly all manned aircraft flying occurs in the troposphere and the stratosphere.

2.4.3 Isothermal Case

The simplest of the two types of altitude bands or layers of the standard atmosphere is the isothermal layer. All three properties must be specified at the base or lowest altitude of the layer, which we will label h_1. The pressure here will be labeled P_1 and the density ρ_1. The temperature T will be constant throughout the layer. The hydrostatic equation and the perfect gas law are then used to define the variation of the density and pressure. To begin, divide Eq. (2.2) by Eq. (2.1) to yield:

$$\frac{dP}{P} = -\frac{g}{RT}\,dh \qquad (2.4)$$

Integrating Eq. (2.4) from h_1 to any arbitrary altitude h in the isothermal layer,

$$\ln \frac{P}{P_1} = -\frac{g}{RT}(h - h_1)$$

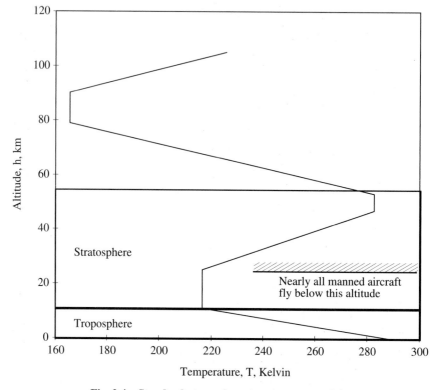

Fig. 2.4   Standard atmosphere temperature model.

or

$$\frac{P}{P_1} = \exp[-(g/RT)(h - h_1)] \tag{2.5}$$

Then, applying the perfect gas law to the numerator and denominator of the left side of Eq. (2.5),

$$\frac{P}{P_1} = \frac{\rho RT}{\rho_1 RT} = \frac{\rho}{\rho_1}$$

so

$$\frac{\rho}{\rho_1} = \exp[-(g/RT)(h - h_1)] \tag{2.6}$$

Equations (2.5) and (2.6) define the variation of pressure and density with altitude in an isothermal layer for known values of these quantities at the base of the layer.

Example 2.4

The tropopause, the boundary between the troposphere and the stratosphere, occurs at 36,152 ft in the standard atmosphere. Above this altitude, up to over

82,000 ft, the temperature stays constant at 389.99°R. If the standard atmospheric pressure is 453.9 lb/ft^2 at 37,000 ft and the density there is 0.000678 slug/ft^3, what are the standard atmospheric pressure and density at 40,000 ft?

Solution: The conditions at $h = 37,000$ ft are used as the starting point for the calculation. Using, Eq. (2.5),

$$P = P_1 \exp[-(g/RT)(h - h_1)]$$

$$= 453.9 \text{ lb/ft}^2 \exp[-(32.2 \text{ ft/s}^2/1716/389.99°\text{R})(40,000 \text{ ft} - 37,000 \text{ ft})]$$

$$= 393.1 \text{ lb/ft}^2$$

Because temperature and pressure are now known at this altitude, density is easily found using Eq. (2.1):

$$\rho = \frac{P}{RT} = \frac{393.1 \text{ lb/ft}^2}{(1716 \text{ ft lb/slug}°\text{R})(389.99°\text{R})} = 0.000587 \text{ slug/ft}^3$$

2.4.4 Gradient Layers

In altitude bands where temperature varies, the variation with altitude is linear and given by

$$T = T_1 + T_h(h - h_1) \tag{2.7}$$

where T_h is the temperature lapse rate defined as

$$T_h \equiv \frac{dT}{dh} = \frac{T - T_1}{h - h_1} \tag{2.8}$$

The subscript h is a shorthand notation for a derivative with respect to h. This notation will be used for derivatives throughout this text. Solving the definition of T_h for dh, we have

$$dh = \frac{1}{T_h} dT$$

Then, substituting this expression into Eq. (2.4) yields

$$\frac{dP}{P} = -\frac{g}{RT_h} \frac{dT}{T} \tag{2.9}$$

As with the isothermal layer, we integrate Eq. (2.7) from the base to an arbitrary altitude within the layer:

$$\ln \frac{P}{P_1} = -\frac{g}{T_h R} \ln \frac{T}{T_1}$$

or

$$\frac{P}{P_1} = \left(\frac{T}{T_1}\right)^{-[(g/T_h R)]} \tag{2.10}$$

and, once again using Eq. (2.1):

$$\frac{P}{P_1} = \frac{\rho RT}{\rho_1 RT_1} = \frac{\rho T}{\rho_1 T_1}$$

so

$$\frac{\rho T}{\rho_1 T_1} = \left(\frac{T}{T_1}\right)^{-(g/T_h R)}$$

or

$$\frac{\rho}{\rho_1} = \left(\frac{T}{T_1}\right)^{-[(g/T_h R)+1]} \tag{2.11}$$

Equations (2.7), (2.10), and (2.11) define the variation of properties in a gradient layer, once the properties at the base of the layer are known. The definition of the standard atmosphere model starts with the average sea-level conditions. The troposphere is built from these initial conditions using the gradient-layer equations and the value of T_h for the troposphere shown on Fig. 2.2. The properties at the top of the troposphere are used as the base properties for the lower part of the stratosphere. Equations (2.5) and (2.6) and a constant temperature are used to build the isothermal part of the stratosphere. The properties at the top of this isothermal layer are used at the base of the next layer (the upper part of the stratosphere), and so on until the entire model is built.

Example 2.5

In the troposphere portion of the standard atmosphere, temperature decreases lineaarly from 518.69°R at sea level to 389.99°R at the tropopause, an altitude of 36,152 ft. If the standard atmospheric pressure is 499.3 lb/ft^2 at 35,000 ft and the density there is 0.000738 slug/ft^3, what are the standard atmospheric pressure and density at the tropopause?

Solution: The temperature lapse rate in the troposphere T_h is

$$T_h = \frac{T_2 - T_1}{h_2 - h_1} = \frac{389.99°R - 518.69°R}{36,152\,\text{ft} - 0} = -0.00356\frac{°R}{\text{ft}}$$

The temperature at 35,000 ft is

$$T_2 = T_1 + T_h(h_2 - h_1) = 518.69°R - 0.00356\frac{°R}{\text{ft}}(35,000\,\text{ft} - 0) = 394.09°R$$

so, using Eq. (2.9),

$$\frac{P}{P_1} = \left(\frac{T}{T_1}\right)^{-[(g/T_h R)]}$$

$$P = P_1\left(\frac{T}{T_1}\right)^{-[(g/T_h R)]} = 499.3\,\text{lb/ft}^2\left(\frac{389.99°R}{394.09°R}\right)^{-\left[\frac{32.2\,\text{ft/s}^2}{(-0.00356°R/\text{ft})(1716\,\text{ft lb/slug}°R)}\right]}$$

$$= 472.5\,\text{lb/ft}^2$$

As in Example 2.4, density is found using Eq. (2.1):

$$\rho = \frac{P}{RT} = \frac{472.5 \text{ lb/ft}^2}{(1716 \text{ ft lb/slug}^\circ\text{R})(389.99^\circ\text{R})} = 0.000706 \text{ slug/ft}^3$$

Example 2.6

On Mars, the atmosphere is composed mainly of carbon dioxide. The value of the gas constant for the Martian atmosphere is 192 J/kg K, and the acceleration of gravity there is 3.72 m/s^2. At the average level of the Martian surface, the average temperature is 228 K, the pressure is 774 Pa, and the density is 0.01768 kg/m^3. At an altitude of 1 km above the surface, the average temperature is 225 K. What are the pressure and density at this altitude?

Solution: Equation (2.10) applies on Mars as well as on Earth. The temperature lapse rate between the surface and 1 km altitude is

$$T_h = \frac{T_2 - T_1}{h_2 - h_1} = \frac{225 \text{ K} - 228 \text{ K}}{1000 \text{ m} - 0} = -0.003 \frac{\text{K}}{\text{m}}$$

$$P = P_1 \left(\frac{T}{T_1}\right)^{-[(g/T_h R)]} = 774 \text{ N/m}^2 \left(\frac{225}{228}\right)^{-\left[\frac{3.72 \text{ m/s}^2}{(-0.003 \text{ K/m})(192 \text{ J/kg K})}\right]} = 710.5 \text{ N/m}^2$$

As in Example 2.5, density is found using Eq. (2.1):

$$\rho = \frac{P}{RT} = \frac{710.5 \text{ N/m}^2}{(192 \text{ J/kg K})(225 \text{ K})} = 0.01645 \text{ kg/m}^3$$

2.4.5 Standard Atmosphere Tables

For convenience in using the standard atmosphere, the values of the various properties in the model have been tabulated.[1] Examples of standard atmosphere tables in both English and SI units are included in Appendix B at the end of this text.

2.5 Common Uses of the Standard Atmosphere

2.5.1 Density, Pressure, and Temperature Altitudes

The properties of the standard atmosphere are used frequently as reference conditions for aircraft performance predictions. It is common to refer to these conditions in terms of the altitude within the standard atmosphere model at which those conditions occur. For example, aircraft takeoff performance depends in part on the air density. The density for a given takeoff calculation is often given as a *density altitude*. Density altitude, which is given the symbol h_ρ, is that altitude in the standard atmosphere that has the corresponding density. If the density is, for instance, $0.00199 \text{ slug/ft}^3$, then the density altitude would be given as 6000 ft. Likewise, *pressure altitude* h_P is that altitude in the standard atmosphere with the corresponding pressure, and *temperature altitude* h_T is the standard atmosphere altitude with the appropriate temperature. If the pressure altitude is given as 4200 m, then the atmospheric pressure is $60,072 \text{ N/m}^2$. A temperature altitude of 10,000 ft specifies a temperature of 483.1°R or 23.1°F.

Example 2.7

Instruments on an aircraft measure an air temperature of 80°F and a pressure altitude of 7000 ft. What is the density altitude?

Solution: 80°F = 540°R. From the standard atmosphere table in Appendix B, at a pressure altitude of 7000 ft, the air pressure is 1633 lb/ft². Then, using Eq. (2.1)

$$\rho = \frac{P}{RT} = \frac{1633 \text{ lb/ft}^2}{1716 \text{ (ft lb)/(slug}^\circ\text{R)} \, 542^\circ\text{R}} = 0.001756 \text{ slug/ft}^3$$

and from the standard atmosphere table the density altitude h_ρ = 10,000 ft.

2.5.2 Altimetry

A standard aircraft altimeter is simply a pressure gauge, connected to an orifice exposed to the atmospheric pressure around the aircraft. The gauge is calibrated in units of altitude instead of pressure, however. This is possible because the standard atmosphere provides a model for the variation of pressure with altitude. Of course, the actual variation of pressure with altitude on a given day is seldom exactly as prescribed in the standard atmosphere model. As a means of partially correcting for this, most altimeters have a knob on them that allows the pilot to adjust the reference pressure on which their altitude indications are based. When the reference pressure is adjusted to the standard sea-level pressure of 29.92 in. of mercury, then the altitude indicated by the altimeter is pressure altitude. Weather services use radios to provide information to pilots on the current reference pressure which they should use to get accurate altitude indications in the area in which they are operating. Figure 2.5 shows a schematic of an aircraft altimeter.

2.6   Chapter Summary

2.6.1 Vocabulary

The condition at any point in a fluid is described by four *point properties:*

1) Pressure
2) Temperature
3) Density
4) Velocity

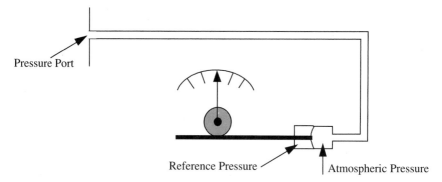

Fig. 2.5   Altimeter schematic.

2.6.2 Equation of State

The equation of state governs the relationship between three of the four properties:

$$P = \rho RT \tag{2.1}$$

The proportionality constant R is called the *gas constant*:

$$R = 287 \frac{J}{(kg)(K)} = 1716 \frac{(ft)(lb)}{(slug)(°R)} \quad \text{for air}$$

2.6.3 Hydrostatic Equation

This equation governs the variation of pressure with altitude in a still fluid:

$$dP = -\rho g \, dh$$

2.6.4 Manometry Equation

The manometry equation is the hydrostatic equation integrated with respect to altitude assuming density is constant:

$$P_2 - P_1 = -\rho g(h_2 - h_1)$$

1) It is valid for liquids, which are incompressible (constant density).
2) It governs manometers—liquid-filled tubes used to measure pressure.

2.6.5 Standard Atmosphere

1) Standard atmosphere is a model of how properties vary with altitude in the atmosphere:
    a) It allows all engineers to use the same atmospheric conditions when calculating aircraft performance.
    b) It reduces confusion and misrepresentation.
2) The hydrostatic equation is integrated with respect to altitude and assumes a particular variation of temperature with altitude:
    a) It assumes a linear decrease of temperature in troposphere.
    b) It maintains a constant temperature in stratosphere.
3) Instead of solving these equations when you need a value, use published tables of solutions to these equations at selected altitudes, like the one in Appendix B:
    a) Common tables list values every 1000 or 500 ft, 100 or 500 m.
    b) Tables list values for pressure, density, temperature, speed of sound, and viscosity.

2.6.6 Density, Pressure, and Temperature Altitudes and Altimetry

1) Density, pressure, and temperature altitudes are altitudes in the standard atmosphere at which a density, pressure, or temperature in question is found:
    a) For example, if the density is 0.002 slug/ft^3, then the density altitude is 5000 ft.
    b) It is an easy way to communicate to pilots how their aircraft performance will be influenced by nonstandard conditions because they know how their aircraft performs at a given altitude on a "normal" or standard day.

2) An altimeter is a pressure gauge calibrated in altitude instead of pressure according to the way pressure varies in the standard atmosphere.

a) It is adjustable to correct reference pressure for nonstandard conditions.

2.7   More Details—Molecular Collisions and Aerodynamics

2.7.1   Solids, Liquids, and Gases

Air does not act like our experience with solids and liquids would lead us to expect. Because we cannot see air, we are not aware of these peculiarities, but they are there, and understanding them is key to understanding aeronautics. To do so requires learning about the behavior of solids, liquids, and gases on a molecular level.

Figure 2.6. illustrates a conceptual drawing of the arrangement of molecules in a solid. The circles represent molecules, and the lines between them represent molecular bonds. The density of a solid is a measure of the number of molecules in a unit volume of the matrix. These bonds are primarily electrical in nature, the result of sharing or exchanging electrons. The bonds are very strong, but not rigid. When a force is applied to the solid, the bonds allow the molecules to move relative to each other and move back when the force is removed, like springs. If the force moves the molecules too far from their original orientation, the solid can permanently deform or break. The temperature of a solid is the measured by the energy of the molecules vibrating or oscillating about their average location in the matrix. As temperature increases, the energy of the molecules causes them to stretch their bonds and move further apart, causing the solid to increase in size and decrease in density. If this vibrational energy becomes great enough, the molecules are able to move more freely from their rigid location in the matrix, and the solid melts and becomes a liquid.

A conceptual drawing of a liquid would look more or less like the solid, except that the arrangement of molecules is more random and the bonds are not rigid. In fact they are just forces between molecules. The forces hold the liquid together, as evidenced by the surface tension of water. But the molecules have enough energy to move fairly freely relative to each other. The resistance that the forces exert against this motion is called *viscosity*. A very viscous liquid, like honey or syrup at low temperature, has fairly strong intermolecular forces, and its molecules barely have enough vibrational energy to move relative to each other. As the temperature increases, the molecules have more and more energy to move, and the liquid

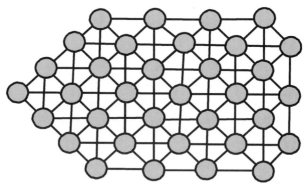

Fig. 2.6   Conceptual drawing of molecules in a solid.

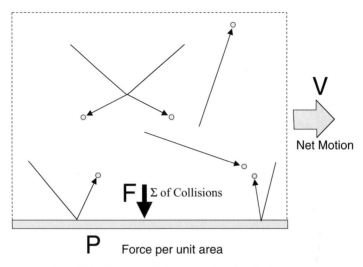

Fig. 2.7   Conceptual drawing of molecules in a gas.

becomes less viscous. Also, as the energy of the molecules increases, it causes them to move further apart, causing the liquid to increase in size and decrease in density. If temperature increases enough, the energy of liquid molecules increases to the point that they can break completely free from the forces exerted by the other molecules, and the liquid boils and becomes a gas.

As shown in Fig. 2.7, a gas is completely different from solids and liquids, because its molecules have broken completely free from each other, flying around randomly, bouncing off of each other in perfectly elastic collisions. When a molecule bounces off another gas molecule or a molecule on the surface of a solid or liquid, it transfers momentum to that molecule. The sum total of momentum transferred by millions and billions of these collisions is measured as pressure. The temperature of a gas is a measure of the average speed of the molecules between collisions. As the temperature of a gas increases, the molecules move faster and transfer more momentum when they collide, and so the pressure of the gas increases. These more energetic collisions also tend to move the molecules further apart, and so the gas expands and its density decreases. Viscosity of a gas is a measure of its ability to equalize differences in momentum among different groups of molecules. Because the only way momentum can be equalized (transferred) is through collisions and these collisions occur more often as the temperature of the gas and speed of its molecules increase, a gas becomes more viscous as it gets hotter. This is just the opposite of the relationship between temperature and viscosity for a liquid.

Velocity is the fourth property, after temperature, density, and pressure, that describes the condition of a substance. The velocity of a solid is just the velocity of the matrix. The velocity of a liquid is the average velocity of its molecules. The velocity of a gas is also the average of the velocity of its molecules, but this average velocity is usually small compared to the random velocities of the molecules. A gas that is not moving, for which the net or average motions of its molecules are zero, still has molecules that move at very high speeds between collisions. This is not true of a still liquid.

2.7.2 Equation of State

The equation of state for a gas, also known as the perfect gas law, is a relationship among the gas's density, pressure, and temperature. These properties all depend on the speed, mass, and number of molecules colliding with each other at a given point in the gas. If the molecules are heavier, or there are more of them, or they are moving faster between collisions, they will transfer more momentum as they collide with other molecules. These other molecules might be at the edge of the matrix of a solid, on the surface of a liquid, or they might just be other gas molecules. In any case, this transfer of momentum is measured as pressure. Heavier molecules or more molecules (more density) or higher molecular velocities (more temperature) will mean stronger collisions, more momentum transfer and more pressure. Therefore, either higher temperature or higher density results in more pressure, and, as it turns out, the relationship is linear. The relationship among pressure, temperature, and density is called the equation of state or perfect gas law:

$$P = \rho R T \tag{2.1}$$

where R is a proportionality constant, unique for each gas. For air the value of R is

$$R = 287 \frac{J}{(kg)(K)} = 1716 \frac{(ft)(lb)}{(slug)(^{\circ}R)}$$

This is not the form of the perfect gas law that you learned in chemistry. The one you learned was

$$Pv = nRT$$

where v is molar volume (volume occupied by one mole of molecules) and n is the number of moles of molecules. This form is easily converted to the one we use in aeronautics by dividing the number of moles by the molar volume; to yield the density and Eq. (2.1), the form of the perfect gas law we use.

Now, all of this only works if the collisions between the molecules remain perfectly elastic. If the molecules hit each other too hard, as in situations of high temperature, pressure, and density, the molecules might stick a little, or they might break apart and react chemically. Then, all bets are off. It is no longer a perfect gas, and the perfect gas law does not apply. This situation occurs in hypersonic flight, which we will discuss very briefly in Chapter 4.

2.7.3 Speed of Sound

Sound in air is also different. When sound moves through a solid or a liquid, a disturbance at one part of the substance is transmitted through the various bonds or intermolecular forces to other parts of the substance. This occurs quite rapidly. A more dense solid or liquid transmits sound faster because the bonds between molecules are stronger and shorter.

But remember, air is just a bunch of molecules flying free and bouncing off of each other. So sound is transmitted in air by collisions. The way this works is that a disturbance is transmitted to the molecules of air closest to the disturbance by collisions. But the disturbance causes these collisions to be more or less energetic than normal, transmitting more or less momentum (pressure) to the molecules. These molecules then go off and collide with other molecules, transmitting their high or low momentum to them. The momentum variations move through the gas as a series of atypical collisions, which we call pressure waves. When these pressure waves reach your ear, you hear them as sound.

So, if sound moves by collisions, the rate air carries sound must be proportional to the average speed at which molecules move from one collision to the next. But that speed is a measure of the air's temperature. In fact, air temperature measures the average kinetic energy of the air molecules as they travel from collision to collision. You learned in physics that the equation for kinetic energy of a given molecule is

$$\text{K.E.} = 1/2 \, mV^2$$

where m is the mass of that particular molecule and V is its current velocity. The kinetic energy of a mass of air is just the average of the kinetic energies of all the molecules, so that the temperature is proportional to the square of some average molecular speed, which is also the average speed at which disturbances are transmitted through the air from one collision to another. Therefore, if temperature is proportional to the square of the speed of the molecules and the speed of sound then the speed of sound in air is proportional to the square root of the air's absolute temperature. It turns out that the proportionality constant is the square root of the air's gas constant R multiplied by the square root of its ratio of specific heats $\gamma = c_p/c_v$, which is 1.4 for air.

The expression for the speed of sound in air is therefore

$$a = \sqrt{\gamma R T}$$

where a is the speed of sound, γ is the ratio of specific heats, R is the perfect gas constant, and T is the temperature. Contrary to what most people believe, the speed of sound does not depend on air pressure or density, but only on temperature. And higher temperatures, which usually produce lower densities, result in higher speeds of sound, again contrary to what most folks believe. Air is weird stuff, all because it is just a bunch of molecules banging into each other.

Reference

[1]Minzner, R. A., Champion, K. S. W., and Pond, H. L., "The ARDC Model Atmosphere," AFCRC-TR-50-267, 1959.

Problems

Synthesis Problems

S-2.1 Brainstorm at least five ways to indicate to a balloonist the height of the balloon above mean sea level (MSL).

S-2.2 A university aeronautics laboratory needs a visual indication of the pressures at 100 discrete points on the surface of a wind-tunnel model. The pressures can be displayed on the surface of the model or on any type of device or display outside the tunnel. The model is a hollow shell that is easily opened up to install sensors, wiring, tubing, etc. Brainstorm at least five design concepts for such a display system.

S-2.3 Suggest at least five measures of merit for a surface pressure display system as in Problem S-2.2.

S-2.4 An Earth resources and pollution monitoring agency requires an estimate of the total mass of the Earth's atmosphere. Suggest at least two methods for making this estimate. Can the hydrostatic equation and standard sea-level pressure be used in this process?

S-2.5 In your own words, why do aeronautical engineers and pilots need a standard atmosphere?

Analysis Problems

A-2.1 Calculate the mass of air (in kg) contained within a room 10 m long, 8 m wide, and 3 m high. Assume atmospheric pressure and temperature of 79,501 N/m^2 (Pa) and 20°C, respectively.

A-2.2 One of the design requirements for the multirole jet fighter listed in Table 1.2 requires a certain performance at 15,000 ft MSL. What atmospheric conditions (P, T, and ρ) would you use in calculations that determine if a proposed design can meet this requirement?

A-2.3 Flight-test course students are preparing for a flight-test sortie in a light aircraft. As part of the systems check, they note that the altimeter indicates a pressure altitude of 7000 ft. The student flight-test engineer reads an air temperature of 65°F. What is the density altitude for these conditions?

A-2.4 A NASA Learjet has a section of one wing instrumented with devices capable of measuring temperature and pressure at discrete points on the wing. If at one of these points, the instruments measure pressure = 10 psi and temperature = −20°F, what is the density of the air at that point?

A-2.5 An aircraft flying at a geometric altitude of 20,000 ft has instrument readings of $P = 900$ lb/ft^2 and $T = 460°$R.
   (a) Find h_P, h_T, and h_ρ to the nearest 1000 ft.
   (b) If the aircraft were flying in a standard atmosphere, what would be the relationship among h_P, h_T, and h_ρ?

A-2.6 (a). If you are standing at the edge of a pool at sea level, standard day conditions, what pressure would you feel?
   (b). What is the pressure a swimmer would feel at the bottom of a 10-ft pool? The density of water is 1.94 slugs/ft^3.
   (c). If the air pressure remained the same and the air temperature was 70°F, what is the density of the air near the surface of the water?

A-2.7   A fairing for an optically perfect window for an airborne telescope is being tested in a wind tunnel. A manometer is connected to two pressure ports, one on the inner side of the window and one on the outside. The manometer is filled with water. During testing at the maximum design airspeed for the window, the column of water in the tube connected to the outside pressure port is 12 in. higher than the column in the tube connected to the inside port.

(a) What is the pressure difference between the inner and outer surfaces of the window?

(b) If the window has a total area of 4 ft^2, how much total force is on it in this situation?

A-2.8   If the hydrostatic equation is integrated from the Earth's surface to the top of the atmosphere, the resulting equation states that the force on 1 m^3 of the Earth's surface is equal to the total weight of the air in a column, which is 1 m^2 and as high as the atmosphere. If sea-level pressure is 101,325 N/m^2 and the area of the Earth's surface is 5.09×10^{14} m^2, then what is the total weight and mass of the Earth's atmosphere?

3
Aerodynamics and Airfoils

"Isn't it astonishing that all these secrets have been preserved for so many years just so that we could discover them!!"

Orville Wright

3.1 Design Motivation

3.1.1 Physics of Aerodynamic Forces

Figure 3.1 shows a cross-section view of an aircraft wing. A wing cross section like this is called an *airfoil*. Lines called *streamlines* drawn above and below the airfoil indicate how the air flows around it. The shape of the airfoil and the pattern of airflow around it have profound effects on the lift and drag generated by the wing. Aircraft designers choose a particular airfoil shape for a wing in order to optimize its lift and drag characteristics to suite the requirements for a particular mission. It is essential that an aircraft designer understand how the changes that occur in air as it flows past a wing create lift and drag and how airfoil shape influences this process.

3.1.2 Basis for Airspeed Indication

The changes that occur in the properties of moving air as it encounters obstructions provide the basis for the airspeed indicating systems used on most aircraft. An understanding of how these systems work is essential to anyone who designs, builds, or operates aircraft.

3.2 Basic Aerodynamics

3.2.1 Language

A number of terms must be defined to facilitate a discussion of aerodynamics. The lines in Fig. 3.1 that indicate how the air flows are known as *streamlines*. Each streamline is drawn so that at every point along its length the local velocity vector is tangent to it. A tube composed of streamlines is called a *stream tube*. In a steady flow, each streamline will also be the path taken by some particle of air as it moves through the *flowfield* (a region of airflow). A *steady flow* is defined as one in which the flow properties (pressure, temperature, density, and velocity) at each point in the flowfield do not change with time. If, as in Fig. 3.2, a streamline runs into an obstruction, the airflow along the streamline comes to a stop at the obstruction. The point where the flow stops is called a *stagnation point*, and the streamline leading to the stagnation point is called a *stagnation streamline*.

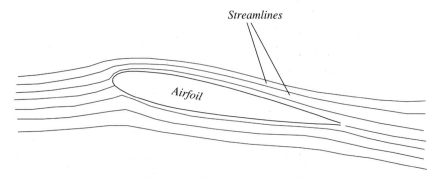

Fig. 3.1 Flowfield around an airfoil.

If, at each point along a streamline, there is no variation in the flow properties in a plane perpendicular to the flow direction, the flow is said to be *one dimensional*. Figure 3.3 illustrates a flow that is one dimensional at stations 1 and 2.

3.2.2 Continuity Equation

Figure 3.3 depicts a flow in a stream tube. Because the walls of the stream tube are composed of streamlines, the velocity vectors are everywhere tangent to the walls of the tube, so that no air can pass through the tube walls. The rate at which mass is flowing through a plane perpendicular to a steady one-dimensional flow is given by

$$\dot{m} = \rho A V \tag{3.1}$$

where $\dot{m}$ is the mass flow rate and A is the cross-sectional area of the stream tube. In nature, in the absence of nuclear reactions matter is neither created nor destroyed. Therefore, mass that flows into the tube must either accumulate there or else flow out of the tube again. The case where matter is accumulating in the tube like air filling a balloon is an unsteady, time-varying flow. If the flow is a steady flow, then the rate at which mass is flowing into the tube at station 1 must just equal the rate

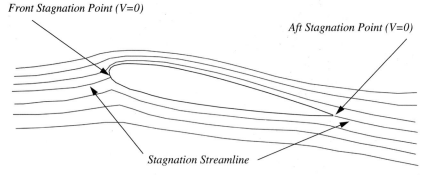

Fig. 3.2 Stagnation point and stagnation streamline.

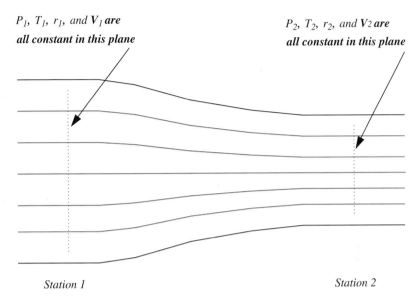

P_1, T_1, r_1, and V_1 *are*
all constant in this plane

P_2, T_2, r_2, and V_2 *are*
all constant in this plane

Station 1   *Station 2*

Fig. 3.3   Flow that is one dimensional at station 1 and station 2.

at which mass is flowing out of the system at station 2:

$$\rho_1 A_1 V_1 = \rho_2 A_2 V_2 \tag{3.2}$$

Equation (3.2) is known as the *continuity equation*. It is a statement of the law of *conservation of mass* for fluid flows. Applying this equation to the flowfield shown in Fig. 3.3 reveals a phenomenon that is very important to the production of aerodynamic forces. If we assume that the flow is *incompressible* (density is constant everywhere in the flowfield) or at least that the changes in air density are small, then Eq. (3.2) makes it obvious that the reduction in stream-tube area at station 2 will produce an increase in the velocity there relative to the velocity at station 1. A simple demonstration of this effect occurs when an obstruction such as a person's thumb is placed over the end of a garden hose that has water flowing out of it. The obstruction of the flow reduces the area of the stream tube and forces the fluid to accelerate in order to maintain the mass flow rate. Figure 3.4 shows a stream tube in a portion of the flowfield around an airfoil. The airfoil is an obstruction to the flow. It reduces the area of the stream tube and forces the flow to speed up as it flows around it. The changes that occur in the properties of the air as it flows past the airfoil produce aerodynamic forces.

Example 3.1

Air flows through a tube, which changes cross-sectional area similar to the one illustrated in Fig. 3.3. At a point in the tube (station 1) where the cross-sectional area is 1 m^2, the air density is 1.2 kg/m^3, and the flow velocity is 120 m/s. At another point in the tube (station 2) the cross sectional area is 0.5 m^2, and the air

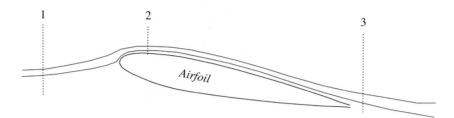

Fig. 3.4   Stream tube in air flowing past an airfoil.

density has decreased to 1.0 kg/m^3. What is the mass flow rate through the tube, and what is the flow velocity at station 2?

Solution: Using Eq. (3.1), the mass flow rate is

$$\dot{m} = \rho A V = (1.2 \text{ kg/m}^3)(1.0 \text{ m}^2)(120 \text{ m/s}) = 144 \text{ kg/s}$$

Then, solving Eq. (3.2) for V_2,

$$V_2 = \frac{\rho_1 A_1 V_1}{\rho_2 A_2} = \frac{\dot{m}}{\rho_2 A_2} = \frac{144 \text{ kg/s}}{(1.0 \text{ kg/m}^3)(0.5 \text{ m/s})} = 288 \text{ m/s}$$

3.2.3 Euler's Equation

To understand the changes that occur in the flow properties of a fluid as its velocity changes, consider an infinitesimally small particle of air moving along a streamline in a steady flow, as shown in Fig. 3.5. A number of forces can act on this particle. Gravity and magnetic fields can exert body forces on it. Viscous shear forces can retard the particle's motion. Pressure imbalances can also exert a net force. If we consider only flows of relatively lightweight gases that do not have large vertical components and no strong magnetic attractions, then the effects of body forces can be ignored. If we consider only *inviscid* (frictionless) flows, then viscous shear forces can also be ignored. For such a situation, the only significant forces remaining are caused by pressure imbalances along the streamline.

Applying Newton's second law to the motion of the particle along the streamline, the sum of the forces in the streamwise direction s is equal to the mass of the fluid

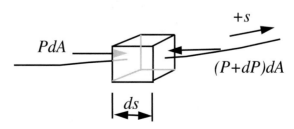

Fig. 3.5   Forces on a fluid element.

particle multiplied by the rate of change of its velocity:

$$\sum F = m\frac{dV}{dt} = P\,dA - (P + dP)\,dA$$

Now the volume of the fluid particle is the infinitesimal streamwise distance ds, multiplied by the area of the perpendicular face, dA, and so

$$m = \rho\,ds\,dA$$

Also, because the velocity vector is everywhere tangent to the streamline, the direction of ds is everywhere parallel to the local velocity, so that

$$\frac{dV}{dt} = \frac{dV}{ds}\frac{ds}{dt} = V\frac{dV}{ds}$$

which yields

$$P\,dA - (P + dP)\,dA = \rho\,dA\,ds\,V\frac{dV}{ds}$$

or

$$-dP = \rho V\,dV \tag{3.3}$$

Equation (3.3) is called *Euler's equation*, after the 18th-century Swiss mathematician who first derived it. The differential equation is a statement of Newton's second law for a weightless, inviscid fluid. It essentially states that for any increase in a fluid's velocity there must be a corresponding decrease in its pressure. Because it relates the rate of change of a fluid's momentum to the forces acting on it, Eq. (3.3) is also known as the *momentum equation*.

3.2.4 Bernoulli's Equation

For many purposes, the integral form of Eq. (3.3) will be more useful to us. For a compressible fluid, the integral of the right-hand side requires a relationship for density. However, many useful flow problems can be solved with reasonable accuracy by assuming density has a constant value throughout the flowfield. This is an extremely accurate assumption for liquids. It also gives reasonable results for air if the velocities throughout the flowfield remain below 100 m/s or 330 ft/s. With ρ assumed constant (incompressible flow) integrate Eq. (3.3) from some arbitrary point along the streamline, station 1, to another point, station 2, to yield

$$\int_1^2 dp = -\rho\int_1^2 V\,dV$$

$$P_2 - P_1 = -\left.\rho\frac{V^2}{2}\right|_1^2$$

or

$$P_1 + \frac{1}{2}\rho V_1^2 = P_2 + \frac{1}{2}\rho V_2^2 = P_0 \tag{3.4}$$

Equation (3.4) is known as *Bernoulli's equation* after another 18th-century Swiss mathematician, Daniel Bernoulli. The two terms on each side of Bernoulli's equation are given descriptive names. The pressure term is called the *static pressure*. The velocity-squared term is called the *dynamic pressure* and is often identified by the symbol q:

$$q \equiv \frac{1}{2}\rho V^2 \equiv \text{dynamic pressure} \tag{3.5}$$

The sum of static pressure and dynamic pressure is called *total pressure*. It is identified by the symbol P_0. Total pressure in a flow governed by Eq. (3.4) is invariant along a streamline:

$$P + \frac{1}{2}\rho V^2 = P + q = P_{\text{static}} + P_{\text{dynamic}} = P_0 = P_{\text{total}}$$

When using Eq. (3.4), remember that it is only valid for the steady flow along a streamline of an inviscid, incompressible fluid for which body forces are negligible. Together with the continuity equation, Bernoulli's equation provides the key to understanding such diverse concepts as how wings generate lift and how airspeed-indicating systems work.

3.3   Basic Aerodynamics Applications

3.3.1   Airspeed Indicators

One of the simpler applications of the aerodynamic equations developed to this point is the analysis and design of common airspeed indicating systems. These systems function by using the relationship between pressure and velocity described by Bernoulli's equation. Figure 3.6 shows a schematic.

The system consists of a *pitot tube*, one or more *static ports*, and a device for indicating differential pressure (a manometer in Fig. 3.6.). The pitot tube is named for its inventor, Henri Pitot, an 18th-century French scientist. It is placed in a flowfield with its opening perpendicular to the flow velocity such that if its opposite end were open air would flow directly through it. Because the opposite end of the pitot tube is blocked by the differential pressure indicator, the air in the tube cannot flow, and a stagnation point exists at the entrance to the tube. We assume that if we

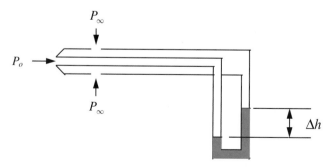

Fig. 3.6   Pitot–static tube and manometer.

look far enough upstream in the flowfield the flow becomes essentially undisturbed by the pitot tube and any shape to which it is attached. The undisturbed flow is called the *freestream*, and the properties of this undisturbed state are called the freestream conditions. Freestream conditions are usually identified by a subscript infinity, for example V_∞. Because total pressure is constant along a streamline, the total pressure for the stagnation streamline leading to the stagnation point at the entrance to the pitot tube is

$$P_0 = P_\infty + \frac{1}{2}\rho V_\infty^2 \qquad (3.6)$$

Velocity is zero at the stagnation point, so that Eqs. (3.4) and (3.6) require that the static pressure there is equal to the total pressure. The pitot tube therefore measures the total pressure of the flow and transmits it to one side of the manometer.

The static ports are oriented parallel to the flow velocity so that no stagnation point develops and the pressure they measure is as close to the freestream static pressure as possible. Aircraft designers use great care in placing static ports, and they often use multiple ports in order to get good approximations to the freestream static pressure. The static ports in Fig. 3.6 are placed on the sides of the pitot tube to form a *pitot–static tube*. The static pressure is transmitted through the connecting tube to the other side of the manometer. Solving Eq. (3.6) for V_∞ yields

$$V_\infty = \sqrt{\frac{2\left(P_0 - P_\infty\right)}{\rho}} \qquad (3.7)$$

Example 3.2

A manometer connected to a pitot-static tube as in Fig. 3.6 has a difference in the height of the two columns of water of 10 cm when the pitot–static tube is placed in a flow of air at standard sea-level conditions. What is the velocity of the airflow?

Solution: In a normally functioning pitot-static tube, the pressure measured at the static port will always be lower than or equal to the total pressure measured at the stagnation point, so that the column of water connected to the static port will be higher than the other. Using the manometry equation, with the subscript 0 identifying total pressure and the subscript ∞ identifying the freestream static pressure approximated at the static port

$$P_0 - P_\infty = -\rho g(h_0 - h_\infty) = -(1000 \text{ kg/m}^3)(9.8 \text{ m/s}^2)(-0.1 \text{ m}) = 980 \text{ N/m}^2$$

Then, substituting the required values into Eq. (3.7),

$$V_\infty = \sqrt{\frac{2(P_0 - P_\infty)}{\rho}} = \sqrt{\frac{2(980 \text{ N/m}^2)}{1.2 \text{ kg/m}^3}} = 40.4 \text{ m/s}$$

The manometer or another differential pressure device measures the difference between the total pressure and the static pressure of the freestream. According to Eq. (3.6), this difference is the dynamic pressure. If the air density is known, then the dynamic pressure is a direct indication of the freestream velocity. In aircraft, a

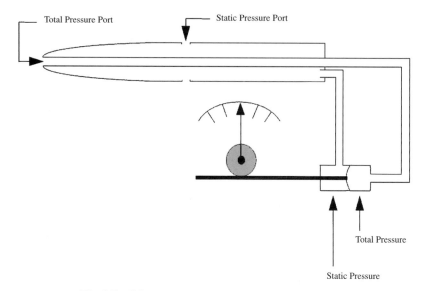

Fig. 3.7   Schematic of an airspeed indicating system.

differential pressure gauge is normally used instead of a manometer. In the differential pressure gauge, the static and total pressure lines are connected to opposite sides of a metal diaphragm. The pressure difference causes the diaphragm to deflect. A linkage connected to the diaphragm moves a needle on the gauge dial when the diaphragm moves. By calibrating the dial scale in terms of velocity instead of pressure, the differential pressure gauge becomes an *airspeed indicator*. Figure 3.7 shows a schematic of an airspeed indicator connected to a pitot–static tube.

3.3.2   Indicated to Calibrated to Equivalent to True (ICeT)

The airspeed that the needle on the airspeed indicator points at for a given set of flight conditions is called the *indicated airspeed*. Indicated airspeed is usually represented by the symbol V_i. Except in very special circumstances, an aircraft's indicated airspeed is *not* the speed at which it is moving through the air. Therefore, pilots make several corrections to indicated airspeed read from their instruments in order to determine their *true airspeed*, the speed at which they are actually moving through the air.

The first correction accounts for errors in the aircraft's airspeed indicating system. These errors mostly result from difficulties in obtaining consistently accurate measurements of freestream static pressure. Because the air speeds up as it flows around any shape, any device placed in the airflow to measure its static pressure disturbs the flow and alters the pressure it is trying to measure. You will learn in Chapter 5 that as an aircraft's speed changes the pattern of airflow around it usually changes. This further complicates the problem of accurately measuring freestream static pressure.

Engineers do their best to design airspeed indicating systems that give accurate measurements over a wide range of speeds and aircraft configurations. In addition

Table 3.1   Compressibility correction f factors

Pressure altitude, ft	Calibrated Airspeed, kn								
	100	125	150	175	200	225	250	275	300
5,000	0.999	0.999	0.999	0.998	0.998	0.997	0.997	0.996	0.995
10,000	0.999	0.998	0.997	0.996	0.995	0.994	0.992	0.991	0.989
15,000	0.998	0.997	0.995	0.994	0.992	0.990	0.987	0.985	0.982
20,000	0.997	0.995	0.993	0.990	0.987	0.984	0.981	0.977	0.973
25,000	0.995	0.993	0.990	0.986	0.982	0.978	0.973	0.968	0.963
30,000	0.993	0.990	0.986	0.981	0.975	0.970	0.963	0.957	0.950
35,000	0.991	0.986	0.981	0.974	0.967	0.959	0.951	0.943	0.934
40,000	0.988	0.982	0.974	0.966	0.957	0.947	0.937	0.926	0.916
45,000	0.984	0.976	0.966	0.956	0.944	0.932	0.920	0.907	0.895
50,000	0.979	0.969	0.957	0.944	0.930	0.915	0.901	0.886	0.871

to errors measuring static pressure, errors can also exist in total-pressure measurements for certain conditions, and in the airspeed indicator itself. Flight testing determines the errors for a range of speeds and configurations. The results are published as a table or graph of *position error* or *installation error*, ΔV_p, as a function of airspeed and configuration. Adding the position error to the indicated airspeed yields *calibrated airspeed V_c*:

$$V_c = V_i + \Delta V_p \tag{3.8}$$

Next, pilots must multiply calibrated airspeed by the f factor to yield *equivalent airspeed V_e*.

$$V_e = f V_c \tag{3.9}$$

The f factor is usually described as a compressibility correction. The details are a little more complicated. Section 3.7.1 explains this correction more completely, but for now let us just call it the f-factor correction. The f factor is usually read from a chart such as Table 3.1. Note that f factor varies with both altitude and calibrated airspeed.

The final correction to obtain true airspeed involves multiplying equivalent airspeed by the square root of the air density ratio:

$$V_\infty = V_e \sqrt{\frac{\rho_{SL}}{\rho_\infty}} \tag{3.10}$$

Because the density ratio ρ_∞ / ρ_{SL} is usually less than or equal to 1, V_∞ is usually $\geq V_e$. When flying at sea level on a standard day, $\rho_\infty / \rho_{SL} = 1$, and $V_\infty = V_e$. Recall that dynamic pressure is given by

$$q = \frac{1}{2}\rho V_\infty^2 \tag{3.5}$$

so that

$$q = \frac{1}{2}\rho_\infty \left(V_e \sqrt{\frac{\rho_{SL}}{\rho_\infty}} \right)^2 = \frac{1}{2}\rho_{SL} V_e^2 \qquad (3.11)$$

Equivalent airspeed can be alternately defined as that airspeed which would produce the same dynamic pressure at sea level as is measured for the given flight conditions. It will become apparent later on in this chapter and in Chapter 4 that, in the absence of compressibility effects, aircraft with identical configurations and orientation to the flow will produce the same aerodynamic forces if the dynamic pressures they are exposed to are the same. Because V_e is a direct measure of dynamic pressure, it is a very useful indicator of an aircraft's force generating capabilities. This fact is very useful to both engineers and pilots.

3.3.3 Ground Speed

It is worthwhile at this point to recapitulate the process for correcting an indicated airspeed. The steps are as follows:

$$V_c = V_i + \Delta V_p \qquad (3.8)$$

$$V_e = f V_c \qquad (3.9)$$

$$V_\infty = V_e \sqrt{\frac{\rho_{SL}}{\rho_\infty}} \qquad (3.10)$$

The result V_∞ is called *true airspeed*. The series of corrections from indicated to calibrated to equivalent to true airspeed is often called an *ICeT* ("ice tee") problem, with the lower case "e" being used as a reminder that equivalent airspeed is usually less than the other airspeeds. However, true airspeed is frequently not very useful until another correction is made. V_∞ is the magnitude of the aircraft's *true velocity* relative to the air mass. However, the air mass itself can be moving relative to the ground. The velocity of the air mass relative to the ground is the *wind velocity*. This must be added vectorially to the true velocity relative to the air mass in order to determine the aircraft's *ground speed* V_g. Ground speed is the magnitude of the aircraft's velocity relative to the Earth's surface. To help distinguish between true airspeed and ground speed, consider the following example.

Example 3.3

An aircraft flying at 300-kn true airspeed has a 50-kn tailwind. What is its ground speed?

Solution: To obtain groundspeed, use vector addition:

$$V_g = V_\infty + V_{wind} \qquad (3.12)$$

$$\overrightarrow{V_\infty = 300 \text{ kn}} + \overrightarrow{V_{wind} = 50 \text{ kn}} = \overrightarrow{V_g = 350 \text{ kn}}$$

This example illustrates the important concept that an aircraft's motion relative to the Earth can be significantly different in both direction and magnitude from its motion relative to the air mass. Whereas motion relative to the air mass is most important for generating sufficient aerodynamic forces to sustain flight, it is usually motion relative to the Earth that allows an aircraft to fulfill its mission. In situations where headwind velocities approach the same magnitude as an aircraft's true airspeed, its usefulness compared to surface transportation can be greatly diminished. The following example gives a complete demonstration of the ICeT (actually ICeTG) method.

Example 3.4

An aircraft flying at 20,000-ft pressure altitude has an indicated airspeed of 205 kn. If the outside air temperature is $-20°$F, position error is -5 kn, and there is a 40-kn headwind, what is the aircraft's ground speed?

Solution: Using Eq. (3.8),

$$V_c = V_i + \Delta V_p = 205 \text{ kn} - 5 \text{ kn} = 200 \text{ kn}$$

Then, from Table 3.1, for this altitude and calibrated airspeed, $f = 0.987$ and using Eq. (3.9)

$$V_e = f V_c = 0.987 \cdot 200 \text{ kn} = 197.4 \text{ kn}$$

Now from the standard atmosphere table in Appendix B the pressure at a pressure altitude of 20,000 ft is 973.3 lb/ft^2, and so solving Eq. (2.1) for the density

$$\rho = \frac{P}{RT} = \frac{973.3 \text{ lb/ft}^2}{[1716 \text{ ft lb/(slug°R)}](-20°\text{F})}$$

$$= \frac{973.3 \text{ lb/ft}^2}{[1716 \text{ ft lb/(slug°R)}](440°\text{R})} = 0.001289 \text{ slug/ft}^3$$

Note that it is not a standard day for the given conditions because the temperature is colder, and therefore the density is higher than in the standard atmosphere at 20,000 ft. Using $\rho = 0.001289$ slug/ft^3 in Eq. (3.10),

$$V_\infty = V_e \sqrt{\frac{\rho_{\text{SL}}}{\rho_\infty}} = 197.4 \text{ kn} \sqrt{\frac{0.002377 \text{ slug/ft}^3}{0.001289 \text{ slug/ft}^3}} = 268 \text{ kn}$$

Now for the "G" step in ICeTG, the correction for wind velocity to determine ground speed. The aircraft has a direct headwind of 40 kn, so that its ground speed is calculated from Eq. (3.12) as

$$V_g = V_\infty + V_{\text{wind}} = 268 \text{ kn} - 40 \text{ kn} = 228 \text{ kn}$$

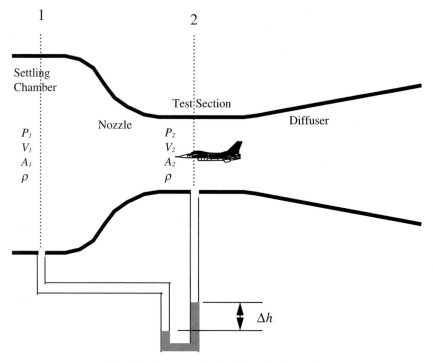

Fig. 3.8   **Low-speed wind-tunnel schematic.**

3.3.4 *Low-Speed Wind Tunnels*

Wind tunnels are devices used to study the aerodynamics of aircraft and other shapes in a laboratory environment. The object to be studied is mounted in the test section of the wind tunnel as shown in Fig. 3.8. A fan or pump at one end of the tunnel creates a flow of air. Air flows into the tunnel through an inlet or settling chamber, accelerates through the nozzle, flows through the test section, and decelerates in the diffuser. The velocity of the air changes as it flows into sections of the tunnel with different cross-sectional areas as required by the continuity equation. The pressure of the air changes with changing velocity in accordance with Bernoulli's equation. Of course, the velocities and pressures predicted by these equations will only be correct if the assumptions made in deriving them are satisfied. For wind tunnels that operate at maximum test-section velocities below 100 m/s or 330 ft/s (so the incompressible assumption is valid), these predictions are reasonably accurate.

The velocity of the air in a wind-tunnel's test section is usually measured either by a pitot tube placed in the test section or by two static ports, one in the settling chamber and one in the test section. The second method has the advantage that static ports do not intrude into the test section and therefore are less likely to interfere with the mounting of a model to be tested. Assuming incompressible flow, Eq. (3.3) can be solved for V_1 to yield

$$V_1 = V_2 \frac{A_2}{A_1} \tag{3.13}$$

Substituting Eq. (3.13) for V_1 in Eq. (3.4) and rearranging to collect like terms yields

$$P_1 - P_2 = \frac{1}{2}\rho\left(V_2^2 - V_2^2\frac{A_2^2}{A_1^2}\right)$$

which can be solved for V_2 to yield

$$V_2 = \sqrt{\frac{2(P_1 - P_2)}{\rho[1 - (A_2/A_1)^2]}} \tag{3.14}$$

Because the required measurement is a differential pressure, the two static ports can be connected to the two sides of a manometer to create a test-section velocity indicator.

Example 3.5

A low-speed wind tunnel similar to the one shown in Fig. 3.8 has a settling chamber cross-sectional area of 10 m^2 and a test-section cross-sectional area of 1 m^2. When the wind tunnel is run at its maximum velocity in standard sea-level conditions, a manometer connected between static ports in the walls of the settling chamber and the test section as shown in Fig. 3.8 has a difference in the heights of its fluid columns of 50 cm. What is the maximum test-section velocity and the mass flow rate through the test section for this tunnel and these conditions?

Solution: The manometry equation is used to determine the static-pressure difference between the settling chamber and the test section. Because the velocity in the test section must be higher than the velocity in the settling chamber, the pressure in the test section will be lower, and the height of the manometer fluid column, which is connected to the test section, will be higher:

$$P_1 - P_2 = -\rho g(h_1 - h_2) = -(1000 \text{ kg/m}^3)(9.8 \text{ m/s}^2)(-0.5 \text{ m}) = 4900 \text{ N/m}^2$$

Once the pressure difference is known and the air density is obtained from the standard atmosphere table, Eq. (3.14) can be used to determine the test-section velocity:

$$V_2 = \sqrt{\frac{2(P_1 - P_2)}{\rho[1 - (A_2/A_1)^2]}} = \sqrt{\frac{2(4900 \text{ N/m}^2)}{1.225 \text{ kg/m}^3[1 - (1 \text{ m}/10 \text{ m})^2]}} = 89.9 \text{ m/s}$$

Because the test-section velocity is below 100 m/s and the settling chamber velocity must be even slower, the assumption of incompressible flow is confirmed as valid, and the analysis can proceed. The density in the test section is therefore

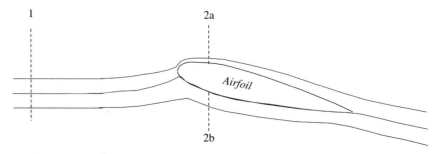

Fig. 3.9   Flow past an airfoil.

approximately the standard sea-level density, and Eq. (3.1) can be used to predict the mass flow rate:

$$\dot{m} = \rho_2 A_2 V_2 = (1.225 \text{ kg/m}^3)(1 \text{ m}^2)(89.9 \text{ m/s}) = 110.1 \text{ kg/s}$$

3.3.5 Airfoils

The continuity equation and Bernoulli's equation also can be used to explain how airfoils generate lift. Consider the steady, inviscid, incompressible flow of air past an airfoil as shown in Fig. 3.9. The entire flowfield is not shown in Fig. 3.9, only two stream tubes: one that passes above the airfoil and one passing below it. At station 1, which is far upstream of the airfoil, the flow is one dimensional. As the flow moves downstream, the orientation of the airfoil causes more of an obstruction to the flow above it than it does to the flow below it. This obstruction to the flow causes the stream tube above the airfoil to be constricted. The stream tube below the airfoil, on the other hand, keeps a nearly constant cross-sectional area all along its length and in fact expands slightly as it approaches the underside of the airfoil leading edge. The continuity equation requires that the flow in the upper stream tube must accelerate to get past the airfoil while the flow in the lower stream tube does not and can even decelerate.

Because the flow is one dimensional far upstream of the airfoil, the same flow conditions, and therefore the same total pressure, will exist on every streamline at station 1. We have made the appropriate assumptions so that Bernoulli's equation will apply along each streamline. Therefore, total pressure will be the same everywhere in the flowfield. Because, to satisfy continuity, the air will be moving faster at 2a than at 2b, the static pressure will be lower at 2a than at 2b. This pressure difference produces lift.

3.3.6 Pressure, Shear, Lift, and Drag

There are only two ways in which a fluid can impart forces to a body immersed in it. The first way, as just described, is by exerting pressure perpendicular to the body's surface. If the pressures on opposite sides of a body are not equal, then a net force such as lift is exerted on the body. A portion of the drag on a moving body likewise results from pressure imbalances, but a significant portion also results from *shear* stresses exerted parallel to the body surface caused by the *viscosity*

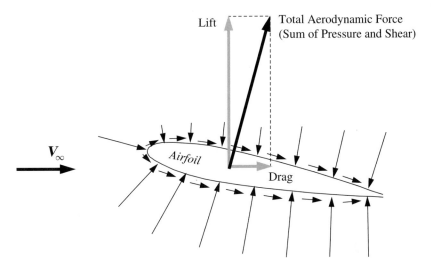

Fig. 3.10   Pressure, shear, and total aerodynamic force on an airfoil.

(resistance to flowing) of the fluid. In reality, lift and drag are components of a *total aerodynamic force* on the body, which is a sum of the net force caused by pressure imbalances and the net force caused by shear stresses. We have arbitrarily chosen to define lift as that component of the total aerodynamic force that is perpendicular to the freestream velocity direction and drag as that component that is parallel to the freestream. Figure 3.10 illustrates pressure, shear stresses, lift, drag, and the total aerodynamic force on an airfoil.

3.3.7   Pressure and Lift

A more detailed analysis of Fig. 3.9 gives further insight into the distribution of the pressure over the surface of the airfoil. If the continuity equation is applied at many points along the stream tubes in Fig. 3.9, a plot can be generated similar to Fig. 3.11, which shows the variation of velocity with chordwise distance in each tube. Note that in Fig. 3.11 zero velocity is assumed to exist at the front and rear stagnation points on the airfoil, even though the stream tubes do not have infinite area at those points. This is possible because the stagnation points are on the side walls of the stream tubes. Applying Bernoulli's equation to these velocity plots yields plots of surface-pressure distribution such as Fig. 3.12.

Note that Fig. 3.12 is for an airfoil with a chord length of 1 m. If the airfoil span is also 1 m, then because the pressure distributions are plotted vs chordwise location the area between the upper and lower surface-pressure curves is the net force caused by pressure perpendicular to the airfoil chord line, the *normal force*. Figure 3.13 shows the relationship between normal force and lift. The angle between the chord line and the freestream direction is called *angle of attack* and is given the symbol α.

Figure 3.14 shows the pressure distribution as arrows drawn perpendicular to the surface of the airfoil. Arrows drawn outward from the surface indicate pressures

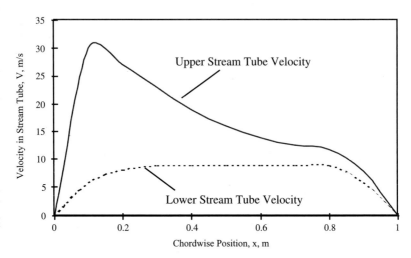

Fig. 3.11   Velocity distributions in stream tubes above and below airfoil.

lower than freestream static pressure, whereas arrows drawn in toward the surface indicate pressures higher than freestream static.

The net normal force on a portion of the airfoil surface is the pressure on that portion multiplied by its area. Because the airfoil surface is not, in general, parallel to the chord line, then if ds is the length of an infinitesimally small portion of the surface and dx is the length of the component of ds along the chord line (see Fig. 3.15), the contribution of its surface normal force to the total force normal to the chord line for an airfoil of unit span is

$$ dn = Pds\frac{dx}{ds} = Pdx \tag{3.15} $$

So the magnitude of the total normal force on the airfoil is

$$ n = \int_0^c (P_l - P_u)\, dx \tag{3.16} $$

which is exactly the same as the area between the two pressure lines on Fig. 3.12. As shown in Fig. 3.13, the lift on the airfoil is the component of normal force perpendicular to the freestream velocity vector (plus a negligible component of the chordwise force on the airfoil, which will be ignored):

$$ l = n \cos \alpha \tag{3.17} $$

Figure 3.14 shows an interesting situation that is commonly achieved by many airfoils. The very low pressures on the rounded leading edge of the airfoil produce a net force in the chordwise direction which is positive *forward*. This effect is known as *leading-edge suction* or *leading-edge thrust*. On airfoils that are fairly thick and that have relatively large leading-edge radii, leading-edge suction frequently has a significant component opposite the drag direction for a range of useful angles of attack. This reduces the net drag on these airfoils, making it a very desirable

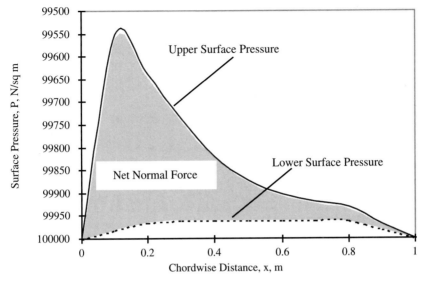

Fig. 3.12 **Typical airfoil surface-pressure distribution.**

feature. One of the advantages of the relatively thick airfoil used by the Fokker DVII in World War I over the thinner airfoils on fighters of the Allies was greater leading-edge suction and therefore less drag.

3.4 Viscous Flow

Viscosity is the tendency for a fluid to resist having velocity discontinuities in it. Viscosity in a liquid results from strong intermolecular forces that resist the

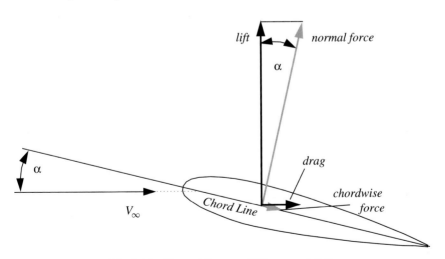

Fig. 3.13 **Normal force and lift on an airfoil.**

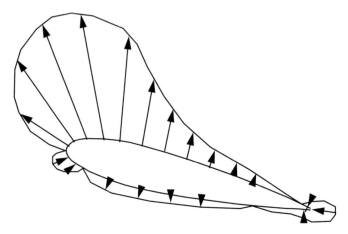

Fig. 3.14   Surface pressures on an airfoil.

motion of molecules relative to each other. The intermolecular forces between faster-moving molecules and slower ones cause velocity differences to be quickly equalized in a viscous liquid. As a liquid heats up, the individual molecules have more energy relative to the intermolecular forces, and so the viscosity of the liquid decreases. In a gas, on the other hand, viscosity results from the diffusion of momentum. Because a gas is composed of free-moving molecules with relatively weak intermolecular forces, the excess velocity of a faster-moving portion of a flowing gas is spread to the slower portions by collisions between faster and slower molecules and by actual movement of the higher-energy molecules

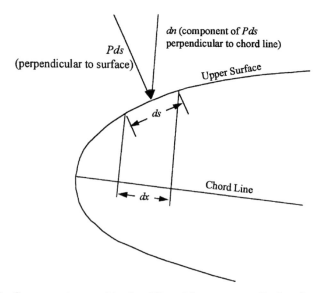

Fig. 3.15   Component normal to chord line of force as a result of surface pressure.

into the slower-moving regions. As a result, when a gas heats up the average speed of its molecules increases, and the rate at which momentum diffuses does also. Hence, a gas becomes *more* viscous as its temperature increases. But aside from these differences, the actions of viscosity in gases and liquids are quite similar.

Viscous effects are most important when a fluid is in contact with and moving relative to a solid body such as an aircraft. That portion of the fluid that is in direct contact with the solid body cannot move relative to it. This is because on a molecular scale even the smoothest polished surface is very rough and full of peaks and valleys. The sides of these peaks and valleys are barriers to the motion of the fluid molecules that are flowing next to the surface. The molecules strike these barriers and impart their excess momentum to the body, so that the fluid closest to the body must move at the same speed as the body. The exchange of momentum between the fluid and the body is the actual mechanism of viscous shear stress. Viscosity causes the velocities of fluid layers further from the body to also be reduced. This reduction in velocity decreases with increasing distance from the body.

3.4.1 Boundary Layer

The region next to a body in which the flow velocities are less than the freestream velocity is know as the *boundary layer*. Figure 3.16 shows a velocity profile for a typical boundary layer. The edge of the boundary layer is normally defined as the point where the velocity reaches 99% of the freestream velocity. Boundary layers on modern aircraft can be from a few millimeters to several meters thick. Table 3.2 indicates typical values of boundary-layer thickness for a variety of objects. Virtually all important viscous effects occur in the boundary layer. As a result, the rest of the flowfield can be treated as inviscid. This greatly simplifies the aerodynamic analysis task.

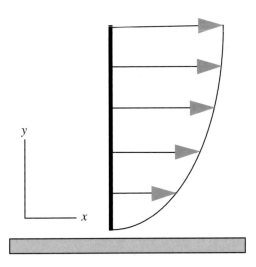

Fig. 3.16 Boundary-layer velocity profile.

Table 3.2   Typical boundary-layer thicknesses

Object	Flowing fluid	Flow velocity, m/s	Order of boundary-layer thickness
Supersonic fighter aircraft wing	Air	500	a few millimeters
Glider wing with 1-m chord length	Air	20	a few centimeters
Ship 200 m long	Water	10	1 m
Smooth ocean	Air (wind)	10	30 m
Land	Air (wind)	10	100 m

3.4.2 Skin-Friction Drag

Several viscous effects in the boundary layer are very important to the aircraft designer. The first is the production of viscous drag, which is also called *skin-friction drag*. Skin-friction drag typically comprises about 50% of the total drag on a commercial airliner at its cruise condition. Because drag must be overcome by thrust, reducing viscous drag will reduce the amount of thrust needed and hence the fuel burned. A designer has several methods for reducing viscous drag. One method is to reduce the surface area of the aircraft that is in contact with the air. This area is called the *wetted area*, a term borrowed from ship designers. Design engineers pay a great deal of attention to minimizing an aircraft's wetted area while keeping enough internal volume so that everything that the airplane must carry will fit.

A second method for minimizing skin-friction drag is controlling the shape of the boundary-layer profile. Figure 3.17 shows the changes a boundary layer undergoes as it flows over a surface. The initial boundary layer that forms at the front or *leading edge* of the surface is very orderly, with all velocity vectors parallel and only the velocity magnitudes decreasing with proximity to the surface. This is known as a *laminar* boundary layer because it is composed of orderly layers. As the flow moves further down the body, the orderly flow breaks down and *transitions* into a swirling, mixing flow known as a *turbulent* boundary layer. The turbulent boundary layer is thicker than the laminar boundary layer.

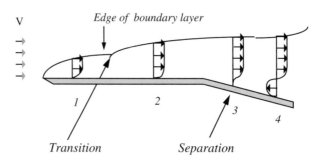

Fig. 3.17   Boundary-layer transition and separation.

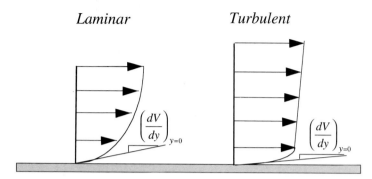

Fig. 3.18   Laminar and turbulent boundary-layer velocity profiles.

Figure 3.18 compares the profiles of the turbulent and laminar boundary layers. Note that, though it is thicker for the same conditions than the laminar boundary layer, velocities in the turbulent boundary layer are higher closer to the surface. This is because the swirling flow in the turbulent boundary layer allows large quantities of faster-moving air to travel en masse down close to the surface, a much more effective way of transferring momentum than diffusion in the orderly laminar boundary layer. Because the velocities in the turbulent boundary layer are higher close to the surface, they transfer more momentum to the body, hence creating more skin-friction drag.

The actual mathematical expression for the shear stress τ is

$$\tau = \mu \left(\frac{dV}{dy} \right)_{y=0} \tag{3.18}$$

where μ is the fluid viscosity and y is the direction perpendicular to the body surface. The rate of change of velocity with y distance (dV/dy) is called the *velocity gradient*, and the subscript $y = 0$ indicates that the gradient of interest is the one at the body surface. The skin-friction drag for a body is given by

$$D_f = \int_0^{S_{\text{wet}}} \tau \, dS$$

where D_f is the skin-friction drag, dS is a differential surface area, and S_{wet} is the total wetted area of the body. The skin-friction drag is often expressed as a dimensionless coefficient C_f, which is defined as

$$C_f = \frac{D_f}{q_\infty S_{\text{wet}}} \tag{3.19}$$

where q_∞ is the freestream dynamic pressure.

Equation (3.18) shows the same difference between laminar and turbulent boundary layers in the shear stress they produce as was just described. Because the turbulent boundary-layer profile has a higher velocity gradient at the body surface than the laminar boundary layer, it produces greater shear stress and hence more skin-friction drag. Smooth body surfaces tend to delay transition from laminar to turbulent flow. If pressure in the flow is gradually decreasing with distance along the surface (corresponding to a gradual increase in flow velocity outside the boundary layer), this also tends to delay transition. The condition of decreasing pressure with distance is called a *favorable pressure gradient* because such a pressure field will help the flow accelerate. Designers can achieve favorable pressure gradients over a large part of a body by placing the point of maximum thickness of the body as far aft (to the rear) as possible.

Of course, a body must eventually end, and the part of the body downstream of the point of maximum thickness will necessarily have an *adverse pressure gradient* as the pressure returns from its low values to freestream pressure. Figures 3.12 and 3.14 both show this effect on the upper surfaces of airfoils at moderate angles of attack. The region of adverse pressure gradient begins upstream of the point of airfoil maximum thickness.

The sloping part of the surface in Fig. 3.17 represents a region of adverse pressure gradient. The flow around the body reaches its maximum speed as it passes the body's point of maximum obstruction to the flow. The adverse pressure gradient on the rear of the body is just enough to slow the flow back down to freestream velocity at the rear end of the body. The flow in the boundary layer has lost momentum compared to that outside the boundary layer. However, the boundary-layer flow still faces the same adverse pressure gradient. Therefore, at some point prior to the *trailing edge* (rear) of the body the flow in the boundary layer slows to a stop and then reverses. Stagnant or reverse flow acts like an obstruction to the rest of the normal forward flow, and so it must detour around the obstruction. Because the reverse boundary layer flow is next to the body surface, the detouring flow moves away from the body, a condition called *separation* or separated flow.

Notice the third boundary-layer profile in Fig. 3.17, the one just downstream of the beginning of the sloped part of the surface. The velocities in the boundary layer close to the surface at this point are zero, but no reverse flow has started. The velocity gradient at the wall for this profile is also zero, and so there is no skin-friction drag. This condition signals the beginning of separation. However, for very controlled conditions a carefully designed airfoil can maintain a zero-gradient velocity profile from its point of maximum thickness all of the way to its trailing edge. Because the pressure on the rear of the airfoil is returning to freestream values, airfoil designers call this area the pressure recovery region. The zero velocity gradient, zero shear stress pressure recovery is called a Stratford recovery after B. S. Stratford, the first engineer to study such a phenomenon.[1]

3.4.3 Pressure Drag

The static pressure at the forward stagnation point on a body is freestream total pressure. There is an *aft stagnation point* on the body as well. For inviscid flow, the static pressure at the aft stagnation point would also be freestream total pressure, and there would be no net drag. When the flow in the boundary layer loses momentum, it also loses total pressure. The static pressure in the flow outside the boundary layer is transmitted to the boundary layer and through it to the body

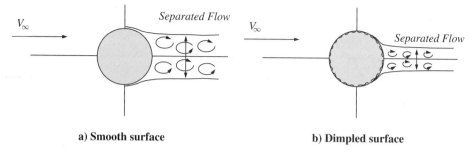

a) Smooth surface   b) Dimpled surface

Fig. 3.19   Effect of dimpled surface on separation point on golf balls.

surface. Therefore, when the boundary layer separates its pressure is generally less than or equal to freestream static pressure. This is always less than total pressure at the front stagnation point. The difference in pressures at the front and rear of the body produces a net force in the drag direction, which is called *pressure drag*. This is also called *drag due to separation*.

Pressure drag can be reduced by delaying separation. The turbulent boundary layer has higher velocities close to the wall and a more effective mechanism for replacing low momentum fluid with faster-moving molecules from outside the boundary layer. A turbulent boundary layer is therefore more resistant to separation and more able to maintain forward velocity for a longer distance against an adverse pressure gradient. Therefore, designers will sometimes use a bumpy surface near the front of a body to force boundary-layer transition. The higher-energy turbulent boundary layer that results, although it has greater skin-friction drag, will separate further aft on the body, reducing pressure drag. A golf ball is a good example of this design decision. The round shape of the golf ball results in very high adverse pressure gradients on the rear surfaces, compared to a more tapered, streamlined, rear section. The high adverse pressure gradient causes separation to occur very early, just aft of the point of maximum thickness, for a laminar boundary layer. This results in higher pressure drag. Figure 3.19 illustrates how the bumpy surfaces of golf balls cause earlier transition to delay separation, reducing pressure drag, and allowing the balls to fly farther.

Separation on a smooth golf ball occurs so early partly because the momentum of the air flowing past the ball is relatively low compared with the viscous shear forces, which tends to slow it down. A nondimensional parameter called the *Reynolds number* is used as a measure of these relative magnitudes of momentum and viscous forces. It is named for Osborne Reynolds, a pioneer researcher in viscous flow phenomena. The parameter is given the symbol Re and defined as

$$Re \equiv \frac{\rho V x}{\mu} \qquad (3.20)$$

where x is a characteristic reference length or distance (such as the chord length of a wing or the distance from the leading edge of a surface to a particular point in a boundary layer) that describes a particular body or surface. The terms in the numerator of the expression for Reynolds number indicate the magnitude of the momentum of the flow, whereas viscosity in the denominator is a measure of the viscous forces present.

Research has shown that the characteristics of a boundary layer can be described as functions of Reynolds number. This means that two bodies with the same shape and orientation to the flow, but with different sizes and in different flow conditions, will have the same type and shape of boundary-layer profile and the same transition and separation characteristics if they have the same Reynolds number. This type of relationship is called a *similarity rule*. It provides an important basis for wind-tunnel testing, because the flowfields around small wind-tunnel models will match those around large aircraft if the Reynolds numbers and other relevant *similarity parameters* are matched. Wind-tunnel testing of this sort inspired and proved design concepts such as the Stratford pressure recovery.

The *critical Reynolds number* is used to predict transition. Critical Reynolds number is defined using the distance from the start of a boundary layer as the reference length. When a distance (e.g., x coordinate) rather than a characteristic length (e.g., chord length) is used to define a Reynolds number, it is sometimes referred to as a *local Reynolds number*. To see how critical Reynolds number is used, consider the boundary layer for air flowing over a flat plate, similar to the left half of the surface in Fig. 3.17. The critical Reynolds number for such a body might be around 5×10^5, depending on the surface roughness. If the flow velocity and density are high and the viscosity is low, critical Reynolds number will be reached, and transition will occur only a short distance from the start of the boundary layer. On the other hand, if the flow is slow moving, more viscous, and less dense, it will take a much larger value of the distance from the start of the boundary layer before local Reynolds number equals the critical Reynolds number. Look again at the equation defining the Reynolds number to see why this is so. In this second case, the boundary layer will remain laminar much further along the surface. This will have a profound effect on drag and separation characteristics of the boundary layer. This is one of the primary reasons why engineers conducting wind-tunnel tests attempt to match Reynolds numbers with the full-scale flight conditions they are modeling. Laminar boundary layers cover only approximately the first 5–15% of a typical aircraft's wing.

Example 3.6

An airfoil in a wind-tunnel test section has a critical Reynolds number of 6×10^5. If the wind tunnel is operating in standard sea-level conditions with a test section velocity of 90 m/s, how far aft of the airfoil's leading edge will transition occur?

Solution: Solving Eq. (3.20) for x (in this case $x_{\text{transition}}$) and substituting in the test-section velocity and standard sea-level values of ρ and μ obtained from the standard atmosphere table,

$$x_{\text{transition}} = \frac{Re_{\text{crit}}\,\mu}{\rho V} = \frac{600,000(0.00001789 \text{ kg/m s})}{(1.225 \text{ kg/m}^3)(90 \text{ m/s})} = 0.097 \text{ m} = 9.7 \text{ cm}$$

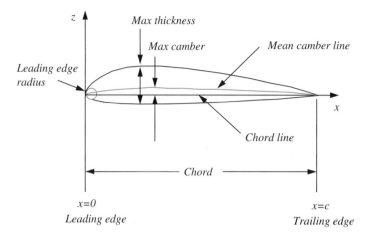

Fig. 3.20   Airfoil shape parameters.

3.5   Airfoil Characteristics

3.5.1   Shape

The differences in velocities and pressures that produce aerodynamic forces on an airfoil and also its boundary-layer profiles, transition, and separation characteristics are caused by the airfoil's shape and orientation. Aircraft designers spend a great deal of effort finding just the right shape for the airfoils they use on a particular design. Currently, many of these airfoil shapes are generated and optimized by computer programs. However, for many applications airfoil shapes can be chosen from geometry and performance data published by airfoil designers. Airfoils of this sort are often grouped into families of similar shapes, distinguished from each other by gradual variation of one or more of the parameters that describe their shape. Figure 3.20 illustrates a typical airfoil shape and the parameters that describe it.

The *chord line* shown in Fig. 3.20 is defined as a straight line drawn from the airfoil's leading edge to its trailing edge. The length of this line is called the *chord* or chord length and is given the symbol c. A curved line drawn from the leading edge to the trailing edge so as to be midway or equidistant between the upper and lower surfaces of the airfoil is called the *mean camber line*. The maximum distance between the airfoil's chord line and mean camber line is called the airfoil's *maximum camber* or just camber. An airfoil whose lower surface is a mirror image of its upper surface is said to be *symmetrical* or *uncambered*, and its mean camber line is coincident with its chord line. The airfoil is described by a *thickness envelope* wrapped around a mean chamber line. Thickness envelope is usually described by parameters, which include the *maximum thickness* as a fraction of the chord length, the point where this maximum thickness occurs, and the *leading-edge radius*.

3.5.2   Lift and Drag Coefficients

The lift and drag generated by an airfoil are usually measured in a wind tunnel and published as coefficients, which are dimensionless. Lift-and-drag coefficients are defined as follows:

$$c_l \equiv \frac{l}{\frac{1}{2}\rho V^2 S} \qquad (3.21)$$

$$c_d \equiv \frac{d}{\frac{1}{2}\rho V^2 S} \qquad (3.22)$$

where l and d are the lift and drag measured on the airfoil and S is the airfoil's *planform area*. Planform area is the area of a projection of the airfoil's shape onto a horizontal surface beneath it, similar to the airfoil's shadow when the sun is directly overhead. Now, we originally defined the airfoil as a slice of a wing, and as such it would have no planform area. When airfoils are tested in a wind tunnel, a section of wing is used that is frequently long enough to reach from one side of the test section to the other, as illustrated in Fig. 3.21.

The length of the section of the wing, that is, the distance that it must reach across the test section, is called its *span*. The wing has the same airfoil shape and size everywhere along its span, so that the same amount of lift and drag per unit

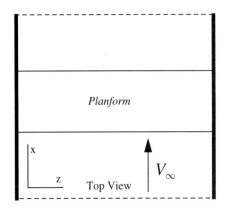

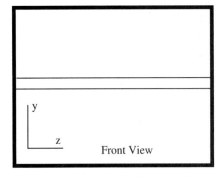

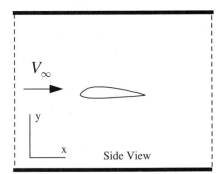

Fig. 3.21   Three-view drawing of rectangular wing section in wind-tunnel test section.

span will be generated by any slice of the wing. A wing section such as this has a finite rectangular planform area, which is used in defining the airfoil lift-and-drag coefficients. The flow around such a wing section is said to be two dimensional because flow properties vary in the *streamwise x* and vertical *y* directions, but not in the *z* or *spanwise* direction. Airfoil lift and drag coefficients are said to be two-dimensional coefficients.

3.5.3 Angle of Attack

Figure 3.22 shows streamlines around an airfoil as its angle of attack is changed. In the first drawing, the airfoil is at zero angle of attack. Because the airfoil is symmetrical, the flowfield above it is a mirror image of the flowfield below it, so no net lift is produced. Note that as angle of attack increases, the stream tubes above the airfoil become more constricted, so that the velocities above the airfoil must increase. This will produce lower static pressure there and consequently more lift. The lower static pressure above the middle of the airfoil will also produce a

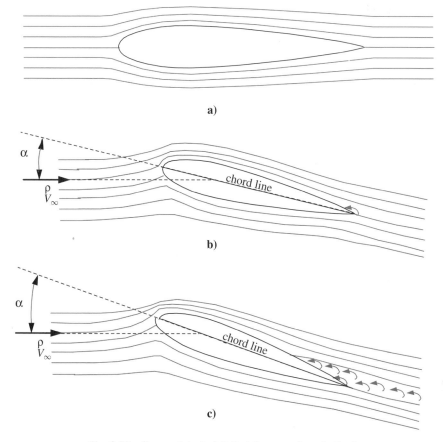

Fig. 3.22 Symmetrical airfoil at three angles of attack.

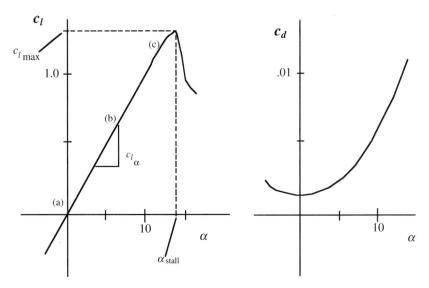

Fig. 3.23   Symmetrical airfoil lift- and drag-coefficient curves.

stronger adverse pressure gradient on the rear portion of the airfoil's upper surface. Note that the second drawing shows flow separation on the airfoil upper surface just ahead of the trailing edge. In the third drawing, the point of separation has moved upstream as a result of the stronger adverse pressure gradient.

3.5.4   Lift- and Drag-Coefficient Curves

Figure 3.23 shows plots of lift coefficient and drag coefficient as functions of angle of attack for the airfoil shown in Fig. 3.22. The letters in parentheses on the lift-coefficient curve correspond to the letters in Fig. 3.22. For smaller angles of attack, the lift coefficient increases linearly, and drag changes very gradually with increasing angle of attack. The rate of change of lift coefficient with angle of attack on this part of the curve is called the *lift-curve slope*:

$$c_{l_\alpha} \equiv \frac{\partial c_l}{\partial \alpha} \tag{3.23}$$

At higher angles of attack, the point of separation on the upper surface of the airfoil moves forward so far that it spoils some of the extra lift created by the additional constriction of the stream tubes. This causes the lift coefficient to increase more slowly with angle of attack and eventually reach a maximum. The earlier flow separation also produces more pressure drag. This causes the drag coefficient to increase much more rapidly at higher angles of attack. At the point on the lift curve where maximum lift coefficient is reached, further increases in angle of attack result in less lift. This phenomenon is called *stall* and the angle of attack for maximum lift coefficient is called the *stall angle of attack* or α_{stall}.

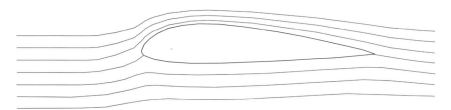

Fig. 3.24   Cambered airfoil at zero angle of attack.

3.5.5   Cambered Airfoils

Figure 3.24 shows the flowfield around a cambered airfoil for an angle of attack of zero. Notice that even though the airfoil is not inclined relative to the freestream direction ($\alpha = 0$), its shape causes the stream tubes above the airfoil to be more constricted than those below. This, of course, causes faster flow velocities and lower pressures above the airfoil. As a result, a cambered airfoil produces lift at zero angle of attack. As angle of attack increases, it has the same effect as for a symmetrical airfoil. However, because lift was already being generated at zero angle of attack the cambered airfoil's lift curve remains above the symmetrical airfoil's curve. Adverse pressure gradients and flow separation also develop sooner for the cambered airfoil, and so its stall angle of attack is lower.

Figure 3.25 shows lift- and drag-coefficient curves for a cambered airfoil and a symmetrical one. Note that $c_{l\alpha}$ is approximately the same for both airfoils. Also note that $c_{l\max}$ is higher for the cambered airfoil, even though it occurs at a lower angle of attack. The angle of attack for which the cambered airfoil generates zero

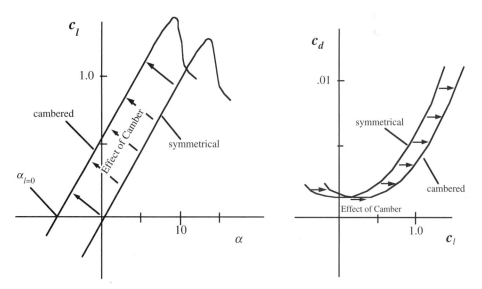

Fig. 3.25   Lift- and drag-coefficient curves for cambered and symmetrical airfoils.

lift is negative. It is called the *zero-lift angle of attack* and is given the symbol $\alpha_{l=0}$. The drag coefficient curves of Fig. 3.20 are plotted against lift coefficient instead of angle of attack in order to facilitate the comparison. Note that, unlike the symmetrical airfoil, the cambered airfoil has its minimum drag at a nonzero value of c_l.

3.5.6 Moment Coefficient and Aerodynamic Center

The distribution of pressure and shear stresses around an airfoil often produces net lift-and-drag forces, and it can also produce a net torque or moment. This is referred to as *pitching moment* and is given the symbol m. Pitching moment tends to rotate the nose or leading edge of the airfoil either up or down. A nose-up pitching moment is normally defined as positive. A pitching-moment coefficient c_m is defined as follows:

$$c_m \equiv \frac{m}{\frac{1}{2}\rho V^2 Sc} \qquad (3.24)$$

where c is the airfoil chord length. Note that the equation defining c_m differs from those for c_l and c_d in having an additional variable, the chord length, in the denominator. This extra quantity in the denominator is required to make c_m dimensionless because moment has dimensions of force multiplied by distance.

The magnitude and sense of the moment generated by the airfoil will be different depending on what point is chosen as the moment reference center. In most cases, it is possible to choose a moment reference center for which the moment is zero. Such a point is called the *center of pressure*. The center of pressure is not very useful, however, because its location must shift with changes in angle of attack in order to keep the moment zero. A more useful moment reference center is the aerodynamic center. The *aerodynamic center* is a fixed moment reference center on the airfoil for which the moment does not vary with changes in angle of attack, at least for that range of angles of attack where the lift curve is linear. Figure 3.26 shows the variation with c_l of c_m for a single airfoil using three different moment reference

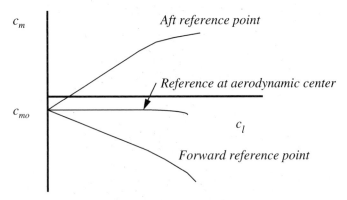

Fig. 3.26   Variation of cambered airfoil pitching-moment coefficient with lift coefficient for three choices of moment reference center.

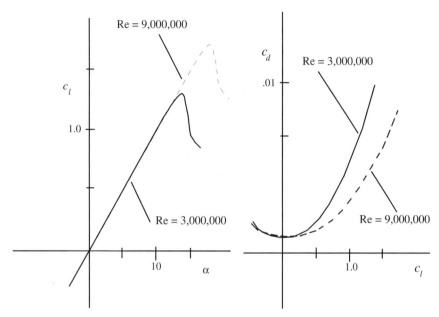

Fig. 3.27   Airfoil lift- and drag-coefficient curves for two different Reynolds numbers.

centers. Note that when summing moments about the aerodynamic center, the value of c_m is not zero for cambered airfoils, but it remains constant for most of the range of lift coefficients.

3.5.7   Reynolds-Number Effects

Figure 3.27 shows lift-and-drag-coefficient curves for an airfoil at two different Reynolds numbers. As Reynolds number increases, transition from a laminar to a turbulent boundary layer occurs closer to the leading edge of the airfoil. This causes more skin-friction drag, but delays separation and reduces pressure drag. At lower angles of attack, this change in the relative magnitudes of skin friction and pressure drag can result in either higher or lower total drag at higher Reynolds numbers. At higher angles of attack, where separation and pressure drag dominate, the reduction in pressure drag caused by delayed separation generally results in less total drag at higher Reynolds numbers. Figure 3.27 shows an airfoil that for higher Reynolds numbers has less drag at higher α and a higher α_{stall}.

3.5.8   Reading Airfoil Data Charts

Figure 3.28 shows a typical page of wind-tunnel airfoil data charts. Data such as these are published in a variety of books[2,3] and technical papers.[4–6] Appendix B in this book contains several similar data pages. Reading one of these charts is easy, if you pay attention to the details. First, note the airfoil designation at the bottom of the chart. NACA is the acronym for the National Advisory Committee for Aeronautics, a U.S. government agency, forerunner of NASA, which performed

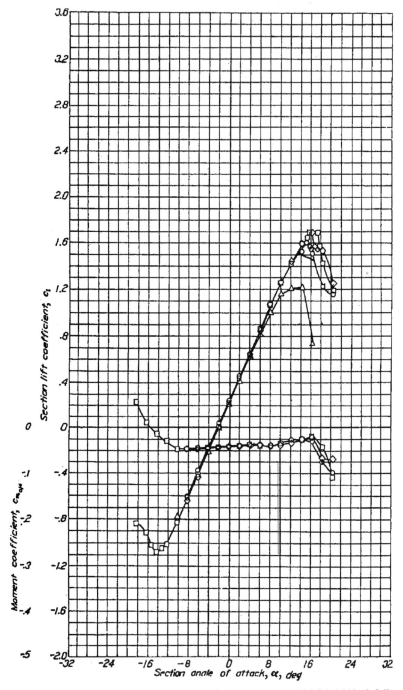

Fig. 3.28a   Lift-, drag-, and moment-coefficient data for a NACA 2412 airfoil.

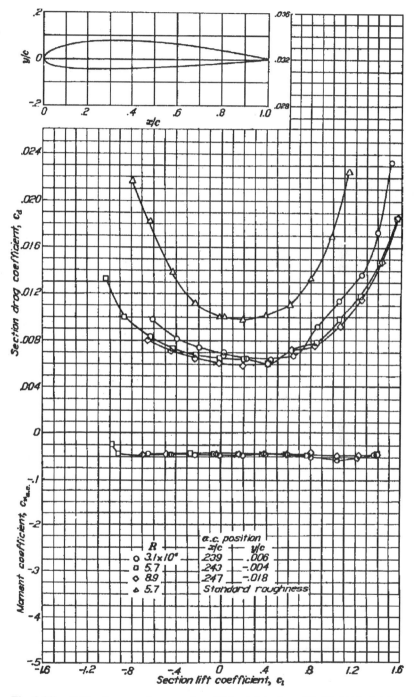

Fig. 3.28b Lift-, drag-, and moment-coefficient data for a NACA 2412 airfoil.

many wind-tunnel tests of airfoils and other shapes in the 1920s and 1930s. The four-digit code identifies the particular airfoil shape.

NACA used a series of codes with four, five, and more digits to systematically classify the airfoils it tested. For instance, the first digit in the four-digit series identifies the airfoil's maximum camber in percent of the chord. The second digit indicates where on the airfoil the point of maximum chamber occurs in tenths of the chord length aft of the airfoil leading edge. The third and fourth digits indicate the airfoil's maximum thickness in percent of the chord length. Thus, a NACA 2412 airfoil has 2% camber, its point of maximum chamber is located at its 40% chord point, and its maximum thickness is 12% of its chord length. If an airfoil with a NACA 2412 section had a chord length of 4 m, its maximum thickness would be 48 cm (see Ref. 2 for more details of the various NACA airfoil series and designations).

A drawing of the airfoil is depicted in Fig. 3.28b. The airfoil-section lift coefficient vs angle-of-attack curves are in Fig. 3.28a. Curves for drag coefficient and the coefficient of pitching moment about the aerodynamic center are plotted against lift coefficient in Fig. 3.28b. A legend on Fig. 3.28b identifies curves for three different Reynolds numbers. The curves for standard roughness are for airfoils that have a surface texture like sand paper on their leading edges. Generally, the data for smooth airfoils (not standard roughness) for an appropriate Reynolds number are used when designing an aircraft.

Example 3.7

A NACA 2412 airfoil with a 2-m chord and 5-m span is being tested in a wind tunnel at standard sea-level conditions and a test-section velocity of 42 m/s at an angle of attack of 8 deg. What is the airfoil's maximum thickness, maximum camber, location of maximum camber, and zero-lift angle of attack? Also, how much lift, drag, and pitching moment about its aerodynamic center is the airfoil generating?

Solution: Airfoil maximum thickness, camber, and location of maximum camber depend only on the NACA 2412 airfoil shape and the length of the airfoil chord. The first digit of the 2412 designation specifies a maximum camber that is 2% of the 2-m chord = 0.04 m. The second digit indicates that the location of the point of maximum chamber is $0.4 c = 0.8$ m aft of the leading edge. The last two digits specify a 12% thick airfoil, so that the maximum thickness is

$$t_{max} = 0.12 \cdot 2 \text{ m} = 0.24 \text{ m}$$

The aerodynamic properties of the airfoil might depend on the Reynolds number, which for standard sea-level conditions and a test-section velocity of 42 m/s is

$$Re = \frac{\rho V c}{\mu} = \frac{1.225 \text{ kg/m}^3 (42 \text{ m/s})(2 \text{ m})}{0.00001789 \text{ kg/m s}} = 5,751,817$$

so that the airfoil data curves for $Re = 5.7 \times 10^6$ (not standard roughness) will be used. The values of $\alpha_{L=0}$ and the c_l at $\alpha = 8$ deg do not, in fact, vary with

Reynolds number. Their values can be read from Fig. 3.28 as

$$\alpha_{L=0} = -2 \text{ deg}, \quad \text{at} \quad \alpha = 8 \text{ deg}, \quad c_l = 1.05$$

Also from Fig. 3.28, for $c_l = 1.05$ and $Re = 5.7 \times 10^6$,

$$c_d = 0.0098 \quad \text{and} \quad c_{m_{ac}} = -0.05$$

The dynamic pressure for the test is

$$q = \frac{1}{2}\rho V_\infty^2 = \frac{1}{2}(1.225 \text{ kg/m}^3)(42 \text{ m/s})^2 = 1080 \text{ N/m}^2$$

The airfoil's planform area is its chord multiplied by its span:

$$S = b \cdot c = 5 \text{ m} \cdot 2 \text{ m} = 10 \text{ m}^2$$

The lift, drag, and moment about the aerodynamic center are then given by

$$l = c_l q S = 1.05(1080 \text{ N/m}^2)(10 \text{ m}^2) = 11,340 \text{ N}$$

$$d = c_d q S = 0.0098(1080 \text{ N/m}^2)(10 \text{ m}^2) = 105.8 \text{ N}$$

$$m_{ac} = c_{m_{ac}} q S c = -0.05(1080 \text{ N/m}^2)(10 \text{ m}^2)(2 \text{ m}) = -1080 \text{ Nm}$$

3.6 Chapter Summary

3.6.1 Terms

(Including expanded definitions of some terms defined in Chapter 2)
1) Fundamental aerodynamic properties: P, ρ, T, V
   a) Intensive point properties
   b) Unsteady: $f(x, y, z, t)$
   c) Steady: $f(x, y, z)$
2) Flowfield: specific aero properties in the region of interest
3) Two sources of aerodynamic forces (on a body)
   a) Pressure surface distribution
   b) Shear-stress surface distribution (friction)
4) Pressure: $P \frac{\text{normal force}}{\text{unit area}}$ $\left(\frac{\text{N}}{\text{m}^2} = \text{Pa}, \frac{\text{lb}}{\text{ft}^2} = \text{psf}, \frac{\text{lb}}{\text{in}^2} = \text{psi, atm}\right)$
   a) Index of momentum exchange between molecules and surface
   b) Not a vector (magnitude and direction), but a scalar (magnitude only)
5) Shear stress: $\tau \frac{\text{tangential force}}{\text{unit area}}$ $\left(\frac{\text{N}}{\text{m}^2} = \text{Pa}, \frac{\text{lb}}{\text{ft}^2} = \text{psf}, \frac{\text{lb}}{\text{in}^2} = \text{psi}\right)$
   a) Caused by exchange of momentum parallel to surface, like friction
   b) Multiply by area to get a force
6) Density: $\rho \frac{\text{mass}}{\text{unit volume}}$ $\left(\frac{\text{kg}}{\text{m}^3}, \frac{\text{slug}}{\text{ft}^3}\right)$
7) Temperature: $T (\text{K}, °\text{C}, °\text{F}, °\text{R})$
   a) Index of molecular kinetic energy
   b) Scalar
   c) Always convert to absolute temperature $(\text{K}, °\text{R})$
8) Velocity: $V \frac{\text{distance}}{\text{time}}$ $\left(\frac{\text{m}}{\text{s}}, \frac{\text{ft}}{\text{s}}, \frac{\text{nm}}{\text{hr}} = \text{kn}, \frac{\text{sm}}{\text{hr}} = \text{mph}\right)$
   a) Vector (magnitude and direction)
   b) Speed is a scalar (magnitude only): V

9) One dimensional
   a) Flow properties are constant in a cross section perpendicular to the flow
   b) Example: wind tunnel—properties are constant at any axial location
10) Airspeed where compressible flow effects must be considered (330 ft/s or 100 m/s)

3.6.2 Important Fluids Equations (See also Fig. 3.29)

Perfect gas law:

$$P = \rho R T$$

Dynamic pressure:

$$q = \tfrac{1}{2}\rho V^2$$

One-dimensional mass flow:

$$\dot{m} = \rho A V$$

One-dimensional compressible continuity equation:

$$\rho_1 A_1 V_1 = \rho_2 A_2 V_2$$

Bernoulli's equation:

$$P_0 = P_1 + \frac{1}{2}\rho V_1^2 = P_2 + \frac{1}{2}\rho V_2^2$$

3.6.3 Bernoulli's Equation—Definitions

$$\underset{\substack{\text{Static} \\ \text{pressure}}}{P} \quad + \quad \underset{\substack{\text{Dynamic} \\ \text{pressure}}}{\frac{1}{2}\rho V^2} \quad = \quad \underset{\substack{\text{Total} \\ \text{pressure}}}{P_0}$$

1) Static pressure: P
   a) Random molecular motion
   b) Source of a force on a body
   c) What we mean when we say "pressure"
   d) What we feel when we move with the flow
2) Dynamic pressure: q
   a) Directed molecular motion
   b) $q = \tfrac{1}{2}\rho V^2$
3) Total pressure: P_0
   a) Sum of static and dynamic pressure
   b) Pressure resulting from isentropically (no friction or heat transfer) slowing the flow to zero velocity
4) Stagnation point
   a) Point of zero velocity on the body surface
   b) Static pressure equal to total pressure (because dynamic pressure is zero)

Relation between pressure and velocity for a fluid

Fundamental Statement:
Sum of forces in a given direction is equal to the product of the mass and the acceleration in that direction

Process: Examine forces acting on a fluid element

Assume:
1) Inviscid flow
2) Negligible body forces
3) Steady flow
4) Flow along a streamline

Yields:

$$dP = -\rho V dV$$ Euler Equation

$\Rightarrow$

Assume: Incompressible flow

Process: Integrate Euler's equation along a streamline

Yields:

$$P_1 + \frac{1}{2}\rho V_1^2 = P_2 + \frac{1}{2}\rho V_2^2 \equiv P_0$$

Bernoulli Equation

where P_0 is constant along a streamline.

Relation between area and velocity for a fluid

Fundamental Statement:
Mass is neither created nor destroyed

Assume: Steady flow

Yields:

$$m_{in} = m_{out}$$

ρ_2
A_2
V_2

ρ_1
A_1
V_1

$\Rightarrow$

Assume:
2) Velocity $\perp$ to area
3) 1D flow

Yields:

$$\dot{m} = \rho A V$$ Mass Flow Rate Equation

and

$$\rho_1 A_1 V_1 = \rho_2 A_2 V_2$$ Compressible Continuity Eqn

$\Rightarrow$

Assume: Incompressible flow

Yields:

$$A_1 V_1 = A_2 V_2$$ Incompressible Continuity Eqn

Relation between pressure and depth in a fluid

(P+dP)dA

dh

W

PdA

Process: Sum forces acting on fluid element in vertical direction

Assume: Static fluid element $\Rightarrow \Sigma F = 0$

Yields:

$$dP = -\rho g dh$$ Hydrostatic Equation

$\Rightarrow$

Process: Integrate hydrostatic equation between two points in column of liquid

Assume:
1) Constant density
2) Constant acceleration of gravity

Yields:

$$P_2 - P_1 = -\rho_\ell g(h_2 - h_1)$$ Manometer Equation

Fig. 3.29 Fluid mechanics equations.

5) Freestream
    a) Flow that is far upstream from a body of interest and is unaffected by that body
    b) Normally considered to be one-dimensional, so total pressure is constant along all streamlines
    c) Freestream static pressure is normally the local atmospheric pressure

3.6.4 Equations for Airfoils

Airfoil lift and lift coefficient:

$$l = C_l q S$$

Airfoil drag and drag coefficient:

$$d = C_d q S$$

Airfoil pitching moment and pitching-moment coefficient:

$$m = C_m q S \bar{c}$$

Airfoil lift-curve slope:

$$C_{l_\alpha} \equiv \frac{\partial C_l}{\partial \alpha}$$

3.6.5 NACA Airfoils

Airfoil terminology.
1) Chord line: line connecting leading edge and trailing edge
2) Chord c: length of chord line
3) Leading edge (LE): $x/c = 0$
4) Trailing edge (TE): $x/c = 1$
5) Mean camber line (MCL): curved line connecting points halfway between upper and lower airfoil surfaces (same as chord line for symmetric airfoil)
6) Max camber: maximum $\perp$ distance between chord line and mean camber line
7) Max thickness t: maximum $\perp$ distance between upper and lower airfoil surfaces

NACA four-digit series nomenclature.
1) Four-digit code used to describe airfoil shapes
2) First digit: maximum camber in percent chord
3) Second digit: location of maximum camber along chord line (from leading edge) in tenths of chord
4) Third and fourth digits: maximum thickness in percent chord

Example 3.8

Calculate the maximum camber, location of maximum camber, and maximum thickness of a NACA 2412 airfoil with a chord of 4 ft.

Solution: 2412 means the maximum camber is equal to 2%, the maximum camber is at 40% chord, and maximum thickness is 12%. Maximum camber: 0.08 ft (2% × 4 ft); location of maximum camber: 1.6 ft aft of leading edge (0.4 × 4 ft); and maximum thickness: 0.48 ft (12% × 4 ft).

3.6.6 How to Use NACA Charts

1) Find Reynolds number Re.
2) Enter c_l chart using α.
3) Find c_{l_α} using two points on the lift-curve slope and taking rise/run. *Be sure the two points are in the linear portion of the graph!*
4) Enter c_d chart using c_l.
5) Find $c_{m_{c/4}}$ using α (use outer scale; inner scale is for c_l).
6) Find $c_{m_{ac}}$ using c_l.
7) Read off $\alpha_{L=0}$, $c_{l_{max}}$, and α_{stall}.

Example 3.9

Determine c_l, c_d, $c_{m_{c/4}}$, $c_{m_{ac}}$, $\alpha_{L=0}$, $c_{l_{max}}$, and α_{stall} for a NACA 2412 airfoil at $\alpha = 12$ deg. Assume $Re = 3.1 \times 10^6$.

Solution:
$Re = 3.1 \times 10^6$
$\alpha = 12$ deg $\rightarrow c_l = 1.4$
$c_{l_\alpha} = (\Delta c_1 / \Delta \square) = (1.05 - 0)/[8 \text{ deg} - (-2 \text{ deg})] = 0.105/\text{deg}$
$c_1 = 1.4 \rightarrow c_d = 0.0173$
$\alpha = 12$ deg $\rightarrow c_{m_{c/4}} = -0.025$
$c_l = 1.4 \rightarrow c_{m_{ac}} = -0.050$ (*Because $c_{m_{c/4}}$ does not equal $c_{m_{ac}}$, the aerodynamic center must not be exactly at the quarter-chord point. Looking at the chart, insert* $\rightarrow x_{ac} = 0.239c$ *and* $y_{ac} = 0.006c$.)
$\alpha_{L=0} = -2$ deg
$c_{l_{max}} = 1.68$
$\alpha_{stall} = 16$ deg

Example 3.10

Calculate lift, drag, and pitching moment using the coefficients obtained from the NACA charts in Example 3.9:

$$c_l = 1.4 \qquad c_d = 0.0173 \qquad c_{m_{ac}} = -0.05$$

Solution:

$$l = c_l \cdot q \cdot S \qquad d = c_d \cdot q \cdot S \qquad m_{ac} = c_{m_{ac}} \cdot q \cdot S \cdot c$$

So to calculate l, d, and m_{ac}, we need to know the dynamic pressure, the chord, and the planform area.

Given that we are at sea level on a standard day with $V_\infty = 100$ ft/s,

$$q = \tfrac{1}{2}\rho V^2 = \tfrac{1}{2}(0.002377 \text{ slug/ft}^3)(100 \text{ ft/s})^2 = 11.885 \text{ lb/ft}^2$$

If $c = 4$ ft and $S = 200$ ft^2,

$$l = c_l \cdot q \cdot S = (1.4) \cdot (11.885 \text{ lb/ft}^2) \cdot (200 \text{ ft}^2) = 3300 \text{ lb}$$

$$d = c_d \cdot q \cdot S = (0.0173) \cdot (11.885 \text{ lb/ft}^2) \cdot (200 \text{ ft}^2) = 40 \text{ lb}$$

$$m_{ac} = c_{m_{ac}} \cdot q \cdot S \cdot c = (-0.05) \cdot (11.885 \text{ lb/ft}^2) \cdot (200 \text{ ft}^2) \cdot (4 \text{ ft}) = -401 \text{ ft-lb}$$

3.7 More Details

3.7.1 ICeT—The Complete Story

The airspeed that the needle on the airspeed indicator points at for a given set of flight conditions is called the *indicated airspeed*. If the airspeed indicator is geared and calibrated based on Eq. (3.6), then it is accurate only at speeds below 100 m/s or 330 ft/s, where the flow is incompressible. Aircraft built prior to around 1925 operated exclusively at incompressible airspeeds and had incompressible airspeed indicators. Incompressible flow indicators are inaccurate for high-speed flight and are no longer used. The Euler equation can be integrated without assuming incompressible flow. The details of this integration go beyond the scope of this text, but the result is a compressible form of Bernoulli's equation. Virtually all modern airspeed indicators are geared and calibrated to represent the compressible analog of Eq. (3.7), which is

$$V_\infty = \sqrt{\left(\left(\frac{1}{\rho_\infty}\right)7P_\infty\left\{\left[\frac{(P_0 - P_\infty)}{P_\infty} + 1\right]^{\frac{1}{3.5}} - 1\right\}\right)} \qquad (3.25)$$

Note that Eq. (3.25) is not a simple equation to engineer into a mechanical instrument. In addition, values of ρ are difficult to measure accurately in flight. For these reasons, it is difficult to build a simple and reliable airspeed indicator based on Eq. (3.25). Engineers surmounted this problem, however, by simplifying the equation. Airspeed indicators are manufactured with gears calibrated to use sea-level standard atmospheric values of P and ρ. In effect, an airspeed indicator is calibrated to solve the expression:

$$V_c = \sqrt{\left(\left(\frac{1}{\rho_{SL}}\right)7P_{SL}\left\{\left[\frac{(P_0 - P_\infty)}{P_{SL}} + 1\right]^{\frac{1}{3.5}} - 1\right\}\right)} \qquad (3.26)$$

where V_c is called the *calibrated airspeed*. Yet, this is still not what is indicated on the airspeed indicator. The static ports on the aircraft can be located such that they do not accurately measure the freestream static pressure. This is referred to as *position* or *installation error*. Additionally, there might be small inaccuracies in the machining of the instrument. To account for these discrepancies, errors are quantified during flight testing and equated to a velocity change ΔV_p called *position error.* The relationship between what is displayed on the airspeed indicator (indicated airspeed V_i) and the calibrated airspeed is given as

$$V_c = V_i + \Delta V_p \qquad (3.27)$$

On a perfect airspeed indicator, with zero position error, a pilot reading indicated airspeed also would be reading calibrated airspeed. However, in most cases ΔV_p does not equal zero, and indicated airspeed will be slightly greater or less than calibrated airspeed.

To obtain true airspeed [Eq. (3.25)] from calibrated airspeed [Eq. (3.26)], two corrections must be made, one for the actual existing pressure and the other for the actual existing density. Making the pressure correction yields *equivalent airspeed*, which is defined as

$$V_e = \sqrt{\left(\left(\frac{1}{\rho_{SL}}\right)7P_\infty\left\{\left[\frac{(P_0 - P_\infty)}{P_\infty} + 1\right]^{\frac{1}{3.5}} - 1\right\}\right)} \qquad (3.28)$$

Note that the actual static pressure is used in Eq. (3.28), as opposed to the sea-level values in Eq. (3.26). The ratio between V_e and V_c is generally called the *compressibility correction factor* and is given the symbol f:

$$V_e = fV_c \qquad (3.29)$$

where

$$f = \frac{V_e}{V_c} = \frac{\sqrt{(1/\rho_{SL})7P_\infty\{[(P_0 - P_\infty)/P_\infty + 1]^{\frac{1}{3.5}} - 1\})}}{\sqrt{(1/\rho_{SL})7P_{SL}\{[(P_0 - P_\infty)/P_{SL} + 1]^{\frac{1}{3.5}} - 1\})}} \qquad (3.30)$$

Note that f varies only with $(P_o - P_\infty)$ and P_∞. All other variables in Eq. (3.30) are constant. P_∞ can be obtained by setting a standard sea-level reference pressure in the aircraft altimeter, and $(P_o - P_\infty)$ can be obtained from knowing the calibrated airspeed. In this manner, a table of f factors such as Table 3.1 can be produced, which applies for any aircraft. It is normally more convenient to find a value for f from the table than to evaluate Eq. (3.30).

For the density correction, observe that

$$V_\infty = \sqrt{\left(\left(\frac{1}{\rho_\infty}\right)7P_\infty\left\{\left[\frac{P_0 - P_\infty)}{P_\infty} + 1\right]^{\frac{1}{3.5}} - 1\right\}\right)}$$

$$= \sqrt{\left(\left(\frac{\rho_{SL}}{\rho_\infty}\right)\left(\frac{1}{\rho_{SL}}\right)7P_\infty\left\{\left[\frac{(P_0 - P_\infty)}{P_\infty} + 1\right]^{\frac{1}{3.5}} - 1\right\}\right)}$$

and

$$V_\infty = V_e\sqrt{\frac{\rho_{SL}}{\rho_\infty}} \qquad (3.31)$$

Because the density ratio ρ_∞/ρ_{SL} is usually less than or equal to 1, V_∞ is usually $\geq V_e$. Notice that when flying at sea level on a standard day $\rho_\infty/\rho_{SL} = 1$, and $V_\infty = V_e$. Recall that dynamic pressure is given by

$$q = \frac{1}{2}\rho V_\infty^2 \tag{3.32}$$

So that

$$q = \frac{1}{2}\rho_\infty \left(V_e \sqrt{\frac{\rho_{SL}}{\rho_\infty}}\right)^2 = \frac{1}{2}\rho_{SL} V_e^2 \tag{3.33}$$

Equivalent airspeed can be defined alternately as that airspeed that would produce the same dynamic pressure at sea level as is measured for the given flight conditions. It will become apparent later on in this chapter and in Chapter 4 that, in the absence of compressibility effects, aircraft with identical configurations and orientation to the flow will produce the same aerodynamic forces if the dynamic pressures they are exposed to are the same. Because V_e is a direct measure of dynamic pressure, it is a very useful indicator of an aircraft's force-generating capabilities. This fact is very useful to both engineers and pilots.

3.7.2 Airfoil Compressibility Effects

The lift-curve and drag data in charts like Fig. 3.28 are valid for relatively low speeds. At higher speeds, the large pressure changes that the air undergoes as it flows around an airfoil cause significant changes in the air density. These density changes in turn magnify the effects of the pressure differences that produce lift and pressure drag. These changes in the magnitudes of the lift and drag are called *compressibility effects*, because they result from the fact that the air's density is changing.

3.7.3 Mach Number

Understanding and predicting compressibility effects requires working with a flow parameter called *Mach number M*. Mach number is named for the Austrian scientist and philosopher Earnst Mach, the first person to point out the significance of this parameter. It is defined as the ratio of the flow velocity to the *speed of sound* in the air. *Freestream Mach number* M_∞ is the ratio of the aircraft's flight speed (and therefore the magnitude of the freestream velocity) to the speed of sound:

$$M_\infty = \frac{V_\infty}{a} \tag{3.34}$$

The speed of sound is represented by the symbol a. As discussed in Sec. 2.7.3, its value is given by the expression

$$a = \sqrt{\gamma R T} \tag{3.35}$$

where $\gamma = c_p/c_v$ is the *ratio of specific heats* (see Ref. 7 for more details). For air, $\gamma = 1.4$.

Understanding why the speed of sound should depend on temperature and no other flow properties is useful in understanding other Mach-number effects. The explanation draws on the discussion in Chapter 1 of the origins of pressure and temperature in the random motions of molecules. The phenomenon called sound is actually fluctuations in air pressure, which move through the air much like waves on the surface of a pond. As described in Chapter 1, air pressure has its origins in the collisions of air molecules, which transfer momentum from the moving molecules to a body or to other air molecules. A sharp rise in pressure that moves as a wave through the air is really a surge in the momentum of the molecules, which is transmitted from one part of the air mass to another through a series of collisions. The rate at which the momentum surge can move through the air (in other words, the speed of a sound wave) is limited primarily by the average speed of the molecules between collisions. But recall that temperature is a measure of average molecular kinetic energy, which depends on the average speed of the molecules. So temperature measures average molecule speed, and average molecule speed determines the speed at which sound can be transmitted.

3.7.4 Prandtl–Glauert Correction

Corrections to airfoil lift-coefficient data to account for compressibility effects are made using an expression known as the *Prandtl–Glauert correction*:

$$c_l = \frac{c_{l_{(M_\infty=0)}}}{\sqrt{1 - M_\infty^2}} \tag{3.36}$$

where $c_{l_{(M_\infty=0)}}$ is the lift coefficient read from the airfoil data chart (assuming airfoil data is from a low-speed test), c_l is the airfoil lift coefficient corrected for compressibility, and M_∞ is the flight Mach number for the conditions to which the airfoil data are being corrected. Note that Eq. (3.30) is valid only for $M_\infty < 0.7$ or so. Also, the correction made by Eq. (3.30) becomes trivial for $M_\infty < 0.3$. Also note that because the Prandtl–Glauert correction applies to all lift coefficients on the lift curve, the lift-curve slope can also be corrected:

$$c_{l_\alpha} = \frac{c_{l_{\alpha(M_\infty=0)}}}{\sqrt{1 - M_\infty^2}} \tag{3.37}$$

Example 3.11

A NACA 2412 airfoil with a 0.5-m chord and 2-m span is being tested in a wind tunnel at standard sea-level conditions and a test-section velocity of 168 m/s at an angle of attack of 8 deg. What is the airfoil's lift-coefficient curve slope, and how much lift is it generating?

Solution: The aerodynamic properties of the airfoil might depend on the Reynolds number, which for standard sea-level conditions and a test-section velocity of 168 m/s is

$$Re = \frac{\rho V c}{\mu} = \frac{1.225 \text{ kg/m}^3 (168 \text{ m/s})(0.5 \text{ m})}{0.00001789 \text{ kg/m sec})} = 5,751,817$$

so the airfoil data curves for $Re = 5.7 \times 10^6$ (not standard roughness) will be used. As in Example 3.7 , the values of $\alpha_{L=0}$ and the c_l at $\alpha = 8$ deg do not vary with Reynolds number. Their values can be read from Fig. 3.28 as

$$\alpha_{L=0} = -2 \text{ deg}, \quad \text{and} \quad c_l = 1.05 \quad \text{at} \quad \alpha = 8 \text{ deg}$$

Because the lift-coefficient curve appears linear between $\alpha_{L=0} = -2$ deg and $\alpha = 8$ deg, the lift-curve slope can be estimated as the change in lift coefficient divided by the change in angle of attack:

$$c_{l_\alpha} = \frac{1.05 - 0}{8 \text{ deg} - (-2 \text{ deg})} = 0.105/\text{deg}$$

The test-section velocity is greater than 100 m/s for this test, and so compressibility corrections must be made. The Mach number for the test is calculated by substituting test-section velocity and standard sea-level speed of sound into Eq. (3.28):

$$M_\infty = \frac{V_\infty}{a} = \frac{168 \text{ m/s}}{340.3 \text{ m/s}} = 0.49$$

Then, applying the Prandtl–Glauert correction to both lift coefficient and lift-curve slope

$$c_l = \frac{c_{l_{(M_\infty=0)}}}{\sqrt{1 - M_\infty^2}} = \frac{1.05}{\sqrt{1 - 0.49^2}} = 1.2 \qquad c_{l_\alpha} = \frac{c_{l_{\alpha(M_\infty=0)}}}{\sqrt{1 - M_\infty^2}} = \frac{0.105/\text{deg}}{\sqrt{1 - 0.49^2}}$$

$$= 0.12/\text{deg}$$

The dynamic pressure for the test is

$$q = \frac{1}{2}\rho V_\infty^2 = \frac{1}{2}(1.225 \text{ kg/m}^3)(168 \text{ m/s})^2 = 17{,}287 \text{ N/m}^2$$

The airfoil's planform area is its chord multiplied by its span:

$$S = b \cdot c = 2 \text{ m} \cdot 0.5 \text{ m} = 1 \text{ m}^2$$

The lift is then

$$l = c_l q S = 1.2 (17{,}287 \text{ N/m}^2)(1 \text{ m}^2) = 20{,}745 \text{ N}$$

3.7.5 Viscosity as Molecular Collisions

If air is just a bunch of molecules making perfectly elastic collisions, how can it have viscosity, and how can it create shear stress on a surface? The answer is found in studying how air molecules interact with a solid surface.

Figure 3.30 shows a conceptual drawing of air flowing over a smooth surface, magnified so you can see individual molecules. Even a surface that is polished mirror smooth is rough as shown at the molecular level. So, when molecules of air collide with the surface, the direction of the momentum they impart depends one the direction they are moving and the slope of the surface where they hit it. If they hit a surface that is parallel to the average velocity of the airflow V_∞, like molecule a in Fig. 3.30, then they impart momentum perpendicular to that surface, contributing to what we call pressure. Note that molecule a's velocity component

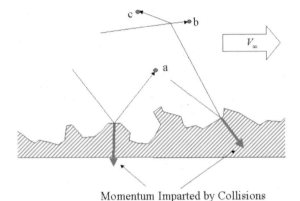

Momentum Imparted by Collisions

Fig. 3.30   Air molecules colliding with solid surface and each other.

parallel to V_∞ is unchanged after the collision. If they hit a part of the surface that is more nearly perpendicular to V_∞, like molecule b in Fig. 3.30, they will impart momentum perpendicular to that surface, which is mostly parallel to V_∞. Note that molecule b lost a lot of momentum parallel to V_∞ from its collision with the surface, which reduces the average velocity in the V_∞ direction of all of the molecules in that area. Molecule b also goes out and collides with molecule c, transferring this low momentum in the V_∞ direction to molecules out away from the body.

The velocity component in the V_∞ direction lost by the molecules when they hit the surface, or when they collided with a molecule that lost its momentum in the V_∞ direction by hitting the surface, etc., is greatest close to the surface. If only molecules next to the surface lost their momentum and molecules just a little way from the surface were unaffected, then we would say the air is nearly inviscid. This would happen if the velocity of the random motion of the molecules were small compared to V_∞. Because the temperature of air is a measure of the random velocities of its molecules, low random velocities occur when the air is very cold.

On the other hand, if the random velocities of the molecules are large compared to V_∞ then the lost momentum close to the surface is propagated out far from above the surface by a series of collisions. This occurs when the air is hot. We would say the air has fairly high viscosity. Another word for the transporting of low momentum outward from the surface is "diffusion" of the momentum deficit.

Another way to look at this is that if the air is cold and not very viscous only molecules very close to the surface are slowed down, so that the boundary layer is fairly thin. If, on the other hand, the air is hot, with high random molecular velocities, and therefore very viscous, molecules relatively far above the surface are slowed down, and the boundary layer is relatively thick.

For laminar boundary layers, molecular collisions are the only way low momentum at the surface is diffused to molecules far above it. That is why laminar boundary layers are thinner. Molecular collisions are also the only way molecules in a laminar boundary layer transport high momentum from outside the boundary down into it. That's why laminar boundary layers have relatively low velocities close to the surface and therefore low skin-friction drag.

Turbulent boundary layers have little swirls, or eddies, which swirl whole groups of low-momentum molecules from near the surface to relatively large distances above the surface, and high-momentum molecules from well above the surface to down close to the surface. As a result, turbulent boundary layers are thicker and have higher velocities parallel to V_∞ near the surface, greater momentum transfer parallel to the surface, and higher drag.

And all that from perfectly elastic collisions!

References

[1]Stratford, B. S., "An Experimental Flow with Zero Skin Friction Throughout Its Region of Pressure Rise," *Journal of Fluid Mechanics*, Vol. 5, May 1959.

[2]Abbott, I. H., and Von Dohenhoff, A. E., *Theory of Wing Sections*, Dover, New York, 1970.

[3]Eppler, R., *Airfoil Design and Data*, Springer-Verlag, Berlin, 1990.

[4]Selig, M. S., Guglielmo, J. J., Broeren, A. P., and Giguere, P., *Summary of Low-Speed Airfoil Data*, SoarTech Publications, Virginia Beach, VA, 1995.

[5]Drela, M., "Low-Reynolds-Number Airfoil Design for the M.I.T. Daedalus Prototype: A Case Study," *Journal of Aircraft*, Vol. 25, No. 8, 1988.

[6]Marsden, D. J., "A High-Lift Wing Section for Light Aircraft," *Canadian Aeronautics and Space Journal*, Vol. 34, No. 1, 1988, pp. 55–61.

[7]Bertin, J. J., and M. L. Smith, *Aerodynamics for Engineers*, Prentice–Hall, Upper Saddle River, NJ, 1989, pp. 299–301.

Problems

Synthesis Problems

S-3.1   Brainstorm at least five concepts for a simple, lightweight means of indicating to the pilot the airspeed of an ultralight aircraft.

S-3.2   State at least three measures of merit for the airspeed indicator in Problem S-3.1.

S-3.3   State at least three measures of merit for an airfoil to be used on a high-performance sailplane.

S-3.4   From the airfoils listed in Appendix B, select an airfoil to be used on a high-performance sailplane. The airfoil will operate on design at a section lift coefficient of 0.4 and a Reynolds number of 3.0×10^6. Justify your choice in terms of the three measures of merit you listed in Problem S-3.3.

S-3.5   The aft end of external fuel tanks carried by the Lockheed-Martin F-16 are truncated rather than tapering to a point. Based on your understanding of skin friction, pressure drag, and boundary-layer separation, why do you think this design decision was made?

Analysis Problems

A-3.1   Define incompressible flow and give airspeeds that allow this assumption to be made for air.

A-3.2   Define steady flow. Give an example of steady flow and one of nonsteady flow.

A-3.3   A wind tunnel has the following flow properties at the inlet {Hint: Draw a sketch!}: $P = 101,000$ N/m^2; $A = 1$ m^2; $T = 288$ K; $V = 200$ m/s and in the test section $A = 0.25$ m^2 and $V = 900$ m/s.
   (a) What is the mass flow rate through the wind tunnel?
   (b) What is the density in the test section?
   (c) Is the flow compressible or incompressible? Why?

A-3.4   Define one-dimensional flow and give an example.

A-3.5   Air flows into the inlet of a low-speed jet with the following properties: $P = 14.5$ psi, $A = 2.65$ ft^2, $V = 300$ ft/s, and $\rho = 0.0024$ slug/ft^3. At the end of the inlet is the compressor face where $P = 15$ psi. Assume the flow is steady, incompressible, one-dimensional, and inviscid.
   (a) What is the mass flow rate?
   (b) Determine the area at the compressor face.

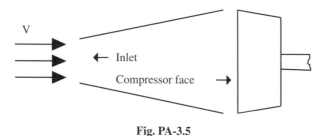

Fig. PA-3.5

A-3.6   A generator on a research aircraft requires 0.2 slugs per second of cooling air. A cooling air scoop for the generator is designed to be operated while the aircraft is cruising at 200 mph at 20,000 ft above sea level in standard atmospheric conditions. The scoop connects to an opening on the generator casing that has an area of 2 ft^2. Determine the required area of the inlet of the scoop. Also determine the air velocities and pressures at the scoop inlet and at the connection with the generator casing.

A-3.7   A high-pressure washing system's pump supplies 1 gallon of water per second at 100 psi pressure. The nozzle should be designed so that the static pressure of the water as it exits the nozzle is the same as the atmospheric pressure. If the inside diameter of the hose leading from the pump to the nozzle is 2 in., what should the diameter of the nozzle exit be? What will be the velocity of the water at the exit of the nozzle?

A-3.8   You are concerned about the moon roof on your new sports car. It seems to flex when driving at high speeds. Calculate how much net force the moon roof must withstand and in what direction. Assume the moon roof is flat with an area of 0.5 m^2 and the pressure and velocity over the moon roof is constant. Your driving speed is 20 m/s, and the velocity over the moon roof is 30 m/s. The pressure inside the car is $90,500 \text{ N/m}^2$ and the freestream pressure and density in front of the car are $90,000 \text{ N/m}^2$ and 1.1 kg/m^3. (Hint: Draw a sketch!)

A-3.9   An F-106 is flying from Minneapolis, Minnesota, to Charleston, South Carolina, a distance of 1400 n miles. The pilot reads an indicated airspeed of 300 kn at a pressure altitude of 50,000 ft MSL. If position error is negligible, what is the aircraft's true airspeed? If it is flying in a 50-kn tailwind, what is the plane's ground speed, and how long will the 1400-n mile flight take?

A-3.10   During a flight test, an indicated airspeed of 98 kn is observed at a pressure altitude of 5000 ft MSL. If position error for these conditions is +2 kn, what is the dynamic pressure for this test? Hint: Remember the alternate definition of equivalent airspeed.

A-3.11   Consider an airplane flying with a velocity of 40 m/s in conditions identical to those at an altitude of 2 km in the standard atmosphere. At a point on the wing, the airflow velocity is 50 m/s. Calculate the pressure at this point. Assume inviscid, incompressible flow.

A-3.12   You are on a bombing run in your F-117 with an indicated airspeed of 304 kn. Your altimeter reads 20,000 ft (pressure altitude), and your stealthy thermometer reads $-33°$F. You have a 40-kn tailwind and know that the position error for the aircraft is -4 kn. Calculate your aircraft's ground speed.

A-3.13   A C-130 pilot desires to fly a low-level navigation mission at 4 n miles/ min (240 kn ground speed). Winds are forecast out of the north (blowing south) at 20 kn, the pressure altitude is 10,000 ft, and the temperature is 80°F. What indicated airspeed should the pilot fly on a northbound leg if the position error is negligible?

A-3.14   A Cessna Citation has an equivalent airspeed of 200 kn at an unknown altitude. What is its dynamic pressure?

A-3.15   An instrument used to measure the airspeed on many early low speed airplanes during the 1910–1930 time period was the venturi, sketched in Fig. PA-3.15. This simple device is mounted on the airplane where the inlet velocity is essentially the same as the freestream velocity. With a knowledge that $A_1/A_2 = 4$ and $P_1 - P_2 = 4000 \text{ N/m}^2$, Find the airplane's velocity at sea level.

A-3.16   A pitot tube is mounted in the test section of a low-speed open-circuit subsonic wind tunnel. Air is flowing through the test section at 100 mph. The air pressure in the test section is 1 atm, and the temperature is 65°F. Calculate the pressure measured by the pitot tube.

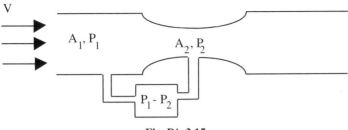

Fig. PA-3.15

A-3.17 A new wing design is being tested in a low-speed wind tunnel and in a flight test. Both tests take place in standard day conditions at an altitude of 2000 m. The velocity in the wind-tunnel test section is 90 m/s, and the true velocity of the flight test aircraft is 90 m/s. (Draw a sketch!)

(a) What are the atmospheric conditions for these tests in SI units?

(b) Calculate the total pressure at the leading-edge stagnation point for the airfoil in the wind tunnel. Assume no total pressure loss.

(c) Calculate the total pressure at the leading-edge stagnation point of the flight-test wing. (Hint: Use a reference attached to the aircraft.)

(d) Will the aerodynamic forces (such as lift) on the two wings be the same or different? Why?

A-3.18 A B-2A is flying at 43,000 ft in standard day conditions at 400 KTAS. The mean chord length is 39.6 ft.

(a) Find the overall or total Reynolds number for the aircraft.

(b) Find the percentage of the wing chord where laminar flow is present, if Reynolds number is 500,000.

A-3.19 A wing section with a rectangular planform and a chord of 2 ft spans the entire 10-ft-wide test section of a low-speed wind tunnel. The wing section uses a NACA 2412 airfoil. It is mounted at an angle of attack of 4 deg. If the tunnel is operated at a test-section velocity of 200 ft/s in standard sea-level conditions, how much lift and drag will the wing section generate?

A-3.20 Calculate the local Reynolds number at a point 2 ft aft of the leading edge of a wing, which is being tested at sea level, standard day conditions, and at 800 ft/s true airspeed. What type of boundary layer exists at this point? Assume $Re_{crit} = 5 \times 10^5$.

A-3.21 Figure PA-3.21 illustrates velocity profiles of the boundary layer at a few locations along a plate with distances in the y direction greatly exaggerated.

(a) What type of boundary layer exists at point A?

(b) What type of boundary layer exists at point B?

(c) Transition occurs between what two profiles?

(d) Reverse flow is evident in which profiles?

(e) Where does separation occur?

(f) Which direction would the transition point move if 1) pressure gradient is made more favorable 2) surface roughness is increased, and 3) freestream turbulence is reduced?

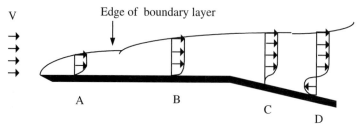

Fig. PA-3.21

A-3.22 a) What is an adverse pressure gradient, and where does it occur on an airfoil?
  b) What causes the flow to separate from an airfoil?
  c) What are the two major consequences of flow separation?
  d) Why do golf balls have dimples?
  e) What is the difference between flow transition and separation?

A-3.23 (a) A NACA 2415 airfoil has a chord of 1.4 m. Give the maximum camber, its location, and the maximum thickness of the airfoil in units of meters.
  (b) A NACA 0015 airfoil has a 0.152-m chord. Give the camber, its position, and the thickness of the airfoil in units of meters.

A-3.24 Draw a typical c_l vs α curve for a positively cambered airfoil and label the following features:
  (a) The axes (c_l and α)
  (b) Alpha at lift $= 0$
  (c) The $c_{l\alpha}$
  (d) The stall angle of attack
  (e) The $c_{l\,max}$

A-3.25 Consider a NACA 2415 airfoil section in a low-speed wind tunnel. Assume $Re = 9 \times 10^6$.
  (a) What is the zero-lift angle of attack?
  (b) Does this airfoil have negative or positive camber?
  (c) What is the stalling angle of attack?
  (d) What is the maximum value of c_l?
  (e) What is the lift-curve slope?

A-3.26 (a) What is the center of pressure (c.p.) of an airfoil?
  (b) What is always true about the sum of aerodynamic force moments about the center of pressure?

A-3.27 (a) What is the aerodynamic center (a.c.) of an airfoil?
  (b) A NACA 4412 airfoil has a chord of one meter. For $Re = 3 \times 10^6$, how far from the leading edge is the aerodynamic center located? Does the location change with changes in Reynolds number?

A-3.28 Consider a rectangular wing mounted in a wind tunnel. The wing model completely spans the test section so that the flow sees essentially an infinite wing. The wing has a NACA 4412 airfoil section, a chord of 3.0 m, and a span of 20 m. The tunnel is operated at the following test conditions: $P = 101,000$ N/m^2; $T = 30°C$; $V = 48$ m/s; and $\mu = 1.86 \times 10^{-5}$ kg/(m s).

(a) Determine the operating Reynolds number.

(b) Calculate the lift, drag, and moment about the aerodynamic center for an angle of attack of 8 deg and $Re = 9 \times 10^6$.

(c) At a Reynolds number of 3×10^6, find the following: 1) what is the stalling angle of attack for this airfoil? 2) what is the angle of attack for zero lift? and 3) what is the lift-curve slope?

A-3.29 A rectangular two-dimensional wing (airfoil) is placed in a low-speed wind tunnel with a test-section velocity of 80 m/s at an altitude of 2500 m on a standard day. The wing has a span of 5 m, a chord of 1.34 m, and a NACA 2415 airfoil cross section. If the wing is at an angle of attack of 4 deg, calculate the following: (a) Reynolds number based on the chord length (b) lift coefficient c_l, drag coefficient c_d, and moment coefficient about the aerodynamic center $c_{m_{ac}}$; (c) lift l, drag d, and moment about the aerodynamic center m_{ac}; and (d) max lift coefficient $c_{l_{max}}$, zero-lift angle of attack $\alpha_{l=0}$, zero-lift drag coefficient c_{d_0}, and angle of attack for minimum drag.

4
Wings and Airplanes

"After running the engine and propellers a few minutes to get them in working order, I got on the machine at 10:35 for the first trial. The wind, according to our anemometers at this time was blowing a little over 20 miles; 27 miles according to the government anemometer at Kitty Hawk. On slipping the rope the machine started off increasing in speed to probably 7 or 8 miles. The machine lifted from the truck just as it was entering the fourth rail."

From the Diary of Orville Wright for 17 December 1903

4.1 Design Motivation

4.1.1 Lift and Drag of Wings

The study of airfoils in Chapter 3 gave insight into how wings generate lift, but it did not tell the whole story. The flow over a wing near the wing tips is very different from the two-dimensional flow around an airfoil. The differences have profound effects on the lift and drag generated by a wing. Understanding these effects is crucial to the aircraft designer who must shape an aircraft's wing to optimize its performance. Section 4.2 discusses wing lift-and-drag theory and analysis methods.

4.1.2 Whole Aircraft Lift Curve

Other components besides the wing contribute to an aircraft's lift. The lift contributions of the aircraft's fuselage, control surfaces, high-lift devices, strakes, etc. all must be considered to predict an aircraft's lifting capability accurately. The aircraft's maximum lift coefficient is one of the governing factors in an aircraft's instantaneous turn capability, landing speed and distance, and takeoff speed and distance. Section 4.3 describes a variety of devices for increasing an airplane's maximum lift coefficent, while Section 4.8 presents methods for estimating the lift-curve slope and maximum lift coefficient of a complete airplane, including the effects of strakes, high-lift devices, control surfaces, etc.

4.1.3 Whole Aircraft Drag Polar

The drag of all aircraft components must also be included when estimating whole aircraft drag. The variation of an aircraft's drag coefficient with its lift coefficient is called the aircraft's *drag polar*. The drag polar is the key information about an aircraft needed to estimate most types of aircraft performance. Aircraft maximum speed, rate and angle of climb, range, and endurance depend so heavily on an aircraft's drag polar that a 1% change in drag can make a huge difference in a jet fighter's combat effectiveness or an airliner's profit potential. Section 4.5 presents a

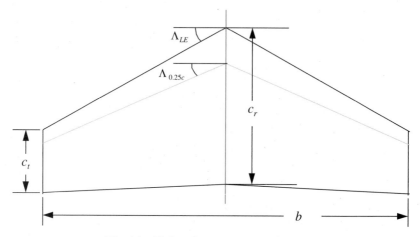

Fig. 4.1   Finite wing geometry definitions.

simple method for predicting an airplane's drag polar at low speeds, while Section 4.6 describes how high flight Mach numbers change an airplane's aerodynamics. Section 4.8 presents methods for estimating a complete airplane's lift-curve slope and drag polar at high flight Mach numbers. Section 4.9 is an example of an aerodynamic analysis for a supersonic jet fighter aircraft. The analysis predicts aircraft lift and drag characteristics for Mach numbers ranging from 0 to 2.0.

4.2   Wings

4.2.1   Terminology

Figure 4.1 illustrates a view of a wing planform with some of the important dimensions, angles, and parameters used to describe the shape of an aircraft wing. The wing span b is measured from wing tip to wing tip. The symbol c is used for the chord length of an airfoil at any point along the wing span. The subscript r indicates the chord length at the wing root or the aircraft centerline. The subscript t denotes the wing-tip chord. The overbar denotes an average value of chord length for the entire wing. The symbol R is used for a parameter called aspect ratio. Aspect ratio indicates how short and stubby or long and skinny the wing is. The symbol Λ is used for wing sweep angle with the subscript LE denoting the wing leading edge. The subscript $0.25c$ denotes the line connecting the 25% chord positions on each airfoil of the wing. The symbol λ is used for the wing taper ratio, or ratio of tip chord to root chord.

$$\lambda = \frac{c_t}{c_r} \tag{4.1}$$

$$R = \frac{b^2}{S} \tag{4.2}$$

$$S = b \cdot \bar{c} \tag{4.3}$$

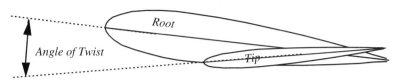

Fig. 4.2   Wing twist.

Figure 4.2 shows a side view of the wing to illustrate the angle of twist. Wings that are twisted so that the wing-tip airfoil is at a lower angle of attack than the wing-root airfoil are said to have wash out. Wing twist in the opposite sense from wash out is wash in. Wing twist of this sort is also called *geometric twist.*

An effective twist of the wing also can be achieved by changing the airfoil shape along the wing span. If the airfoil at the wing tip has less camber than the airfoil at the root, this has much the same affect on the wing lift as if the airfoils were the same but the wing tip airfoil was at a lower angle of attack than the root. Changing airfoils along the wing span in this way is called *aerodynamic twist.*

4.2.2 Wing-Tip Vortices

The flow around a wing section that spans the test section of a wind tunnel approximates the flow around a wing with an infinite span, no twist, and a constant chord length along its span. In Chapter 3, this type of flow was labeled two dimensional because flow properties did not vary in the spanwise direction. The flowfield around a finite wing, or wing with a finite span is not two dimensional. The majority of differences between the flow around a finite wing and that around an infinite wing result from flow phenomena, which occur at the wing tips. Figure 4.3 shows a front view of the flowfield around a finite wing. Note that the differences between the pressures above and below the wing that produce lift also produce a strong flow around the wing tip. The arrows in Fig. 4.3 are intended to illustrate a front view of flow streamlines in the plane of the 50% chord point on the wing. The lengths of the tails of the arrows do not indicate the magnitude of the velocity vectors. Of course, the actual magnitudes of the velocity vectors must be such that there is no flow through the surface of the wing.

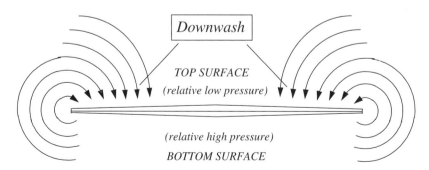

Fig. 4.3   Front view of wing with flow around the wing tips.

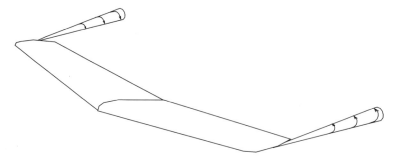

Fig. 4.4   Trailing vortices.

As shown in Fig. 4.4, these circular flow patterns around the wing tips become concentrated into very strong tornado-like swirling flows known as wing tip vortices or trailing vortices. The trailing vortices generated by large aircraft persist for many miles behind them and can pose serious hazards to smaller aircraft that encounter them. Air traffic controllers must allow sufficient spacing between aircraft so that the action of air viscosity and turbulence can dissipate a preceding plane's trailing vortices before the arrival of the next one. This spacing requirement to allow vortex dissipation is the limiting factor on traffic density at most commercial airports.

4.2.3   Downwash

Also note in Fig. 4.3 that the circular flow pattern around the wing tips results in a downward component to the flow over the wing. This downward flow component is called *downwash*. Figure 4.5 shows that downwash adds vectorially to the freestream velocity to change the direction of the flow velocity. Note that the resulting total velocity vector still results in flow parallel to the wing surface, but the orientation of the effective freestream velocity direction relative to the airfoil is altered.

The change in flow direction as a result of downwash is called the downwash angle and is given the symbol ε. The angle between the airfoil chord line and the local flow velocity vector is called the effective angle of attack α_{eff}. Each individual

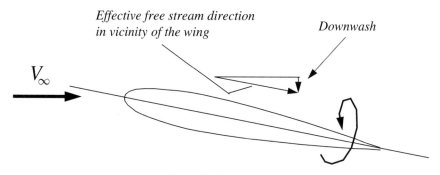

Fig. 4.5   Downwash.

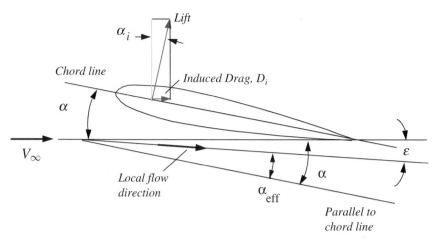

Fig. 4.6 Downwash angle and induced drag.

wing section's lift, drag, and angle of attack vary with the airfoil's orientation to this local flow direction, but the whole wing's lift, drag, and angle of attack must still be defined relative to the freestream direction. Figure 4.6 reveals that, as a consequence of the change in effective flow direction caused by the downwash, the effective angle of attack of the airfoil is reduced, and the lift generated by each airfoil has a component in the wing's drag direction. This component of lift in the drag direction is called *induced drag*. The reduction in effective angle of attack caused by the downwash causes the wing to produce less lift than it would if there were no downwash.

Figure 4.7 illustrates lift-coefficient curves for an airfoil and for a finite wing with the same airfoil section shape. Note that c_l denotes a two-dimensional airfoil lift coefficient, whereas

$$C_L = L/qS \qquad (4.4)$$

is used for the three-dimensional finite-wing lift coefficient. This same convention will be followed for c_d and

$$C_D = D/qS \qquad (4.5)$$

The reduction in effective angle of attack as a result of downwash decreases lift at any given α and delays stall to higher values of α. As in Chapter 3, slopes of the lift curves are defined as

$$c_{l_\alpha} \equiv \frac{\partial c_l}{\partial \alpha} \quad \text{and} \quad C_{L_\alpha} \equiv \frac{\partial C_L}{\partial \alpha} \qquad (4.6)$$

4.2.4 Spanwise Lift Distribution

Unlike the two-dimensional flow around an airfoil in a wind tunnel, the flow around a finite wing varies in the spanwise direction. This spanwise variation is primarily caused by the inability of the wing to support a pressure difference at

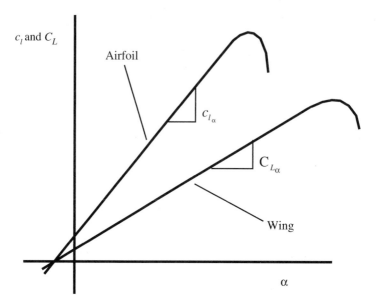

Fig. 4.7   Two-dimensional and three-dimensional lift-coefficient curves.

its tips (the cause of trailing vortices). It can be influenced by wing taper, wing twist, or even differences in airfoil shape at different spanwise positions on the wing. Spanwise variation of airfoil shape is called aerodynamic twist. But even an untapered, untwisted, unswept wing still has spanwise variation of the flowfield around it. This is because the trailing vortices on such a wing have a stronger effect and produce more downwash near the wing tips than they do far from the tips. As a result, even though the wing is not twisted increasing downwash reduces effective angle of attack and therefore lift near the wing tips. Tapering the wing or giving it wash out can help reduce this effect. In fact, a wing that is tapered and/or twisted to give an elliptical spanwise distribution of lift will have a constant downwash at every spanwise position. Figure 4.8 shows an elliptical

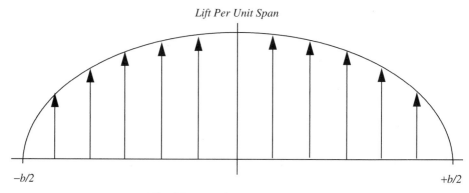

Fig. 4.8   Elliptical lift distribution.

Fig. 4.9   Supermarine Spitfire (courtesy National Air and Space Museum, SI 91-1443, Smithsonian Institution) and Republic P-47 Thunderbolt (courtesy Fairchild via National Air and Space Museum, SI 89-21538, Smithsonian Institution) fighter aircraft of World War II both had elliptical wing planforms.

spanwise lift distribution. An untwisted wing with an elliptical planform will have an elliptical lift distribution. As shown in Fig. 4.9, the famous Supermarine Spitfire and Republic P-47 Thunderbolt fighter aircraft of World War II both used elliptical wing planforms. Such wings are relatively complex and expensive to build, and so straight-tapered wings are much more common.

4.2.5   Finite-Wing Induced Drag

Figure 4.6 shows that induced drag is a component of the three-dimensional lift in the drag direction:

$$D_i = L \sin \varepsilon \quad \text{or} \quad C_{D_i} = C_L \sin \varepsilon \qquad (4.7)$$

It can be shown that the induced angle of attack everywhere along the span of wings with elliptical lift distributions is given by

$$\varepsilon = \frac{C_L}{\pi \mathcal{R}} \text{ radians} = \frac{57.3 \, C_L}{\pi \mathcal{R}} \text{ deg}$$

For ε small, $\sin \varepsilon \approx \varepsilon$ (in radians), and

$$C_{D_i} = C_L \varepsilon = C_L \frac{C_L}{\pi \mathcal{R}} = \frac{C_L^2}{\pi \mathcal{R}} \qquad (4.8)$$

4.2.6   Span Efficiency Factor

Equation (4.8) applies only to wings with elliptical lift distributions. However, it is possible to modify Eq. (4.8) slightly to make it apply to any wing by using a span efficiency factor e such that

$$C_{D_i} = \frac{C_L^2}{\pi e \mathcal{R}} \qquad (4.9)$$

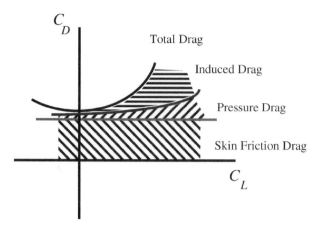

Fig. 4.10   Finite-wing total drag.

The value of e is 1 for elliptical wings and between 0.5 and 1 for most common wing shapes.

4.2.7 Finite-Wing Total Drag

The total drag of the wing is the sum of profile drag and induced drag:

$$C_D = c_d + \frac{C_L^2}{\pi e \mathcal{R}} \tag{4.10}$$

Recall, however, from Chapter 3 that profile drag is composed of skin-friction drag and pressure drag. Figure 4.10 illustrates the variation of each type of drag with lift coefficient.

4.2.8 Winglets and Tip Plates

A variety of devices have been used on aircraft to reduce induced drag. Figure 4.11a shows three such devices. Of the three, the winglet is the most effective and most widely used. In addition, jet fighter aircraft that carry fuel tanks or air-to-air missiles on their wing tips experience a small reduction in induced drag when such wing-tip stores are in place. All of these devices inhibit the formation of the wing-tip vortices and therefore reduce downwash and induced drag. Figure 4.11b shows a winglet on the wing tip of a McDonnell–Douglas C-17.

Of course, just extending the wing to increase its span and aspect ratio will have a similar effect. However, the increased lift far out at the end of the wing will increase the bending moment at the wing root and create greater loads on the wing-root structure. The winglet increases wing span only slightly. It is preferred because it achieves an effective increase in aspect ratio without significantly increasing wing-root structural loads.

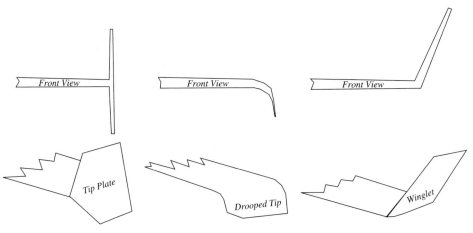

a) Three induced-drag-reducing wing-tip devices

b) Winglets on a C-17 (U.S. Air Force Photo)

Fig. 4.11   Wing-tip devices for reducing induced drag.

4.2.9   Finite-Wing Lift

Because the induced angle of attack for a wing with an elliptical lift distribution is constant everywhere along the span, it is relatively easy to determine the lift of such a wing. If the wing has an elliptical planform and no geometric or aerodynamic twist, it will have an elliptical lift distribution for a wide range of angles of attack. A twisted rectangular or tapered wing will normally achieve a true elliptical lift distribution at only one angle of attack. The elliptical planform wing's zero-lift angle of attack will be the same as for its airfoil section. At an arbitrary positive angle of attack below stall, an elliptical wing's effective angle of attack will be given by

$$\alpha_{\text{eff}} = \alpha - \varepsilon = \alpha - \frac{57.3\, C_L}{\pi \, \mathcal{R}} \tag{4.11}$$

As shown in Fig. 4.7, the airfoil and finite-wing lift-curve slopes can be represented as

$$c_{l_\alpha} = \frac{c_l}{(\alpha - \alpha_{l=0})} \qquad C_{L_\alpha} = \frac{C_L}{(\alpha - \alpha_{L=0})} \tag{4.12}$$

where α is any arbitrary angle of attack in the linear range of the lift curves. C_L and c_l are the lift coefficients at that arbitrary value of α. From Fig. 4.7 we recognize that

$$C_L = C_{L_\alpha}(\alpha - \alpha_{L=0}) = c_{l_\alpha}(\alpha_{\text{eff}} - \alpha_{L=0}) = c_{l_\alpha}\left(\alpha - \frac{57.3\, C_L}{\pi \, \mathcal{R}} - \alpha_{L=0}\right) \tag{4.13}$$

Combining Eqs. (4.12) and (4.13), the expression for C_{L_α} becomes

$$C_{L_\alpha} = \frac{c_{l_\alpha}}{1 + 57.3\, c_{l_\alpha}/\pi \, \mathcal{R}}$$

Following the same convention as in Eq. (4.7) for nonelliptical wings, the expression can be written

$$C_{L_\alpha} = \frac{c_{l_\alpha}}{1 + 57.3\, c_{l_\alpha}/\pi \, e \mathcal{R}} \tag{4.14}$$

Note that in general for a given wing, the value of e required for Eq. (4.14) is not the same as that required for Eq. (4.7). The two values are typically quite close to each other, however.

Example 4.1

A wing with a rectangular planform, a NACA 2412 airfoil, a span of 5 m and a chord of 2 m is operating in standard sea-level conditions at a free stream velocity of 42 m/s and an angle of attack of 8 deg. If the wing's span efficiency factor is 0.9, how much lift and drag is it generating?

Solution: The aerodynamic properties of the airfoil might depend on the Reynolds number, which for standard sea-level conditions and a freestream velocity of 42 m/s is

$$Re = \frac{\rho V c}{\mu} = \frac{1.225\ \text{kg/m}^3\,(42\ \text{m/s})\,(2\ \text{m})}{0.00001789\ \text{kg/m s}} = 5{,}751{,}817$$

so that the airfoil data curves for $Re = 5.7 \times 10^6$ (not standard roughness) will be used. The values of $\alpha_{L=0}$ and the c_l at $\alpha = 8$ deg do not, in fact, vary with Reynolds number. Their values can be read from Figs. 3.28 as

$$\alpha_{L=0} = -2\ \text{deg}, \qquad \text{at} \qquad \alpha = 8\ \text{deg}, \quad c_l = 1.05$$

Because the lift-coefficient curve appears linear between $\alpha_{L=0} = -2$ and 8 deg the lift-curve slope can be estimated as the change in lift coefficient divided by the change in angle of attack:

$$c_{l_\alpha} = \frac{1.05 - 0}{8 \text{ deg} - (-2 \text{ deg})} = 0.105/\text{deg}$$

Also from Figs. 3.28, for $c_l = 1.05$ and $Re = 5.7 \times 10^6$, $c_d = 0.0098$. The dynamic pressure for the test is

$$q = \frac{1}{2}\rho V_\infty^2 = \frac{1}{2}(1.225 \text{ kg/m}^3)(42 \text{ m/s})^2 = 1080 \text{ N/m}^2$$

The wing's planform area is its chord multiplied by its span:

$$S = b \cdot c = 5 \text{ m} \cdot 2 \text{ m} = 10 \text{ m}^2$$

Its aspect ratio is determined using Eq. (4.2):

$$\mathcal{R} = \frac{b^2}{S} = \frac{(2 \text{ m})^2}{10 \text{ m}^2} = 2.5$$

and the finite-wing lift-curve slope is predicted by Eq. (4.14):

$$C_{L_\alpha} = c_{l_\alpha} \left/ 1 + \frac{57.3\, c_{l_\alpha}}{\pi e \mathcal{R}} \right. = 0.105/\text{deg} \left/ 1 + \frac{(57.3 \text{ deg/rad})(0.105/\text{deg})}{\pi\,(0.9)(2.5)} \right.$$

$$= 0.0567/\text{deg}$$

The lift coefficient is then calculated using Eq. (4.13):

$$C_L = C_{L_\alpha}(\alpha - \alpha_{L=0}) = 0.0567/\text{deg}[8 \text{ deg} - (-2 \text{ deg})] = 0.567$$

If the wing had an elliptical planform, the airfoil lift coefficient everywhere on the wing would equal the finite-wing lift coefficient, and a different value of c_d could be read from the airfoil chart for this lower c_l value. However, for a rectangular planform c_l varies, and as a conservative estimate of the average value of c_d the value of c_d read from the airfoil data chart for $c_l = 1.05$ is used. The finite-wing drag coefficient is then calculated using Eq. (4.10):

$$C_D = c_d + \frac{C_L^2}{\pi e \mathcal{R}} = 0.0098 + \frac{0.567^2}{\pi\,(0.9)(2.5)} = 0.055$$

The lift and drag are then given by

$$L = C_L\, qS = 0.567(1080 \text{ N/m}^2)(10 \text{ m}^2) = 6124 \text{ N}$$

$$D = C_D\, qS = 0.055(1080 \text{ N/m}^2)(10 \text{ m}^2) = 597 \text{ N}$$

It is interesting to compare these results with the forces generated by an airfoil in a wind tunnel with the same geometry and freestream conditions, but purely two-dimensional flow around it, as calculated in Example 3.7. The decrease in lift and increase in drag caused by the three-dimensional flow around the finite wing's tips is significant.

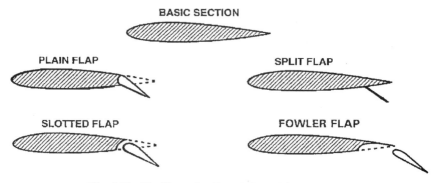

Fig. 4.12   Trailing-edge flaps (adapted from Ref. 1).

4.3 High-Lift Devices

Relatively thin airfoils with low camber generally give low drag at high speeds. Unfortunately, these airfoils also typically have relatively low values of maximum lift coefficient. Most aircraft are equipped with devices that can be used to increase lift when needed, at the expense of additional drag. These devices are of several types.

4.3.1 Trailing-Edge Flaps

Moveable surfaces on the rear portion of the wing that can be deflected downward to increase the wing's camber are called trailing-edge flaps or simply flaps. Figure 4.12 shows four different types of flaps. The plain flap changes camber to increase lift, but its effect is limited by additional flow separation, which occurs when it is deflected. The additional separation occurs because the upper surface of the deflected flap experiences a stronger adverse pressure gradient. The split flap deflects only the underside of the trailing edge so that, while it creates a great deal of pressure drag, it avoids the strong adverse pressure gradient on its upper surface and therefore keeps the flow attached slightly longer. This gives the split flap slightly greater lift.

Slotted flaps have a gap or slot in them to allow faster-moving air from the lower surface to flow over the upper surface. The higher-energy air from the slot gives the boundary layer more energy to fight the adverse pressure gradient and delay separation. A single-slotted flap creates the slot by moving away from the wing slightly when it is deflected. Double- and triple-slotted flaps are also used. Each slot admits more high-energy air onto the upper surface to further delay separation and increase lift. The *Fowler flap* moves aft to increase the wing area before deflecting downward to increase camber. Fowler flaps usually have one or more slots to increase their effectiveness.

Figure 4.13 shows airfoil lift-and-drag-coefficient curves for a typical trailing-edge flap. Note that in general the effect of flaps is to increase camber, moving the lift-curve up and to the left. For flaps other than Fowler flaps the lift-curve slope is unchanged. The angle of attack for zero lift is made more negative. With the flap extended, the wing generates more lift at all angles of attack below stall. The maximum lift coefficient is greater, but it occurs at a lower angle of attack. The amount of this shift in $\alpha_{l=0}$ and increase in $C_{L\max}$ is different for each type of flap.

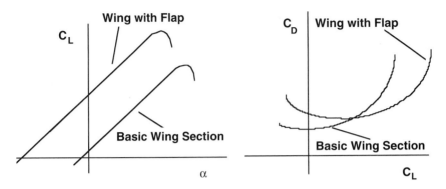

Fig. 4.13   Lift-and-drag-coefficient curves for wings with flaps.

Slots in flaps help delay the stall to higher angles of attack and higher values of C_{Lmax}. The lift-curve slope increases when Fowler flaps are used. This is because Fowler flaps increase the actual lifting area of the wing when they are extended, but the lift coefficient is defined using the same reference planform area as when the flaps are retracted.

4.3.2   Strakes and Leading-Edge Extensions

Figure 4.14 shows a strake on an F-16. A similar device on the F-18 is referred to as a leading-edge extension (LEX). The strake has a sharp leading edge. When the aircraft operates at high angles of attack, the flow cannot stay attached as it flows over the sharp strake leading edge, and it separates. Because the leading edge of the strake is highly swept, the separated flow does not break down into turbulence, but instead rolls up into a tornado-like vortex. The vortex generates an intense low-pressure field, which, because it is on the upper surface of the strake and wing, increases lift. The presence of the vortex gives the rest of the wing a more favorable pressure gradient, so that stall is delayed. The strake also increases the total lifting area, but it is usually not included in the reference planform area. Therefore, the strake increases lift-coefficient-curve slope even at low angles of attack when the vortex does not form. Figure 4.15 shows lift-and-drag-coefficient curves for a wing with and without strakes. Note that at relatively high angles of attack the lift curve for the wing with strakes is actually above the dotted line that

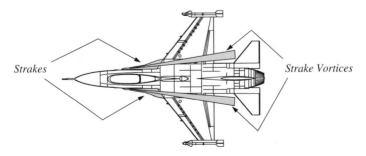

Fig. 4.14   F-16 strakes.

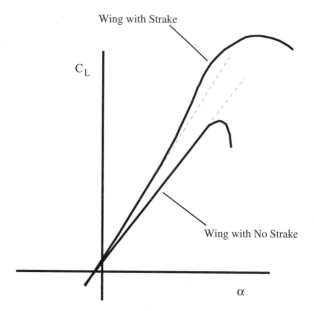

Fig. 4.15   Lift-coefficient curves for wing alone and wing with strakes.

is an extension of the linear region of the curve. It is at these angles of attack where strakes are most effective.

4.3.3   Leading-Edge Flaps and Slats

Figure 4.16 shows several devices that are used on wings to increase lift. Plain leading-edge flaps deflect to increase wing camber and move the point of minimum pressure further aft on the upper surface of the airfoil at high angles of attack. The aft movement of the point of minimum pressure extends the region of favorable pressure gradient and delays separation. A fixed slot can be used to admit higher-speed air onto the upper wing surface to reenergize the boundary layer and delay separation. A slat is a leading-edge flap that, when it is extended, opens up a slot as well. All three leading-edge devices delay stall and extend the lift curve to higher angles of attack and higher maximum lift coefficients. Because angle of attack is defined using the chord line of the airfoil with no high-lift devices extended, extending a leading-edge device can actually decrease the lift coefficient at a particular angle of attack. Some slats increase the lifting area when they are deployed, so that they increase the lift-curve slope like Fowler flaps. Figure 4.17 illustrates lift-coefficient curves for a wing with and without a typical leading-edge slot, slat, or flap. The magnitude of the increase in maximum lift coefficient and stall angle of attack is different for each type of leading-edge device.

4.3.4   Boundary-Layer Control

Because flow separation and stall are caused by depletion of flow velocity in the boundary layer, several methods can be used to remove or reenergize this

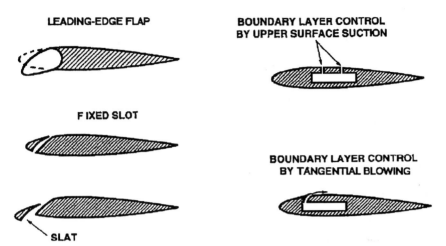

Fig. 4.16   Leading-edge flaps and boundary-layer control devices (adapted from Ref. 1).

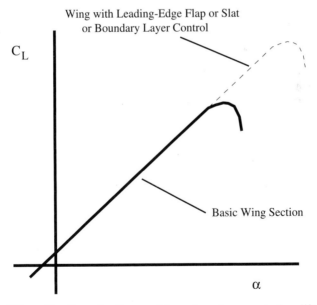

Fig. 4.17   Effect of leading-edge flaps and boundary-layer control on lift-coefficient curves.

low-energy air and delay separation. One method is to drill thousands of tiny holes in the wing surface and use *suction* to pull the low-energy air inside the wing. Another method is to use *blowing* of high-velocity air tangent to the wing surface to reenergize the boundary layer and delay separation. Air for tangential blowing is normally obtained as *bleed air* from a jet engine's compressor. Both of these boundary-layer control devices delay separation and stall to higher angles of attack. Their lift curves look similar to those for leading-edge devices shown in Fig. 4.17. Examples of boundary-layer suction and blowing are illustrated in Fig. 4.16.

4.3.5 Powered Lift and Vectored Thrust

An internally blown flap or jet flap has bleed air directed onto its leading edge and upper surface from the rear of the wing. The high-velocity air delays separation and increases lift. Figure 4.18a shows a typical internally blown flap configuration. Engine exhaust also can be used to increase or assist lift. Figures 4.18b to 4.18d show three ways this can be done. The exhaust can be directed at the leading edge of a flap as on the McDonnell–Douglas C-17, or at the wing and flap's upper surface, as on the Boeing YC-14. In either case, the vastly increased airflow over the flap increases lift. The engine nozzle can also be moveable to redirect or *vector* the engine exhaust downward. This reorients the engine thrust vector so that it has a component in the lift direction to assist the lift generated by the wing. Also note in Fig. 4.18 the multiple slots in each Fowler flap. Several high-lift devices are often used together on an aircraft. Each device adds to the total $C_{L\max}$. In some cases the devices complement each other so that the total increase in $C_{L\max}$ for several devices used together is greater than the sum of the $C_{L\max}$ increments for each device used alone.

4.4 Whole Aircraft Lift

The total lift generated by a whole aircraft for a given set of flight conditions and angle of attack depends on much more than just the lift generated by the wing. A typical aircraft's fuselage, engine nacelles, horizontal tail, strakes, and other components can add significantly to the total lift the aircraft generates. Section 4.8 describes a simple method for predicting these effects. However, for many common aircraft configurations the lift of the whole aircraft is very nearly the same as the lift of the wing alone. Therefore, predicting the lift of the wing alone using the methods discussed in Sec. 4.2 will also for these configurations predict the lift of the whole aircraft with reasonable accuracy. Note that this approximation is quite good for all-wing aircraft like the B-2 and Avro Vulcan and reasonably accurate for airliners like the Boeing 707/727/737/747/757/767/777 and transports like the Lockheed C-5. However, it is a poor approximation for fighter aircraft configurations like the F-16 and F-22, for which the strakes and wide fuselages contribute a very significant part of the aircraft's total lift.

4.5 Whole Aircraft Drag and Drag Polar

Like lift, the total drag of an aircraft depends on the drag of all its components. However, for nearly all aircraft configurations drag of the whole aircraft is usually many times greater than the drag of the wing alone. For this reason, whole aircraft

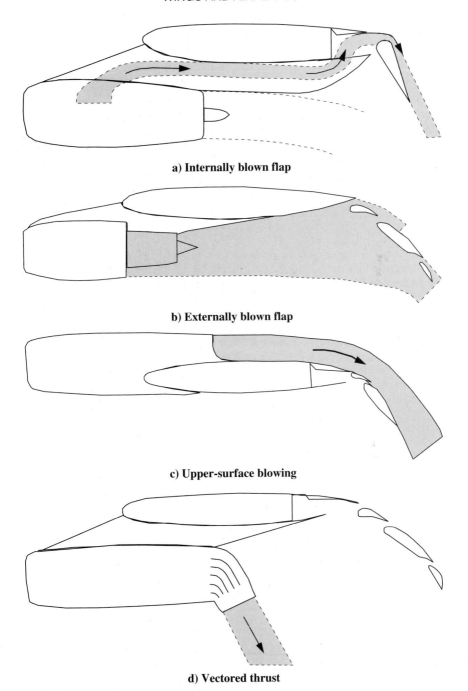

a) Internally blown flap

b) Externally blown flap

c) Upper-surface blowing

d) Vectored thrust

Fig. 4.18   Four powered lift configurations (adapted from Ref. 1).

drag predictions must include the effects of all parts of the aircraft, not just the wing alone.

Engineers typically represent the drag of a whole aircraft in a slightly different fashion from the way Eq. (4.10) represents the total drag of a wing. Because whole aircraft drag depends on so many factors, engineers typically group these effects into those that vary with lift and those that do not. They call the nondimensional coefficient form of this representation the aircraft's *drag polar*. A simple drag polar representation has the following form:

$$C_D = C_{D_0} + kC_L^2 \qquad (4.15)$$

where

$$k = 1/(\pi e_O R) \qquad (4.16)$$

C_{D_0} is called the *parasite drag coefficient*. It is a constant that represents all drag generated by the aircraft when it is not generating lift. The variable e_O in the expression for k is called *Oswald's efficiency factor*. It is not the same as span efficiency factor e used in Eqs. (4.9) and (4.14) because it includes all other types of drag due to lift.

C_{D_0} in Eq. (4.15) is a constant, but both terms in Eq. (4.10) vary with lift coefficient. This is because the airfoil two-dimensional drag coefficient c_d varied with c_l as shown in the NACA data charts. Having C_{D_0} be a constant in an aircraft's drag polar is very convenient, but this requires that the kC_L term model much more drag than does the induced drag term of Eq. (4.10).

In addition, when an aircraft changes its angle of attack in order to change its lift this changes the airflow patterns around its fuselage, engine nacelles, etc., as well as its wing. The changing orientation of these components to the freestream velocity vector changes the location and extent of flow separation on these components. As a result, the variation of an aircraft's drag with its lift is much greater than that of its wing alone. Therefore, in order to model all of this extra drag the value of e_O for an aircraft is typically less than e for its wing alone. Section 4.8 presents a simple method for predicting the drag polar of a whole aircraft.

Example 4.2

The Avro Vulcan jet bomber is nearly an all-wing aircraft, with only a small fuselage section sticking forward of its large triangular-shaped or *delta* wing. Its wing uses a NACA 0009 airfoil and has a span of 111 ft and planform area of 3964 ft². It is flying in standard day 5000-ft conditions at a true airspeed of 300 ft/s and an angle of attack of 8 deg. Its C_{D_0} is 0.009, its span efficiency factor is 0.9, its Oswald's efficiency factor is 0.7, and its average wing chord length is 36 ft. Estimate how much lift and drag it is making for these conditions.

Solution: As stated in Sec. 4.4, the methods used in Example 4.1 to predict the lift of a wing alone are reasonable approximations of the whole aircraft lift of the Vulcan. The aerodynamic properties of the airfoil might depend on the Reynolds number, which for standard day 5000-ft conditions and a freestream velocity of 300 ft/s is

$$Re = \frac{\rho V c}{\mu} = \frac{0.002048 \text{ slug/ft}^3 \, (300 \text{ ft/s}) \, (36 \text{ ft})}{0.0000003637 \text{ slug/ft s}} = 60{,}817{,}926$$

Therefore, the airfoil data curves for $Re = 9 \times 10^6$ (not standard roughness) will be used. The values of $\alpha_{L=0}$ and the c_l at $\alpha = 8$ deg do not, in fact, vary with Reynolds number. Their values can be read from the NACA 0009 data chart in Appendix B as

$$\alpha_{L=0} = 0 \text{ deg}, \qquad \text{at} \qquad \alpha = 8 \text{ deg}, \quad c_l = 0.85$$

Because the lift-coefficient curve appears linear between $\alpha_{L=0} = 0$ and 8 deg, the lift-curve slope can be estimated as the change in lift coefficient divided by the change in angle of attack:

$$c_{l_\alpha} = \frac{0.85 - 0}{8 \text{ deg} - (0 \text{ deg})} = 0.106/\text{deg}$$

The dynamic pressure is

$$q = \frac{1}{2}\rho V^2 = \frac{1}{2}(0.002048 \text{ slug/ft}^3)(300 \text{ ft/s}^2) = 92.16 \text{ lb/ft}^2$$

The Vulcan's aspect ratio is determined using Eq. (4.2):

$$R = \frac{b^2}{S} = \frac{(111 \text{ ft})^2}{3964 \text{ ft}^2} = 3.1$$

and the wing lift-curve slope is predicted by Eq. (4.14):

$$C_{L_\alpha} = c_{l_\alpha}\bigg/ 1 + \frac{57.3 \, c_{l_\alpha}}{\pi e R} = 0.106/\text{deg}\bigg/ 1 + \frac{(57.3 \text{ deg/rad})(0.106/\text{deg})}{\pi \, (0.9)(3.1)}$$

$$= 0.0626/\text{deg}$$

The lift coefficient is then calculated using Eq. (4.13):

$$C_L = C_{L_\alpha}(\alpha - \alpha_{L=0}) = 0.0626/\text{deg}[8 \text{ deg} - (0 \text{ deg})] = 0.500$$

Then using Eqs. (4.15) and (4.16)

$$C_D = C_{D_0} + kC_L^2 = 0.009 + \frac{0.5^2}{\pi(0.7)(3.1)} = 0.0457$$

The lift and drag are then given by

$$L = C_L q S = 0.5 \, (92.16 \text{ lb/ft}^2)(3964 \text{ ft}^2) = 182{,}989 \text{ lb}$$

$$D = C_D q S = 0.0457 \, (27.8 \text{ lb/ft}^2)(3964 \text{ ft}^2) = 16{,}697 \text{ lb}$$

4.6   Mach-Number Effects

Mach number was defined in Sec. 3.73 as

$$M_\infty \equiv \frac{V_\infty}{a} \qquad (3.34)$$

where a is the speed of sound. Speed of sound is determined either from a standard atmosphere table like the one in Appendix B or is calculated using the expression

$$a = \sqrt{\gamma R T} \qquad (3.35)$$

In Eq. (3.35), T is the absolute temperature, R is the gas constant, and γ is the ratio of specific heats, with $\gamma = c_p/c_v = 1.4$ for air.

Even though they are nondimensionalized to remove the effect of velocity, an aircraft's lift coefficient curve and drag polar still change as the aircraft flies at very high speeds. As with airfoils (discussed in Sec. 3.7), pressure changes around an aircraft are magnified by density changes at higher Mach numbers. As Mach number increases to near unity and above, additional changes occur to the flow. These changes have profound effects on an aircraft's lift and drag.

4.6.1   Mach Waves

Consider an infinitesimally small body moving in the atmosphere. The body is making small pressure disturbances that are transmitted as sound waves. The body's Mach number indicates the relative speed between it and the sound waves it creates. If $M = 0$, then the sound waves radiate outward in concentric circles from the body like ripples from the point where a stone lands in a pond. Figure 4.19a illustrates this situation.

If the body is moving then the sound waves upstream of it are closer together because each successive wave is generated from a point further upstream. The speed relative to the body at which each wave moves upstream is $a - V$ because the body is moving the same direction as the wave. Downstream of the body just the reverse is true. The spacing between the waves is greater and the waves are moving at $a + V$ relative to the body. The closer spacing of the waves upstream of the body causes the sound to have a higher frequency or pitch, while the sound downstream has a lower pitch. This is why the sound of an automobile horn or train whistle shifts to a lower frequency as the vehicle passes. The effect is called the Doppler shift. Similar shifts in the frequencies of reflected radio and light waves are the basis for radar and laser speed detectors. The situation is illustrated in Fig. 4.19b.

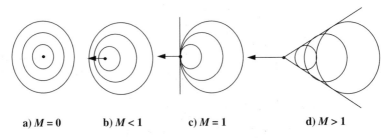

a) $M = 0$ b) $M < 1$ c) $M = 1$ d) $M > 1$

Fig. 4.19   Sound waves generated by a moving body.

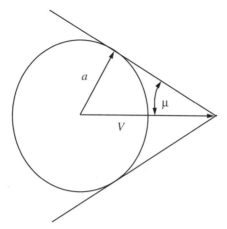

Fig. 4.20   Mach wave geometry.

The situation when $M = 1$ is illustrated in Fig. 4.19c. Note that the body is moving at the same speed as the sound waves it emits, so that all of the sound emitted by the body reaches a point ahead of it at the same time it does. The sound waves collect into a single pressure wave, known as a *Mach wave*, which is perpendicular to the direction of movement of the body.

When $M > 1$, the Mach wave trails back from the body at an angle, as shown in Fig. 4.19d. An expression for μ, the angle of the Mach wave (also known as the *Mach angle*), can be derived from the relationship between the velocity of the body and the velocity at which the sound waves move out from their point of origin, as shown in Fig. 4.20.

On the basis of the geometry of the Fig. 4.20, the expression for μ is

$$\mu = \sin^{-1}\frac{a}{V_\infty} = \sin^{-1}\frac{1}{M_\infty} \qquad (4.17)$$

4.6.2 Shock Waves

The pressure waves caused by a body moving through the air likewise influence the flowfield ahead of the body. Consider now a large body such as an aircraft or missile moving through the air. The influence of the high pressure at a stagnation point on the front of the body is transmitted upstream at the speed of sound, so that the flow slows down gradually rather than suddenly when encountering it. However, as the speed of the body through the air exceeds the speed of the sound waves this process of "warning" the air ahead that the body is approaching becomes impossible. In such a situation, the pressure change occurs suddenly in a short distance. This sudden pressure change is called a *shock wave*. Air flowing through a shock wave undergoes a rapid rise in pressure, density, and temperature, a rapid decrease in velocity, and a loss of total pressure. The angle of a shock wave is usually different than the Mach angle. It depends on the Mach number and the angular change of the flow direction as it goes through the shock wave. Figure 4.21 shows shock waves around a model of the space shuttle in the U.S. Air Force Academy's trisonic (high subsonic, transonic, and supersonic) wind tunnel. The

Fig. 4.21   Shock waves around a model of the space shuttle at $M = 1.7$ and two different angles of attack in the U.S. Air Force Academy's trisonic wind tunnel.

waves are made visible by the bending of the light waves as they pass through the regions of rapidly changing air density.

4.6.3   Critical Mach Number

Shock waves can occur around a body even when it is flying at speeds below the speed of sound. This happens because the air accelerates as it flows around the body. An airfoil, for instance, might be moving at $M = 0.8$ relative to the freestream, but it was shown in Chapter 3 that the shape of the airfoil causes the flow to be moving much faster over its upper surface. The local flow velocity over the upper surface of the airfoil might be greater than the speed of sound. This situation is described by saying the local Mach number is greater than one ($M > 1$), and the flow in this region is said to be *supersonic*. The freestream Mach number at which the local Mach number first equals unity is called the *critical Mach number* M_{crit}. Figure 4.22a illustrates this situation.

At $M_\infty = M_{crit}$ no shock wave forms because the local Mach number only equals 1.0 at one point. As M_∞ increases above M_{crit} however, the region where $M > 1$ grows. As shown in Fig. 4.22b, pressure waves from decelerating flow downstream of the supersonic region cannot move upstream into that region, and so they "pile up" into a shock wave. This shock wave at the downstream end of the supersonic region is called a *terminating shock* because it terminates the supersonic region and slows the flow abruptly to below the speed of sound. The strong adverse pressure gradient in the shock wave that slows the supersonic flow also slows the flow in the

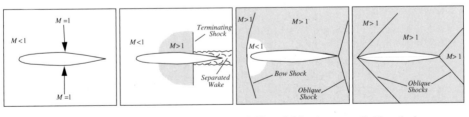

a) $M_\infty = M_{crit}$ b) $M_{crit} < M_\infty < 1$ c) $M_\infty > 1$, blunt nose   d) $M_\infty > 1$, sharp nose

Fig. 4.22   Flowfields at transonic and supersonic speeds.

boundary layer and often causes it to separate. This phenomenon is called *shock-induced separation*. It causes a significant increase in drag and decrease in lift. The sudden rise in drag as M_∞ approaches 1 was once thought to be an absolute barrier to higher speeds. It was called the sound barrier. The Mach number at which this rapid rise in drag occurs is called the *drag-divergence Mach number* M_{DD}.

As M_∞ exceeds 1.0, another shock wave forms a short distance in front of bodies with blunt or rounded leading edges. As shown in Fig. 4.22c, air flowing through this shock wave, called the *bow shock*, is abruptly decelerated to $M < 1$. The subsonic flow downstream of the bow shock can accelerate again to be supersonic as it flows around the body, but it will exert a significantly lower pressure on the rear part of the body because it has lost so much total pressure. This low pressure on the rear of the body produces a great deal of pressure drag, which is called *wave drag*. The bow wave is perpendicular or normal (it is also called a normal shock) to the flow directly ahead of the body, but its angle to the flow becomes the same as the Mach angle off to the sides of the body's path. The terminating shock moves to the trailing edge of the body, and no longer slows the flow to subsonic.

If M_∞ is sufficiently greater than unity and if the leading edge of the body is sharp, the bow shock will touch the body's point, as shown in Fig. 4.22d. The shock is said to be *attached*. Except at the point of attachment, the flow no longer decelerates below $M = 1$, but remains supersonic as it flows past the body. The shock wave at the leading edge and the one at the trailing edge trail off at an angle that initially depends on the shape of the body. Further from the body the shock angles become the same as the Mach angle. These shock waves are referred to as *oblique shocks* because they are not perpendicular to the flow. The loss of velocity and total pressure in oblique shocks is less than for normal shocks.

4.6.4 Flight Regimes

The range of Mach numbers at which aircraft fly is divided up into *flight regimes*. The regimes are chosen based on the aerodynamic phenomena that occur at Mach numbers within each regime and on the types of analysis that must be used to predict the consequences of those phenomena. Figure 4.23 shows these regimes. Mach numbers below M_{crit} are grouped together as the *subsonic* flight regime. Within this regime, compressibility effects are usually ignored for $M < 0.3$.

For freestream Mach numbers greater than about 1.3 (depending on aircraft shape), the flow is entirely supersonic (local Mach number remains greater than 1 everywhere in the flowfield), as shown in Fig. 4.22d. This is called the *supersonic* flight regime. Note that while some local Mach numbers in the flowfield are supersonic for speeds below the supersonic flight regime, an aircraft is only considered to be operating in the supersonic regime when all of the flow around it

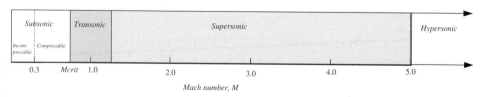

Fig. 4.23   Flight regimes.

(excluding flow entering the engine inlets) remains supersonic. Freestream Mach numbers above about 5.0 are considered *hypersonic*. The hypersonic flight regime is characterized by extreme temperature changes and significant interactions between oblique shock waves and the boundary layer.

Between the subsonic and the supersonic flight regimes lies the *transonic* regime. The transonic regime is characterized by a mixture of supersonic and subsonic flow, and in many cases there are also large areas of separation. Figures 4.22b and 4.22c show two examples of this. Transonic flowfields are too complex for accurate analysis by any but the most advanced methods, and the analysis often requires hours of computing time on the fastest supercomputers for a single flight condition. These methods are beyond the scope of this text.

4.6.5 Lift

All features of the lift curves of most aircraft vary with Mach number. The most important of these effects is the change in $C_{L\alpha}$. The relationship between $C_{L\alpha}$ and $c_{l\alpha}$ does not change significantly with changes in subsonic Mach number, so that the Prandtl–Glauert correction can be applied at subsonic speeds to $C_{L\alpha}$ in the same way it was applied to $c_{l\alpha}$ in Sec. 3.7.4:

$$C_{L_\alpha} = \frac{C_{L_\alpha M_\infty = 0}}{\sqrt{1 - M_\infty^2}} \qquad (4.18)$$

Equation (4.18) is valid only for $M_\infty < M_{\text{crit}}$. Also, the correction made by Eq. (4.18) becomes trivial for $M_\infty < 0.3$. This fact is part of the basis for setting the dividing line between incompressible and compressible flow at $M_\infty = 0.3$. For the supersonic flight regime, if $M_\infty > 1/\cos \Lambda_{LE}$ the lift-curve slope is given by

$$C_{L_\alpha} = \frac{4}{\sqrt{M_\infty^2 - 1}} \text{ per radian} \quad \text{or} \quad C_{L_\alpha} = \frac{4/57.3}{\sqrt{M_\infty^2 - 1}} \text{ per degree} \qquad (4.19)$$

Both Eqs.(4.18) and (4.19) yield an infinite value for lift-curve slope at $M_\infty = 1.0$. In fact, in the transonic regime, with so much shock-induced separation and complex flowfields, $C_{L\alpha}$ is difficult to predict. For a well-designed supersonic aircraft, $C_{L\alpha}$ levels off from the subsonic curve defined by Eq. (4.18) and transitions smoothly to the supersonic curve of Eq. (4.19). Figure 4.24 illustrates a typical variation of $C_{L\alpha}$ vs M_∞.

$C_{L\text{max}}$ and $\alpha_{L=0}$ also vary with Mach number. $C_{L\text{max}}$ initially increases as a result of compressibility effects and then decreases as shock-induced separation causes stall at lower angles of attack. The zero-lift angle of attack for cambered airfoils remains unchanged at subsonic speeds but becomes zero in the supersonic regime. In addition an airfoil's aerodynamic center moves from near an its quarter-chord for subsonic speeds to near its half-chord when supersonic. This can have a profound effect on an aircraft's stability and control, and will be discussed in Chapter 6.

4.6.6 Drag at High Subsonic Mach Numbers

Drag results from a complex set of phenomena, and high Mach numbers only add to the complexity. In the subsonic regime, primary changes in C_{D_0} and k_1 for a given aircraft are caused by increasing Reynolds number as Mach number increases. These changes depend on the relative importance of skin-friction drag and

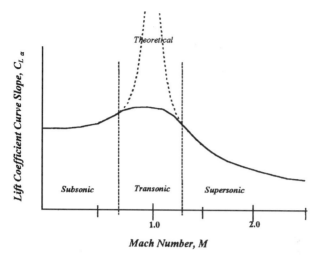

Fig. 4.24   Typical variation of lift-coefficient-curve slope with Mach number.

pressure drag for particular aircraft configurations. For many aircraft the changes are negligible. For early conceptual design, it is frequently acceptable to assume C_{D_0} and k_1 do not vary with Mach number below M_{crit}.

4.6.7 Supersonic Zero-Lift Drag

For the supersonic flight regime, wave drag is added to other types of drag. Theoretical analyses and wind-tunnel tests show that at supersonic speeds slender, pointed bodies whose cross-sectional areas vary as shown in Fig. 4.25 have minimum wave drag for their size. These low-wave-drag shapes are known as *Sears–Haack bodies* after the engineers[2] who initially studied them. The mathematical relationship for the area distribution that produces minimum wave drag is

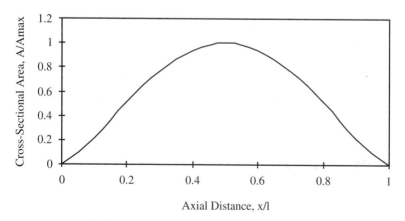

Fig. 4.25   Sears–Haack body area distribution.

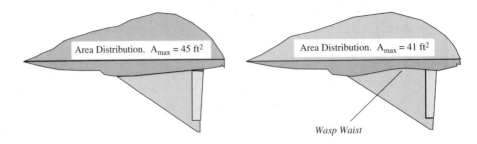

Note: Both aircraft have the same internal volume

Fig. 4.26   Area rule applied to a supersonic fighter aircraft.

called the *area rule*. The magnitude of wave drag for these bodies varies as

$$C_{D_{\text{wave}}} = \frac{4.5\pi}{S}\left(\frac{A_{\max}}{l}\right)^2 \tag{4.20}$$

where $A_{\max}$ is the maximum cross-sectional area of the body and l is its overall length. To achieve minimum wave drag for supersonic aircraft, designers strive to make the cross-sectional areas of their designs vary like Fig. 4.25. The process is called "applying the area rule" or just "area ruling." Area ruling might require reducing the area of the fuselage where the wing is attached to avoid a bump in the area plot. The result is a "wasp waist" as is seen on such aircraft as the T-38 and F-106. Figure 4.26 illustrates an example of this, while Fig. 4.27 illustrates the T-38's wasp waist.

At this point it is interesting to remember the discussion in Chapter 3 about the advantages a thicker airfoil with a larger leading-edge radius gave to the Fokker DVII in World War I. Thick airfoils were popular on all subsequent types of aircraft

Fig. 4.27   Area ruling of the T-38 fuselage.

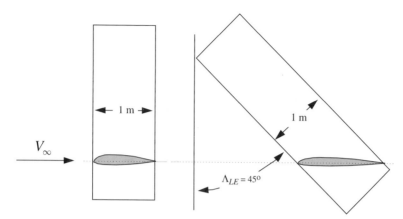

Fig. 4.28   Effect of wing sweep on streamwise thickness-to-chord ratio.

until after the start of World War II, when the fastest piston-engined fighters began reaching or exceeding the critical Mach number for their wing's airfoil. Severe control difficulties frequently resulted. The phenomenon was not well understood at the time, but it was observed that fighter planes with thinner airfoils could fly faster before encountering the problem. Two similar fighter aircraft produced by the same company exemplify the effect. The Hawker Typhoon and Tempest fighters had the same engine and fuselage, but the Typhoon had a smaller wing with a greater thickness-to-chord ratio. The Tempest had a maximum speed that was 50+ kn faster than the Typhoon, though it was nearly identical except for its wing. As maximum speeds of fighter aircraft have continued to increase, their airfoils have gotten progressively thinner, so that the thin, highly cambered airfoil sections of the outer wing panel of the F-15 are similar (though far from identical) to the airfoils of the World War I Sopwith Camel!

4.6.8   Effect of Wing Sweep

In addition to reducing airfoil thickness, aircraft designers can also raise a wing's M_{crit} by sweeping it either forward or aft. To understand how this works, consider the untapered, swept wing in Fig. 4.28. Sweeping the wing without changing its shape increases the effective chord length. Figure 4.28 shows why this is true. Increasing the airfoil chord length without changing thickness lowers the airfoil's thickness-to-chord ratio. This in turn reduces the amount the flow must speed up to get past the airfoil. M_{crit} therefore increases because, if the flow does not accelerate as much, the freestream Mach number can be higher before flow over the airfoil reaches the speed of sound.

4.6.9   Supersonic Drag Caused by Lift

At supersonic speeds, all airfoils, regardless of shape, generate zero lift at zero angle of attack. Practical supersonic airfoil shapes also generate minimum drag at zero lift and zero angle of attack. For well-designed supersonic aircraft, the drag-due-to-lift parameter k increases smoothly to very high supersonic values, sometimes exceeding 0.5.

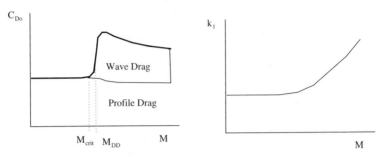

Fig. 4.29   Variation of C_{D_0} and k with Mach number.

4.6.10   Total Drag

In summary, the total drag on an aircraft is the sum of profile drag (the subsonic drag not caused by lift), wave drag, and drag-due-to-lift or induced drag:

$$C_{D_0} = C_{Dp} + C_{Dwave} \qquad \text{and} \qquad C_D = C_{D_0} + kC_L^2 \qquad (4.21)$$

Figure 4.29 shows how these vary with Mach number for a typical supersonic aircraft. Note the plot of C_{D_0} vs Mach in Fig. 4.29 shows the critical Mach number M_{crit} and the drag-divergence Mach number M_{DD}, as discussed in Sec. 4.6.3.

4.6.11   Mach Cone

A further consideration for the shape of supersonic aircraft is the benefit to be gained by keeping the wing inside the shock-wave cone generated by the aircraft's nose, as shown in Fig. 4.30. This practice reduces the aircraft's wave drag because

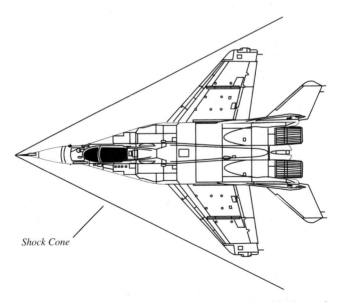

Fig. 4.30   Wings stay within nose shock cone at design Mach number.

the Mach number inside the cone is lower than M_∞, and shock waves are weaker than they would be if the wing were exposed to the full M_∞ flow velocity. Also, if a wing tip or other component projects outside the shock cone it will generate an additional shock wave. This further increases total wave drag on the aircraft and, where the two shock-waves intersect or interfere, causes additional disruption to the flow. In the case of very high Mach numbers, this shock-wave interaction contributes to additional heating of the aircraft's skin, which can lead to structural damage or failure.

Example 4.3

The SR-71 was a mystery plane in the 1970s and 1980s. Its maximum speed was a carefully kept secret. Yet, the practice of designing every part of an aircraft to fit inside the shock cone generated by its nose is especially important for aircraft like the SR-71 that fly at very high Mach numbers. At those speeds, impingement of a shock wave on a wing leading edge could cause excessive heating (i.e., the wing would melt!). The SR-71 designers must have obeyed this rule. So, with a picture of the SR-71 like Fig. 4.31, predict the Blackbird's maximum Mach number.

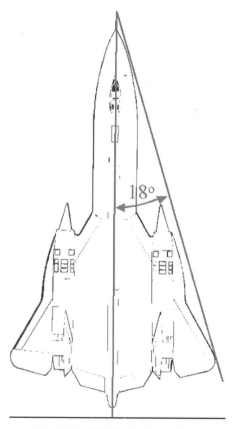

Fig. 4.31   SR-71 planform view.

Solution: From Fig. 4.31, the minimum shock angle for the SR-71 is 18 deg. Assuming this is the same as the Mach angle at its maximum speed, we solve Eq. (4.17) for M:

$$\mu = \sin^{-1}\frac{a}{V} = \sin^{-1}\frac{1}{M}$$

$$M = \frac{1}{\sin\mu} = \frac{1}{\sin(18\ \text{deg})} = 3.23$$

Information on the SR-71, once a military secret, is now available without restriction. Figure 4.32, from the plane's flight manual, verifies our prediction.

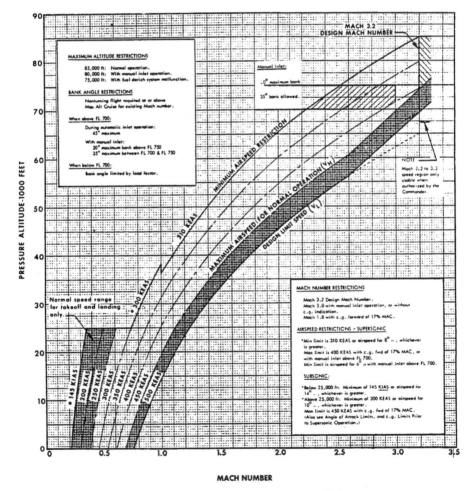

MACH NUMBER

Fig. 4.32   SR-71 flight envelope (data basis: flight test).

4.7 Chapter Summary

4.7.1 *Wing Nomenclature*

$$\lambda = \frac{c_t}{c_r} \qquad (4.1)$$

$$\mathcal{R} = \frac{b^2}{S} \qquad (4.2)$$

$$S = b \cdot \bar{c} \qquad (4.3)$$

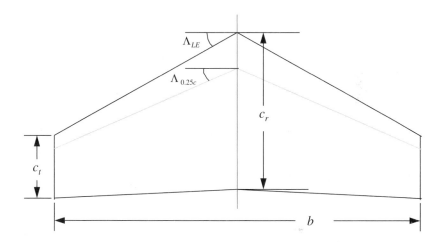

4.7.2 *Wing-Tip Vortices*

When a wing ends in the flow (wing tip), air flows around the tip from high pressure on bottom to low pressure on top.

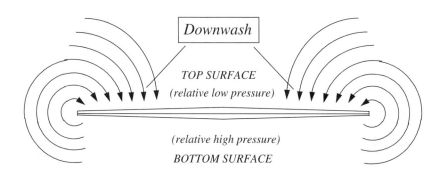

It creates a swirling, tornado-like flow pattern called a wing-tip vortex.

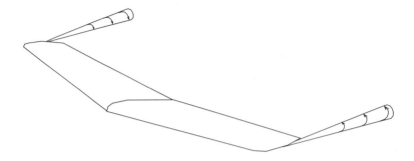

The vortex causes downwash.

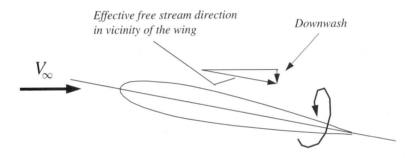

Downwash leads to less lift (because angle of attack is reduced) and more drag (because lift vector tilts).

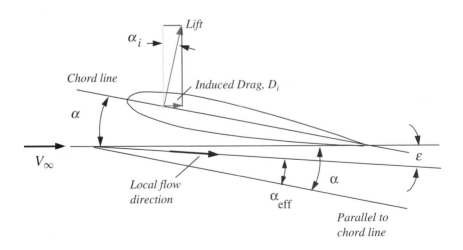

4.7.3 Equation Summary

Mach number:

$$M = \frac{V}{a}$$

Lift and lift coefficient:

$$L = C_L \, q S$$

Drag and drag coefficient:

$$D = C_D \, q S$$

Pitching moment and pitching-moment coefficient:

$$M = C_M \, q S \bar{c}$$

Aspect ratio:

$$R = \frac{b^2}{S}$$

Finite-wing lift-curve slope, where c_{l_α} is a two-dimensional airfoil lift-coefficient-curve slope and e is span efficiency factor:

$$C_{L_\alpha} = \frac{c_{l_\alpha}}{1 + 57.3 c_{l_\alpha}/\pi e R}$$

Finite-wing lift coefficient:

$$C_L = C_{L_\alpha} \, (\alpha - \alpha_{L=0})$$

Finite-wing drag coefficient, where c_d is a two-dimensional airfoil drag coefficient and e is span efficiency factor:

$$C_D = c_d + \frac{C_L^2}{\pi e R}$$

Aircraft drag polar:

$$C_D = C_{D_0} + k C_L^2$$

Aircraft drag-due-to-lift factor, where e_O is Oswald's efficiency factor:

$$k = \frac{1}{\pi e_o R}$$

See Fig. 4.33 for additional details.

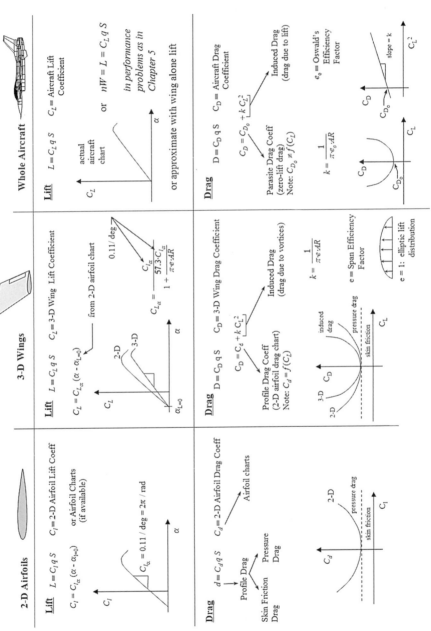

Fig. 4.33 Lift and drag summary.

4.7.4 Mach Effects

Speed of sound. The phenomenon of sound is caused by the propagation of pressure waves through a medium. The speed of propagation depends on the medium (air, water, etc.) and is known as the speed of sound. The abbreviation used is a. The speed of sound in a gas is a function of temperature as follows:

$$a = \sqrt{\gamma R T}$$

where γ is the ratio of specific heats of the gas and $\gamma = 1.4$ for air. Because temperature decreases in the standard atmosphere up to about 37,000 ft, the speed of sound decreases as you ascend in this region. This partially explains why an aircraft can attain a higher Mach number at higher altitudes. The speed of sound at sea level is about 661 kn, whereas it is only about 574 kn at 40,000 ft.

The *Mach number* is defined as the ratio of velocity to the speed of sound:

$$M = \frac{V}{a}$$

A weak pressure wave (called a *Mach wave*) forms in front of an aircraft as it is flying and can give us an indication of the aircraft's design Mach number:

$$M = 1/\sin(\mu)$$

$$\mu = \arcsin(1/M)$$

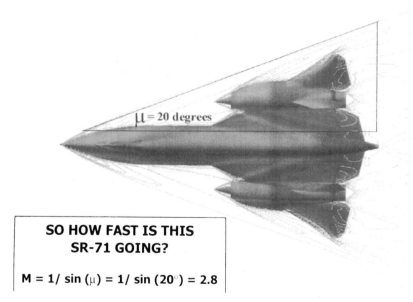

$\mu = 20$ degrees

SO HOW FAST IS THIS SR-71 GOING?

$$M = 1/\sin(\mu) = 1/\sin(20°) = 2.8$$

Critical Mach number M_{crit}. As an aircraft approaches Mach 1, the flow at some point on the aircraft will reach Mach 1 before the freestream velocity V_{∞} does because the flow accelerates over the aircraft's surface. The freestream Mach number M_∞ where this occurs is called the critical Mach number or M_{crit}. M_{crit} is always less than 1.

Shock waves.   As the aircraft accelerates beyond M_{crit}, shock waves form on the wings, fuselage, and other surfaces. A shock wave is a thin, highly viscous region that sharply increases the flow's static pressure but decreases the flow's velocity and total pressure.

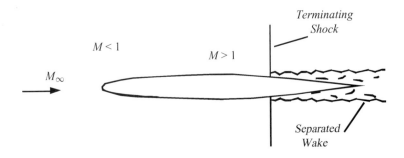

The location of the terminating shock moves aft as the aircraft's Mach number increases until it reaches the trailing edge of the wing. Another shock forms in front of the wing at higher Mach numbers. The type of shock in front depends on the geometry of the wing (or fuselage, etc.).

1) Blunt bodies generally experience normal (or bow) shocks:

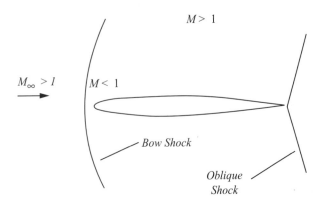

2) Bodies with sharp edges generally experience oblique (or attached) shocks:

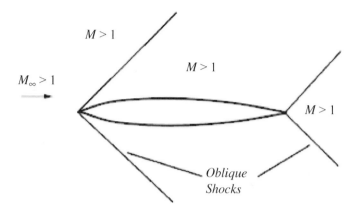

Flight regimes. We can divide the airspeed envelope into different regimes based on Mach effects:

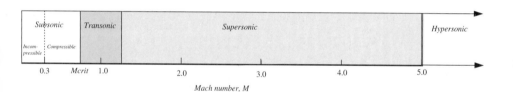

1) For the *subsonic* regime the entire flowfield has a velocity less than Mach 1.

2) The *transonic* regime begins at M_{crit}; the flowfield has velocities both less than and greater than Mach 1.

3) In the *supersonic* regime the entire flowfield has a velocity greater than Mach 1.

4) In the *hypersonic* regime the freestream velocity is greater than Mach 5 (depending on vehicle shape).

Shock waves cause additional drag. Because shock waves cause a loss in the flow's total pressure, there is an increase in pressure drag when they are present. This drag is referred to as *wave drag* because it is caused by the shock waves in particular as opposed to other mechanisms that create pressure drag. Shock waves tend to cause the flow to separate from the aircraft, which greatly increases the wave drag. This is known as shock-induced separation.

Drag-divergence Mach number M_{DD}. The drag rise is sharpest just above M_{crit}. The Mach number where this sudden drag rise occurs is called the *drag-divergence Mach number* or M_{DD}. M_{DD} is always slightly larger than M_{crit} but is still less than Mach 1.

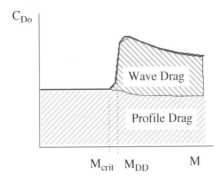

Lift-curve slope changes with Mach number.   Initially, the lift-curve slope increases as the density of the air changes because of the compressibility effects of Mach numbers greater than about 0.3. At higher Mach numbers, shock-induced separation dominates, and the lift-curve slope decreases.

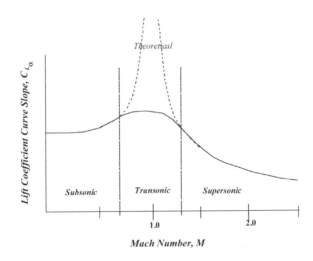

Mach effects on location of aerodynamic center.   Beyond Mach 1, the aerodynamic center of all aerodynamic surfaces moves from approximately the quarter-chord point to approximately the half-chord point. The importance of this will be discussed during the stability and control lessons.

4.8 More Details

4.8.1 Whole Aircraft Lift Calculations

A complete aircraft will frequently generate significantly more lift than its wing alone. An estimate of a whole aircraft's lift can be made by summing the lift contributions of its various components. The following is a simple method for making an initial estimate of an aircraft's lift. The method is suitable for use in the early conceptual phase of design.

4.8.2 Wing Contribution

For most aircraft, the majority of the lift is generated by the wing. The finite-wing lift prediction methods discussed in Sec. 4.2 give good initial estimates of wing lift curve slope, provided an appropriate value of e can be estimated. The results of extensive wind-tunnel testing[3] of a vast variety of wing shapes suggest the following empirical expression for span efficiency factor e

$$e = \frac{2}{2 - \mathcal{R} + \sqrt{4 + \mathcal{R}^2(1 + \tan^2 \Lambda_{t_{max}})}} \tag{4.22}$$

where $\Lambda_{t_{max}}$ is the sweep angle of the line connecting the point of maximum thickness on each airfoil of the wing.

One effect of airfoil camber and wing twist on lift is to shift the zero-lift angle of attack. A way to avoid the need for predicting zero-lift angle of attack early in the design process is to work in terms of *absolute* angle of attack:

$$\alpha_a = \alpha - \alpha_{L=0} \tag{4.23}$$

Because of the way α_a is defined, it always equals zero when lift is zero. Using absolute angle of attack is usually adequate for early conceptual design.

Estimating wing maximum lift coefficient is difficult without more advanced analysis methods. However, a practical constraint of takeoffs and landings leads to a simple way to estimate the maximum *usable* lift coefficient for those two phases of flight. Figure 4.34 shows an aircraft with a tricycle landing gear on a runway. When the aircraft accelerates to takeoff speed, it must rotate to the takeoff angle of attack in order to generate enough lift to become airborne. The aircraft normally tips back on its main landing gear as it rotates. The amount that the aircraft can rotate is limited by the tail striking the ground. This limitation also applies to landing because the aircraft will be at its landing angle of attack when it touches down. For many aircraft this angle is well below the wing's stall angle. Therefore, the maximum usable lift coefficient for takeoff or landing can be estimated as the wing lift-curve slope $C_{L\alpha}$ multiplied by the maximum usable absolute angle of attack $\alpha_{a_{max}} = 15$ deg $- \alpha_{L=0}$, in the case of Fig. 4.34.

$$C_{L_{max}} = C_{L_\alpha} \cdot \alpha_{a_{max}} = C_{L_\alpha} \cdot (\alpha_{max} - \alpha_{L=0}) \tag{4.24}$$

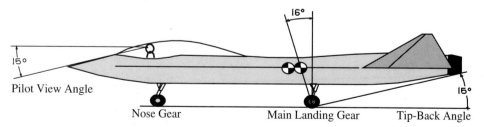

Fig. 4.34   Tip-back angle and pilot view angle.

Figure 4.34 also shows the pilot's downward view angle over the nose. The maximum usable angle of attack of an aircraft might be limited, at least for landing, by the pilot's visibility over the nose. This visibility requirement is particularly important for aircraft that must land on an aircraft carrier. This geometry constraint also can limit the maximum usable lift coefficient. So, as a result of limits on maximum rotation or tip-back angles and pilot view angles, a good rule of thumb for a value for maximum usable angle of attack is about 15 deg.

4.8.3 *Contribution of High-Lift Devices*

An approximate estimate for the effect of trailing-edge flaps on C_{Lmax} can be easily added to the wing C_{Lmax} prediction. Because most flaps change $\alpha_{L=0}$ but not $C_{L\alpha}$, their effect can be represented as an increment to the maximum usable absolute angle of attack. For flaps that span the entire wing, this increment in α_a is the same magnitude but of opposite sign as the increment in $\alpha_{l=0}$ in two-dimensional wind-tunnel data for an airfoil with the flap system mounted on it. If flapped airfoil data are not available, the increment can be approximated by another rule of thumb. Aircraft often use partial extension of flaps for takeoff and full flaps for landing. As a first approximation, a 10-deg increment in α_a for takeoff flap settings and 15 deg for landing flaps is acceptable.

For flaps that do not span the entire wing (a much more common situation), the increment in α_a is scaled by the ratio of flapped area S_f to reference planform area S. S_f is the area of that part of the wing that has the flaps attached to it. Figure 4.35 depicts S_f (shaded gray) for a typical wing. The change in maximum usable absolute angle of attack $\Delta\alpha_a$ is given by

$$\Delta\alpha_a = \Delta\alpha_{a_{2-D}} \frac{S_f}{S} \cos \Lambda_{h.l.} \qquad (4.25)$$

where $\Lambda_{h.l.}$ is the sweep angle of the flap hinge line, as shown in Fig. 4.45. After

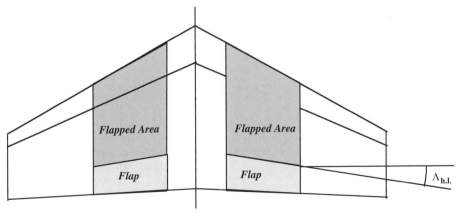

Fig. 4.35   Flapped area and flap hinge line sweep angle.

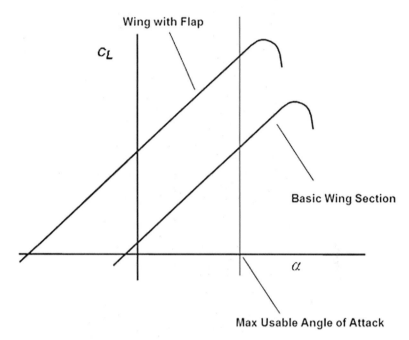

Fig. 4.36   Effect of flaps and maximum usable angle of attack on $C_{L\text{max}}$.

estimating $\Delta\alpha_a$, the maximum usable lift coefficient with flaps is approximated as

$$C_{L_{\max}} \cong C_{L_{\max}(\text{no flap})} + C_{L_\alpha} \cdot \Delta\alpha_a \qquad (4.26)$$

Note that Eq. (4.25) seems to disagree with the relationship between $C_{L_{\max}}$ for flapped and unflapped wings presented in Fig. 4.13. The reason for the difference becomes apparent if the maximum usable angle of attack line is superimposed on the C_L vs α curve, as shown on Fig. 4.36. Although this is not always the case, the situation depicted in Fig. 4.21 is common. The aircraft's maximum usable angle of attack for takeoff and landing is significantly below its clean stall angle of attack. When flaps are deflected, the maximum usable angle of attack is still below α_{stall} with flaps, so that the change in $C_{L_{\max}}$ is correctly predicted by Eq. (4.26).

4.8.4   Contribution of Fuselage and Strakes

An aircraft fuselage is usually relatively long and slender and therefore does not produce much lift. In the region of horizontal lifting surfaces, however, the lift being generated by those surfaces carries over onto the fuselage. This effect is modeled by treating the wing as if it extends all of the way through the fuselage without any change in airfoil, sweep, or taper. In fact, the fuselage shape is significantly different from the wing's airfoil shape and might be less effective at producing lift. However, because the fuselage lifting area is generally larger than the portion of the wing in the fuselage the two effects can be treated as canceling each other out, at least for early conceptual design. This is especially true for all-wing aircraft, like the B-2A, and for airliners like the Boeing 777, which has a long, slender fuselage. It is *not* a good approximation for many jet fighter configurations, like the Lockheed Martin F-16 and F-22, which generate a very significant part of their lift from their fuselages and strakes.

For fuselages with strakes or leading-edge extensions, the effect should be included, even for a first estimate. For angles of attack below 15 deg, the strake vortex is not very strong, and extensive wind-tunnel testing[4] has shown that the lift-curve slope of the wing with strake can be modeled as

$$C_{L_\alpha(\text{with strake})} = C_{L_\alpha(\text{without strake})} \frac{S + S_{\text{strake}}}{S} \tag{4.27}$$

where S_{strake} includes only the *exposed surface area* of the strake, not any portion inside the fuselage. Because $\alpha = 15$ deg is usually the maximum usable α, Eq. (4.27) is adequate for the usable range.

4.8.5 Horizontal Stabilizers and Canards

The purposes of additional horizontal lifting and stabilizing surfaces on an aircraft will be discussed in Chapter 6. For a first estimate of the lift contributions of these surfaces, it is sufficient to treat them as additional wings. However, the *downwash* created by the main wing will change the effective angle of attack of smaller horizontal surfaces in the wing's wake. Figure 4.37 illustrates this effect. Figure 4.37 also shows an *upwash* field, which increases the effective angle of attack of horizontal surfaces ahead of the wing. Of course, these smaller surfaces also create their own upwash and downwash. These upwash and downwash fields as a result of smaller surfaces will be ignored because they are generally much weaker than those of the main wing.

To determine a horizontal surface's contribution to the whole aircraft's lift-curve slope, it is first necessary to determine the rate at which downwash (or upwash as appropriate) ε changes with changing aircraft angle of attack. In the extreme case where the rate of change in the downwash angle equals the rate of change in angle of attack (i.e., $\partial\varepsilon/\partial\alpha = 1$), the rate of change of the effective angle of attack of a surface in that downwash field is zero. That surface would make no contribution to the whole aircraft's lift-curve slope.

Estimates of the rate of change of downwash angle with angle of attack can be made using the following empirical (based on testing rather than theory) curve fit of wind-tunnel[3] data:

$$\frac{\partial\varepsilon}{\partial\alpha} = \frac{21°C_{L_\alpha}}{R^{0.725}} \left(\frac{c_{\text{avg}}}{l_h}\right)^{0.25} \left(\frac{10 - 3\lambda}{7}\right) \left(1 - \frac{z_h}{b}\right) \tag{4.28}$$

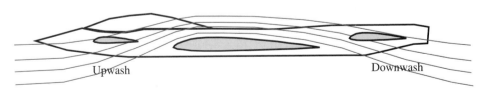

Fig. 4.37   Upwash and downwash.

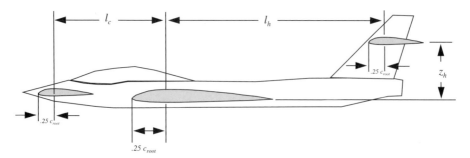

Fig. 4.38   Airplane geometry for downwash prediction.

where c_{avg} is the mean geometric chord of the wing l_h is the distance from the quarter-chord point of the average chord of the main wing to the quarter average chord point on the horizontal surface, as shown in Fig. 4.38; and z_h is the vertical distance of the horizontal surface above the plane of the main wing, as shown in Fig. 4.38.

Once $\partial\varepsilon/\partial\alpha$ is predicted, the horizontal surface's contribution to the aircraft's $C_{L\alpha}$ is approximated as

$$\Delta C_{L_\alpha(\text{due to horizontal tail})} = C_{L_{\alpha t}}\left(1 - \frac{\partial\varepsilon}{\partial\alpha}\right)\frac{S_t}{S} \qquad (4.29)$$

where the subscript t denotes parameters for the horizontal tail. Common $\Delta C_{L_\alpha(\text{due to horizontal tail})}$ values vary from almost zero to 35% or more of C_{L_α}.

For horizontal surfaces ahead of the wing, also known as *canards*, the empirical equation[3] for predicting the rate of change of upwash with angle of attack for wings with $\Lambda_{0.25} < 35$ deg is

$$\frac{\partial\varepsilon_u}{\partial\alpha} = (0.3\,\mathcal{R}^{0.3} - 0.33)\left(\frac{l_c}{c}\right)^{-(1.04 + 6\mathcal{R}^{-1.7})} \qquad (4.30)$$

where ε_u is the upwash angle and l_c is the distance from the wing's quarter-chord to the canard's quarter-chord as shown in Fig. 4.50.

Once $\partial\varepsilon_u/\partial\alpha$ is estimated, the canard's contribution to the aircraft's $C_{L\alpha}$ is approximated as

$$\Delta C_{L_\alpha(\text{due to canard})} = C_{L_{\alpha c}}\left(1 + \frac{\partial\varepsilon_u}{\partial\alpha}\right)\frac{S_c}{S} \qquad (4.31)$$

where the subscript c identifies quantities related to the canard. Contributions of canards to the total aircraft lift-curve slope are typically larger than those for horizontal tails. This is partly because of the canard being in an upwash field rather than the downwash field surrounding most horizontal tails. Once the contributions of canards and horizontal tails are estimated, the whole aircraft lift-curve slope is given by

$$C_{L_\alpha(\text{whole aircraft})} = C_{L_\alpha(\text{wing+body+strake})} + \Delta C_{L_\alpha(\text{due to horizontal tail})} + \Delta C_{L_\alpha(\text{due to canard})} \qquad (4.32)$$

4.8.6 Calculating Whole Aircraft Drag

The drag of a whole aircraft is frequently much greater than that of the wing alone. For this reason, it is important to include all significant contributions to total drag when estimating whole aircraft drag. The following is a simple method for estimating the drag polar of a complete aircraft, suitable for use in conceptual aircraft design.

Recall that the drag polar for a complete aircraft is written somewhat differently than that for a wing alone. For the whole aircraft, drag is identified as either *parasite drag* or *drag due to lift*. The parasite drag is all drag on the aircraft when it is not generating lift. This includes both skin-friction and pressure drag, as well as several additional types of zero-lift drag that are associated with the complete aircraft configuration. The drag due to lift includes all types of drag that depend on the amount of lift the aircraft is producing. Although confusing, it is common to refer to the $k C_L^2$ term in Eq. (4.32) as induced drag, though it is significantly different from the induced drag in Eq. (4.10). Drag due to lift includes induced drag caused by wing-tip vortices and downwash, the pressure drag that increases with lift caused by forward movement of the separation point, induced and pressure drag from canards and horizontal tails, and addition drag such as vortex drag caused by the leading-edge vortices on strakes and highly swept wings. All of these types of drag can be approximated by the following simple expression for drag coefficient:

$$C_D = C_{D_0} + k_1 C_L^2 + k_2 C_L \qquad (4.33)$$

where

$$k_1 = 1/(\pi e_O R) \qquad (4.34)$$

Note that $k_2 C_L$ is an additional term added to the drag polar Eq. (4.33) when compared with Eq. (4.15). The coefficient k_2 is chosen to allow modeling of wings with airfoils that generate minimum drag at some nonzero value of lift. C_{D_0} is still called the parasite drag coefficient or the zero-lift drag coefficient, but it might have a slightly different value when k_2 is nonzero. It still represents all drag generated by the aircraft when it is not generating lift.

To model the common situation where minimum drag occurs at a positive value of lift coefficient, the coefficient k_2 must be negative. This has the effect of shifting the entire C_D vs C_L curve to the right. Figure 4.39 illustrates this effect. The C_L for which C_D is a minimum is called C_{LminD}.

4.8.7 Parasite Drag

Just as lift predictions for the early stages of conceptual design rely heavily on results of wind-tunnel testing of similar configurations, so drag predictions rely heavily on drag data for similar types of aircraft. In later design stages, it is necessary to make very precise predictions of the aircraft's drag because just a 1% difference in the drag at cruise conditions, for instance, can make the difference between success and failure of a design. The methods used in making these precise predictions go far beyond the scope of this textbook and require details of the design that are generally not available early in the conceptual design phase. It is important, however, to understand in a qualitative sense where the drag on an aircraft comes from.

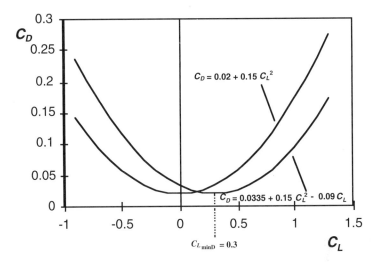

Fig. 4.39 Example of drag polar with minimum drag coefficient at nonzero lift coefficient.

Skin-friction drag on a complete aircraft configuration is generally much greater than that on the wing alone because the wetted area S_{wet} is greater. Wetted area of an aircraft is all of the surface area over which air flows and therefore to which the flowing air imparts shear stress. Pressure drag for the complete aircraft includes drag caused by separation of the airflow around the aircraft fuselage, control surfaces, etc., in addition to the wing. *Interference drag* results from flow interactions between the various components of an aircraft, which cause them to have more drag when assembled together than the sum of their drags when tested in a wind tunnel separately. Miscellaneous drags include drag caused by cooling air flowing through heat exchangers, air that leaks through doors and fairings that do not fit perfectly and around moveable surfaces, plus the drag of antennae, gun barrels, sensors, etc. that protrude from the aircraft. The total of all of these drags is the *profile drag* of the complete aircraft. To this must be added wave drag if the aircraft flies near or above the speed of sound.

A very good initial estimate of subsonic parasite drag can be made from drag data for similar aircraft using the concept of an *equivalent skin-friction drag coefficient*, C_{fe}, which is defined as follows:

$$C_{fe} = C_{D_0} \frac{S}{S_{wet}} \qquad (4.35)$$

Table 4.1 lists average C_{fe} values for several classes of aircraft. These values are based on historical data[5,6] for large numbers of each type of aircraft. C_{fe} is a function of such diverse factors as aircraft skin materials and shape; paint; typical flight Reynolds numbers; number of additional air scoops for ventilation; type, size, number, and location of engine air inlets; and attention to detail in sealing doors, control surface gaps, etc. Naturally, these details vary significantly from aircraft to aircraft, but the data in Ref. 6 suggest that there is enough similarity

Table 4.1   Common C_{fe} values

Type	C_{fe}
Jet bomber and civil transport	0.0030
Military jet transport	0.0035
Air Force jet fighter	0.0035
Carrier-based Navy jet fighter	0.0040
Supersonic cruise aircraft	0.0025
Light single-propeller aircraft	0.0055
Light twin-propeller aircraft	0.0045
Propeller seaplane	0.0065
Jet seaplane	0.0040

among aircraft of a given class that useful average C_{fe} values can be established. Table 4.1 lists the most commonly used values of C_{fe} (Refs. 5 and 6).

Using C_{fe} to predict C_{D_0} for an aircraft that generates minimum drag when it is generating zero lift only requires selecting a C_{fe} for the appropriate category of aircraft and estimating the total wetted area of the aircraft concept. The value of C_{D_0} is then obtained by solving Eq. (4.35):

$$C_{D_0} = C_{fe}\frac{S_{wet}}{S} \tag{4.36}$$

4.8.8 Supersonic Wave Drag

$C_{D_{wave}}$ for aircraft with reasonably smooth area distributions that conform approximately to Fig. 4.25 can be predicted using the following modification of Eq. (4.20):

$$C_{D_{wave}} = \frac{4.5\pi}{S}\left(\frac{A_{max}}{l}\right)^2 E_{WD}(0.74 + 0.37\cos\Lambda_{LE})\left[1 - .3\sqrt{M - M_{C_{D_0}max}}\right] \tag{4.37}$$

where

$$M_{C_{D_0}max} = \frac{1}{\cos^{0.2}\Lambda_{LE}} \tag{4.38}$$

estimates the Mach number where the maximum value of C_{D_0} occurs. Equation (4.37) is only valid for $M_\infty \geq M_{C_{D_0}max}$. E_{WD} is an empirical (based on experimental data) wave drag efficiency parameter. It is a measure of how closely the area distribution for the aircraft approximates the smooth curve of Fig. 4.20 and how free the aircraft is of additional sources of wave drag (antennae, leaks, bulges, wing-body junctions, engine inlets, etc.) The magnitude of E_{WD} averages about 2.0 for typical supersonic aircraft. The modifications in Eq. (4.37) are based on curve fits of wind-tunnel and flight-test data.[3]

Accurate estimation of the variation of C_{D_0} through the transonic regime is extremely difficult. As a simple approximation, a straight line is drawn between the subsonic C_{D_0} at M_{crit} and the C_{D_0} predicted at $M_{C_{D_0}max}$ by adding the $C_{D_{wave}}$

from Eq. (4.38) to the subsonic C_{D_0}. The resulting error typically is acceptable for early conceptual design, provided the aircraft will not cruise in the transonic regime.

M_{crit} is determined either by the shape of the fuselage or the shape of the wing, depending on which component creates the fastest velocities in the air flowing around it. For an unswept wing, the airfoil shape, especially its maximum thickness-to-chord ratio, determines how much the air accelerates as it flows around the wing and therefore how high V_∞ can be before $M = 1$ somewhere in the flowfield. Airfoil designers can expend considerable effort carefully shaping an airfoil to make its M_{crit} as high as possible and to delay the development of strong shock waves above M_{crit}. When actual airfoil data are not available, the following curve fit of M_{crit} data for NACA 64-series airfoils can be used:

$$M_{crit} = 1.0 - 0.065 \left(100 \, \frac{t_{max}}{c} \right)^{0.6} \tag{4.39}$$

where t_{max} is the airfoil's maximum thickness.

4.8.9 Effect of Wing Sweep

As mentioned in Sec. 4.6.8, aircraft designers also can raise a wing's M_{crit} by sweeping it either forward or aft. As shown in Fig. 4.28, sweeping the wing without changing its shape increases the effective chord length. From the geometry of Fig. 4.28, the relationship between the chord of the unswept wing and the chord of the swept wing is

$$c_{(swept\ wing)} = c_{(unswept\ wing)} / \cos \Lambda_{LE} \tag{4.40}$$

so that

$$\left(\frac{t_{max}}{c} \right)_{(swept\ wing)} = (\cos \Lambda_{LE}) \left(\frac{t_{max}}{c} \right)_{(unswept\ wing)} \tag{4.41}$$

Substituting the swept wing chord into Eq. (4.41) yields an expression for critical Mach number for swept wings:

$$M_{crit} = 1.0 - 0.065 \cos^{0.6} \Lambda_{LE} \left(100 \, \frac{t_{max}}{c} \right)^{0.6} \tag{}$$

or, in terms of the unswept wing's M_{crit},

$$M_{crit} = 1.0 - \cos^{0.6} \Lambda_{LE} [1.0 - M_{crit\,(unswept)}] \tag{4.42}$$

For tapered wings, the effect is modeled by using $\Lambda_{0.25c}$, the sweep angle of the line connecting the quarter-chord points of the wing's airfoils, and using the maximum value of (t_{max}/c) on the wing,

$$M_{crit} = 1.0 - \cos^{0.6} \Lambda_{0.25c} [1.0 - M_{crit\,(unswept)}] \tag{4.43}$$

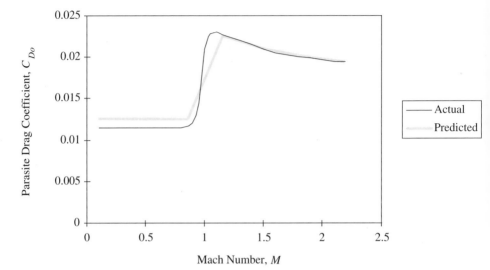

Fig. 4.40   Variation of actual and predicted C_{D_0} with Mach number for the F-106.

4.8.10   Fuselage Contribution

Only fuselages with relatively blunt noses will produce a value for M_{crit} that is lower than the one determined by the shape of the wing. Generally, a fuselage that has a long pointed nose, so that the fuselage reaches its maximum area at least six fuselage diameters downstream of the point of the nose, will ensure that M_{crit} caused by the fuselage is higher than M_{crit} resulting from the wings. The value for M_{crit} for the entire aircraft will be the lowest of the two.

The preceding methods give reasonably accurate predictions for C_{D_0} for a wide variety of existing supersonic aircraft. Figure 4.40 illustrates the actual variation of C_{D_0} with Mach number for the Convair F-106 Delta Dart supersonic fighter, along with C_{D_0} values predicted using the methods just described. Figure 4.50 also shows the type and magnitude of error that can be expected when approximating the transonic variation of C_{D_0} with a straight line.

4.8.11   Drag Due to Lift

Predicting drag due to lift must begin with predicting Oswald's efficiency factor e_O. This is done with a curve fit of wind tunnel data[3] for a variety of wing and wing–body combinations. The equation for this curve fit is

$$e_O = 4.61(1 - 0.045 R^{0.68})(\cos \Lambda_{LE})^{0.15} - 3.1 \qquad (4.44)$$

Note that increasing wing sweep tends to decrease the value of e_O. Also note that increasing R will tend to decrease e_O. This is because, for high-aspect-ratio wings, that part of the airfoil profile drag that varies with lift is a larger part of the total drag due to lift that e_O must model. Because increasing R decreases the value of k_1, e_O must decrease to partially offset this effect.

4.8.12 Effect of Camber

There are a number of reasons why an aircraft might generate its minimum drag at a positive (nonzero) value of lift coefficient. As one example, the profile drag on cambered airfoils is typically at a minimum at some small positive value of lift coefficient. As another example, the shape and orientation of an aircraft's fuselage can cause it to generate the least amount of drag at other than the zero-lift condition. Equation (4.33) has an additional term, the $k_2 C_L$ term, to model this effect. If, for instance, the minimum drag coefficient for an aircraft occurs at a lift coefficient signified by the symbol $C_{L\mathrm{minD}}$, then the necessary value of k_2 is given by

$$k_2 = -2 k_1 C_{L\mathrm{minD}} \tag{4.45}$$

The value of $C_{L\mathrm{minD}}$ is determined by plotting the drag polar for the wing using actual airfoil data and Eq. (4.10). If actual airfoil data are not available, as a crude approximation assume that the airfoil generates minimum drag when it is at zero angle of attack and that the effect of induced drag is to move $C_{L\mathrm{minD}}$ to a value halfway between zero and the value of C_L when $\alpha = 0$. The value of C_L when $\alpha = 0$ is given by

$$C_{L_{\alpha=0}} = C_{L_\alpha}(\alpha_a) = C_{L_\alpha}(-\alpha_{L=0}) \quad \text{because} \quad \alpha_a = \alpha - \alpha_{L=0} \quad \text{and} \quad \alpha = 0 \tag{4.46}$$

and

$$C_{L\mathrm{minD}} = C_{L_\alpha}\left(\frac{-\alpha_{L=0}}{2}\right) \tag{4.47}$$

This value of $C_{L\mathrm{minD}}$ is then used for the entire aircraft. This is done because it is assumed that the aircraft designer will design the fuselage, strakes, etc. so that they also have their minimum drag at the angle of attack that puts the wing at its $C_{L\mathrm{minD}}$. When this is done, the minimum value of C_D, which is given the symbol $C_{D\mathrm{min}}$, must not be any lower than the C_{D_0} predicted by Eq. (4.36). Recall that C_{D_0} is the aircraft's zero-lift drag coefficient. For aircraft with minimum drag at nonzero lift this leads to the following revised predictions:

$$C_{D\mathrm{min}} = C_{\mathrm{fe}}\frac{S_{\mathrm{wet}}}{S} \tag{4.48}$$

$$C_{D_0} = C_{D\mathrm{min}} + k_1 C_{L\mathrm{minD}}^2 \tag{4.49}$$

Remember that, for supersonic flight, wave drag must be added to the subsonic C_{D_0} to get the total supersonic C_{D_0}. Because cambered airfoils generate more wave drag in supersonic flight than do symmetrical airfoils, the additional term in Eq. (4.49) is retained when predicting total supersonic C_{D_0}.

4.8.13 Supersonic Drag Due to Lift

At supersonic speeds, all airfoils, regardless of shape, generate zero lift at zero angle of attack. Practical supersonic airfoil shapes also generate minimum drag at zero lift and zero angle of attack, so that in the supersonic regime $k_2 = 0$. The supersonic value of k_1 is given by

$$k_1 = \frac{R(M^2 - 1)}{(4\,R\,\sqrt{M^2 - 1}) - 2} \cos \Lambda_{\mathrm{LE}} \qquad (4.50)$$

For well-designed supersonic aircraft, the transition from subsonic to supersonic values of k is gradual, so that the variation of these parameters through the transonic regime can be approximated with a smooth curve.

4.9   Whole Aircraft Analysis Example

4.9.1   Lift

Figure 4.41 shows a drawing of an F-16 with the lifting surfaces and high-lift devices labeled. The F-16 uses a NACA 64_A-204 airfoil, which has its maximum thickness at 50% chord. The sweep angle of the line connecting the maximum thickness points of the airfoils, $\Lambda_{t\mathrm{max}} = 20$ deg. The flapped area for the trailing-edge flaps, as defined in Fig. 4.20, is approximately 150 ft². The flap hinge line sweep angle $\Lambda_{\mathrm{HL}} = 10$ deg. Aspect ratios of the wing and horizontal tail are

$$R = \frac{b^2}{S} = \frac{30^2}{300} = 3, \quad R_t = \frac{b_t^2}{S_t} = \frac{18^2}{108} = 3$$

Then, using Eq. (4.22) to estimate e (the same for both surfaces because they have

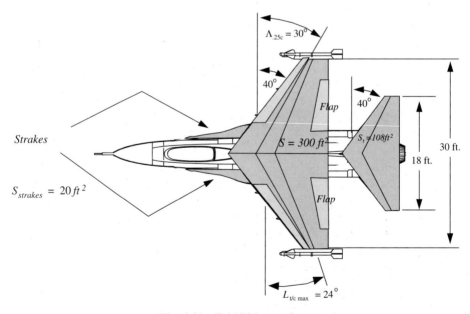

Fig. 4.41   F-16 lifting surfaces.

the same R):

$$e = \frac{2}{2 - R + \sqrt{4 + R^2 \left(1 + \tan^2 \Lambda_{t_{max}}\right)}}$$

$$= \frac{2}{2 - 3 + \sqrt{4 + 9\left(1 + \tan^2 24 \text{ deg}\right)}} = 0.703 = e_t$$

The two-dimensional lift-curve slope for the NACA 64_A-204 is approximately 0.1 per degree, so that

$$C_{L_\alpha} = \frac{c_{l_\alpha}}{1 + 57.3\, c_{l_\alpha}/\pi e R} = 0.0536/\text{deg} = C_{L_{\alpha t}}$$

for this unique situation where $R_t = R$. But this is before the effect of the strakes on the wing is included. For this

$$C_{L_\alpha(\text{with strake})} = C_{L_\alpha(\text{without strake})}\, \frac{S + S_{\text{strake}}}{S} = (0.0536/\text{deg})\, \frac{300 + 20}{300}$$

$$= 0.0572/\text{deg}$$

The distance from the quarter-chord of the main wing's mean chord to the same point on the F-16's horizontal tail $l_h = 14.7$ ft; the wing taper ratio $\lambda = 3.5$ ft / 16.5 ft $= 0.21$; and the distance of the horizontal tail below the plane of the wing z_h averages slightly less than 1 ft; so that using Eq. (4.21),

$$\frac{\partial \varepsilon}{\partial \alpha} = \frac{21°C_{L_\alpha}}{R^{0.725}} \left(\frac{c_{\text{avg}}}{l_h}\right)^{0.25} \left(\frac{10 - 3\lambda}{7}\right)\left(1 - \frac{z_h}{b}\right)$$

$$= \frac{(21\text{deg})\,(0.0572/\text{deg})}{3^{0.725}} \left(\frac{10 \text{ ft}}{14.7 \text{ ft}}\right)^{0.25}\left[\frac{10 - 3\,(0.21)}{7}\right]\left(1 - \frac{1 \text{ ft}}{30 \text{ ft}}\right)$$

$$= 0.64$$

and

$$C_{L_\alpha(\text{whole aircraft})} = C_{L_\alpha(\text{with strake})} + C_{L_{\alpha t}}\left(1 - \frac{\partial \varepsilon}{\partial \alpha}\right)\frac{S_t}{S} = C_{L_\alpha(\text{whole aircraft})}$$

$$= 0.0572/\text{deg} + 0.0536/\text{deg}\,(1 - .64)\,(108/300) = 0.064/\text{deg}$$

It is interesting to compare these predictions with the F-16's actual lift-coefficient-curve slope of 0.065/deg. Despite the complexity of the F-16 configuration, the analysis result agrees quite well with the actual slope. The method does not achieve this degree of accuracy in every case, but its predictions are quite good for a wide variety of aircraft configurations. Figure 4.42 compares actual aircraft-lift-coefficient curve slopes with slopes predicted by the method just described.

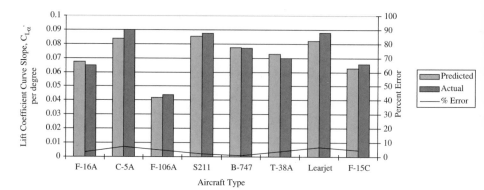

Fig. 4.42   Predicted and actual lift-coefficient-curve slopes for several aircraft.

Once the lift-curve slope is known, and using the fact that the F-16's landing gear limits its maximum usable angle of attack for takeoff and landing to 14 deg, we calculate its $C_{L\max}$ for takeoff and landing:

$$\Delta\alpha_a = \Delta\alpha_{a_{2-D}} \frac{S_f}{S}\cos\Lambda_{h.l.} = 4.9\,\text{deg}, \quad \text{so}\quad C_{L\max} = 0.064/\text{deg}(14\,\text{deg} + 4.9\,\text{deg})$$

$$= 1.21 \quad \text{for takeoff}$$

$$\Delta\alpha_a = \Delta\alpha_{a_{2-D}} \frac{S_f}{S}\cos\Lambda_{h.l.} = 7.36\,\text{deg}, \quad \text{so}\quad C_{L\max} = 0.064/\text{deg}(14\,\text{deg} + 7.36\,\text{deg})$$

$$= 1.37 \quad \text{for landing}$$

Because the F-16 is equipped with plain flaperons that deflect a maximum of 20 deg for both takeoff and landing, rather than slotted flaps that deflect much further for landing, it is no surprise that its actual usable $C_{L\max}$ for takeoff and landing is only 1.2, very close to the takeoff prediction but somewhat less than predicted for landing.

4.9.2   Parasite Drag

The first step in determining the F-16's parasite drag is to estimate its wetted area. This can be done in a variety of ways. If the aircraft is drawn accurately on a CAD system, a function is probably available within the system to determine S_{wet}. A reasonably accurate estimate can be obtained with much less effort than is required to make a detailed drawing, however, by approximating the aircraft as a set of simple shapes as shown in Fig. 4.43. The equations for the surface areas of these simple shapes are well known, and by taking some care to avoid counting areas where two shapes touch, it is relatively easy to determine S_{wet} as shown in Tables 4.2. Note that cross-sectional shapes of the items labeled "cylinders" and "half-cylinders" in Fig. 4.43 can be circular, elliptical, rectangular, or any other shape whose perimeter is easily determined. The surface areas of these cylinders do not include the areas of the ends because these generally butt up against another cylinder and are not wetted. When the longitudinal flat face of a half-cylinder or half-cone touches another body, twice the surface area of that face must be

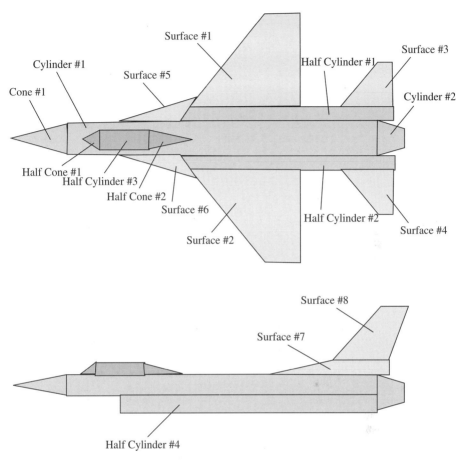

Fig. 4.43   F-16 geometry approximated by simple shapes.

subtracted because it and an equal area of the other body are in contact with each other, and therefore not wetted. It is interesting that the actual wetted area of the F-16 is 1495 ft^2, about 5% more than was estimated by this simple model. With a value for S_{wet} in hand and choosing $C_{\text{fe}} = 0.0035$ for a jet fighter from Table 4.1,

$$C_{D_{\min}} = C_{\text{fe}} \frac{S_{\text{wet}}}{S} = 0.0035 \, (1430 \text{ ft}^2/300 \text{ ft}^2) = 0.0167$$

4.9.3 Induced Drag

The value of the Oswald's efficiency factor e_O is estimated by Eq. (4.44) as

$$e_O = 4.61(1 - 0.045 \mathcal{R}^{0.68})(\cos \Lambda_{\text{LE}})^{0.15} - 3.1 = 0.906$$

Table 4.2a   F-16 wetted area estimation

Surface	Span, ft	c_{root}, ft	c_{tip}, ft	t_{max}/c	Equation[a]	S_{wet}	$-S_{intersections}$	net S_{wet}
1 and 2	12	14	3.5	0.04	(1)	419.5	0	419.5
3 and 4	6	7.8	2	0.04	(1)	117.5	0	117.5
5 and 6	2	9.6	0	0.06	(1)	38.6	0	38.6
7	1.4	12.5	6	0.10	(1)	26.3	0	26.3
8	7	8	3	0.06	(1)	77.3	0	77.3
9 and 10	1.5	5	3	0.03	(1)	23.9	0	23.9

Table 4.2b   F-16 wetted area estimation

Surface	Length, ft	Height, ft	Width, ft	h_2, w_2	Equation[a]	S_{wet}	$-S_{intersections}$	net S_{wet}
			Cylinder					
1	39	2.5	5	——	(2)	551.3	0	551.3
			Cone[b]					
1	6	2.5	5	0, 0	(3)	42.4	0	42.4
2	4	6	6	4, 4	(3)	62.8	0	62.8
			Half-cylinder					
1 and 2	24	0.8	1	——	50% of (2)	67.9	38.4	29.5
3	5	2	2	——	50% of (2)	15.7	10	5.7
4	30	2.5	5	——	50% of (2)	212.1	180	32.1
			Half-cone					
1	2	2	2	——	50% of (3)	3.1	2	1.1
2	4	2	2	——	50% of (3)	6.3	4	2.3
						Total S_{wet}		1418 ft^2

[a]Equations:

1) Wing or stabilizing surface:

$$S_{wet} = S_{exposed} [1.977 + 0.52 (t/c)]$$

where $S_{exposed}$ is the planform area of the surface excluding any portion that is inside another component, (i.e., not exposed). Note that the surfaces for this example were all drawn so that they were not inside any other component.

2) Cylinder sides:

$$S_{wet} = \pi l (h + w/2)$$

for elliptical cross sections and $S_{wet} = 2l (h + w)$ for rectangular cross sections, where l is the length of the fuselage segment, h is its height, and w is its width. The surface areas of the ends of the cylinders are not included because they butt up against the ends of other cylinders or cones. Typically, the only case where the end of a cylinder does not butt up against another surface is at the end of a jet-engine exhaust nozzle, where there is also no surface area. Some fuselages and external fuel tanks also terminate without tapering to a point, but this is generally done because flow separation has already occurred ahead of this position on the body. The equivalent skin-friction coefficient method accounts for this design decision more accurately if the area of the end of the cylinder is not included in the aircraft wetted area.

3) Cone and truncated cone sides:

$$S_{wet} = \pi l \left(\frac{h_1 + w_1 + h_2 + w_2}{4} \right)$$

for elliptical and circular cross sections and $S_{wet} = l(h_1 + w_1 + h_2 + w_2)$ for rectangular cross sections, where l is the length of the fuselage segment, h_1 is the height of the front end, w_1 is the width of the front end, h_2 is the height of the aft end of the segment, and w_2 is its width. The surface areas of the ends of the cones and truncated cones are not included for the same reasons given for cylinders.

[b]Cone and/or truncated cone (tapered cylinder).

Table 4.3   F-16 wing-alone drag-coefficient variation

C_L	c_d	$k_1 C_L^2$	$C_D = c_d + k_1 C_L^2$
−0.2	0.0062	0.0047	0.0109
−0.1	0.006	0.0012	0.0072
0.0	0.0053	0	0.0053
0.1	0.0045	0.0012	0.0057
0.2	0.0042	0.0047	0.0089

so

$$k_1 = 1/(\pi e_O \!R) = 0.117$$

The F-16's average chord is $\bar{c} = b/\!R = 30 \text{ ft}/3 = 10$ ft. For standard sea-level conditions and $M = 0.2$,

$$Re = \rho V \bar{c}/\mu$$

$$= \left(0.002377 \text{ slug/ft}^3\right) (0.2 \cdot 1116.4 \text{ ft/s}) (10 \text{ ft})/(0.3737 \cdot 10^{-6} \text{ slug/ft s})$$

$$= 14{,}200{,}000$$

Using Eq. (4.10) and airfoil data for the NACA 64_A-204 ($Re = 9 \times 10^4$), we generate Table 4.3, drag-coefficient data for the wing alone. From Table 4.3 it is apparent that $C_{L \min D} \cong 0.04$. Assuming a well-designed aircraft will have its minimum drag at approximately the same C_L where the wing alone has its minimum drag

$$C_{D_0} = C_{D_{\min}} + k_1 C_{L_{\min D}}^2 = 0.0167 + 0.117(0.04)^2 = 0.0169$$

$$k_2 = -2 k_1 C_{L_{\min D}} = -2 (0.117)(0.04) = -0.0094$$

$$C_D = 0.0169 + 0.117 C_L^2 - 0.0094 \, C_L$$

4.9.4 Supersonic Drag

The preceding analysis is valid for Mach numbers less than M_{crit}. Using the methods described in Sec. 4.8 to predict M_{crit}, $C_{D\text{wave}}$, and k_1,

$$M_{\text{crit(unswept)}} = 1.0 - 0.065 \left(100 \frac{t_{\max}}{c}\right)^{0.6} = 1.0 - 0.065(4)^{0.6} = 0.85$$

$$M_{\text{crit}} = 1.0 - \cos^{0.6} \Lambda_{0.25c} [1.0 - M_{\text{crit(unswept)}}] = 1.0 - \cos^{0.6} 30 \text{ deg}(1 - 0.85)$$

$$= 0.865$$

$$M_{C_{D_0}\max} = \frac{1}{\cos^{0.2} \Lambda_{\text{LE}}} = \frac{1}{\cos^{0.2} 40 \text{ deg}} = 1.05$$

$$C_{D\text{wave}} = \frac{4.5\pi}{S} \left(\frac{A_{\max}}{l}\right)^2 E_{\text{WD}}(0.74 + 0.37 \cos \Lambda_{\text{LE}}) \left[1 - 0.3\sqrt{M - M_{C_{D_0}\max}}\right]$$

Table 4.4   F-16 drag polar predicted using the
methods of Chapter 4

Mach number	C_{D0}	k_1	k_2
0.3	0.0169	0.117	−0.0094
0.86	0.0169	0.117	−0.0094
1.05	0.0428	0.128	−0.0047
1.5	0.0380	0.252	0
2.0	0.0356	0.367	0

$A_{max} = 25.5$ ft^2 and $l = 48.5$ ft, so that at Mach 1.05, $C_{Dwave} = 0.0261$; at Mach 1.5, $C_{Dwave} = 0.0261$; and at Mach 2, $C_{Dwave} = 0.0189$. And finally, the supersonic k_1 values

$$k_1 = \frac{\mathcal{R}(M^2 - 1)}{(4\mathcal{R}\sqrt{M^2 - 1}) - 2} \cos \Lambda_{LE} = 0.128 \quad \text{at} \quad M = 1.05$$

$$= 0.252 \quad \text{at} \quad M = 1.5$$

$$= 0.367 \quad \text{at} \quad M = 2$$

The drag polar results are summarized in Table 4.4. Once again it is interesting to compare these results with actual values for the F-16 listed in Table 4.5. The predicted values agree reasonably well with the actual values. The 12% lower subsonic C_{D_0} values were to be expected in part because of the 5% lower estimate of wetted area. A more accurate model of the aircraft made from more, smaller simple shapes could be expected to produce better S_{wet} and C_{D_0} estimates. Also, the fixed geometry of the F-16's air inlet produces a great deal of additional wave drag and flow separation at high supersonic Mach numbers. This explains why the F-16's C_{D_0} values actually increase at higher Mach numbers when the model suggests they should decrease. It is best to view the drag values predicted by these methods as goals that can be achieved with careful design. The F-16 is optimized for subsonic maneuvering, not supersonic cruise, and so it should not be expected to achieve the lowest possible supersonic drag. On the other hand, as shown by Fig. 4.45, the methods predict very accurately the wave drag of the F-106, an aircraft optimized for supersonic flight.

Table 4.5   Actual F-16 drag polar

Mach number	C_{D_0}	k_1	k_2
0.3	0.0193	0.117	−0.007
0.86	0.0202	0.115	−0.004
1.05	0.0444	0.160	−0.001
1.5	0.0448	0.280	0
2.0	0.0458	0.370	0

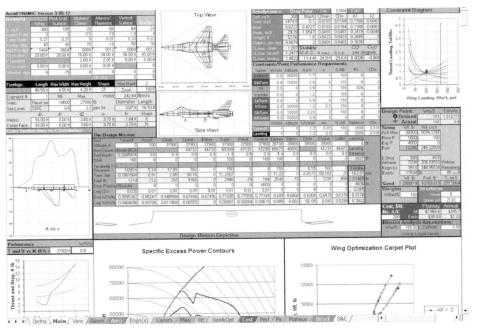

Fig. 4.44   AeroDYNAMIC analysis of the F-16.

4.9.5 AeroDYNAMIC

The computer program called AeroDYNAMIC, which accompanies this text, contains an aircraft design module that allows the user to approximate the shape of a conceptual aircraft design using simple shapes in the same manner as was done in this section. The shapes are described by parameters such as height, length, width, span, thickness, etc., which are entered into a spreadsheet. Aero-DYNAMIC draws the aircraft as the parameters are entered and then performs an aerodynamic analysis similar to the example in this section. This is all accomplished very rapidly, relieving the designer of many tedious calculations and drawing manipulations. The user is then free to explore many alternative design choices and to optimize the design. Similar, though much more capable CAD systems are used by virtually all modern aircraft designers. Figure 4.44 shows an analysis screen from AeroDYNAMIC. Note that more complex shapes have been used to approximate the F-16 shape than in Fig. 4.43. This gives the analysis greater fidelity.

References

[1]Nicolai, L. M., *Fundamentals of Aircraft Design*, METS, Inc. Xenia, OH, 1975.

[2]Sears, W. R., "On Projectiles of Minimum Wave Drag," *Quarterly of Applied Mathematics,* Vol. 4, No. 4, 1947.

[3]Fink, R. D., *USAF Stability and Control DATCOM*, Wright-Patterson AFB, OH, 1975.

[4]Lamar, J. E., and Frink, N. T., "Aerodynamic Features of Designed Strake-Wing Configurations," *Journal of Aircraft*, Vol. 19, No. 8, 1982, pp. 639–646.
    [5]Raymer, D. P., *Aircraft Design: A Conceptual Approach*, AIAA Education Series, AIAA, Washington, DC, 1989.
    [6]Roskam, J., *Airplane Design Part VI*, Roskam Aviation and Engineering Corp., Ottawa, KS, 1990.

Problems

Synthesis Problems

S-4.1 An aircraft design exhibits excessive transonic wave drag and an unacceptably low M_{crit}. Brainstorm five ways to change the design to improve these characteristics.

S-4.2 An aircraft design has excessive induced drag for its high-altitude, low-speed surveillance mission. Brainstorm five ways to reduce its induced drag.

S-4.3 A jet fighter aircraft cannot reach its required maximum speed when carrying its air-to-air missile armament on underwing pylons and wing-tip launch rails. You believe excessive parasite drag is the problem. Brainstorm at least five ways to allow the plane to carry its weapons with less parasite drag.

S-4.4 A large jet transport design is unable to achieve the value of C_{Lmax} it needs for acceptable takeoff and landing performance. Brainstorm five ways to increase its C_{Lmax}.

S-4.5 A jet fighter aircraft design concept achieves its maximum lift coefficient at an angle of attack of 27 deg, but its maximum usable angle of attack for takeoff and landing is only 15 deg. Brainstorm at least five design changes that would allow the aircraft to use more of its maximum lift capability for takeoff and landing.

Analysis Problems

A-4.1 A rectangular wing is placed in the 10-m wide test section of a low-speed wind tunnel with a test-section velocity of 80 m/s at an altitude of 2500 m on a standard day. The wing has a span of 5 m, chord of 1.34 m, and a NACA 2415 airfoil cross section. If the wing is at an angle of attack of 4 deg, calculate the following: (a) Reynolds number based on the chord length, (b) finite-wing lift coefficient C_L and drag coefficient C_D, (c) Lift L and drag D, and (d) Zero-lift angle of attack $\alpha_{L=0}$ and zero-lift drag coefficient C_{D_0}.

A-4.2 (a) Consider a flying-wing aircraft made using a NACA 2412 airfoil with a wing area of 250 ft^2, a wing span of 50 ft, and a span efficiency factor of 0.9. If the aircraft is flying at a 6 deg angle of attack and a Reynolds number of approximately 9×10^6, what are C_L and C_D for the flying wing?

(b) If the flying wing is flying at sea level at $V_\infty = 280$ ft/s, how much lift and drag is it experiencing?

A-4.3   What are two consequences when we have a wing (three-dimensional) instead of an airfoil (two-dimensional)?

A-4.4   (a) Induced drag is also called drag due to _____.

(b) When lift is zero, the induced drag is equal to _____?

(c) When would induced drag be more prevalent: during high-speed flight (such as cruise) or low-speed flight (such as landing)? Why?

A-4.5   (a) Draw a sketch of a typical C_L vs angle-of-attack curve and a C_D vs C_L curve, and show how increasing camber (by putting a flap down) would change these curves.

(b) Draw the same curves, and show how using boundary-layer control or leading-edge devices would change these curves.

(c) Draw the same curves, and show how sweeping the wings back (as on the F-111 or F-14) would change the curves.

A-4.6   (a) The maximum speed achieved by the X-15A-2 rocket-powered research craft was 6630 ft/s at 99,000 ft. Assuming standard day conditions, what was its Mach number?

(b) An SR-71 is traveling at Mach 3.2 in standard day 80,000-ft conditions. What is its velocity in nautical miles per hour?

A-4.7   A straight wing with a critical Mach number of 0.65 is swept back 35 deg. What is its critical Mach number in the swept configuration?

A-4.8   Draw a sketch of a typical C_{D_0} vs Mach-number curve for a jet fighter aircraft and show how sweeping the wings back (as on the F-111 or F-14) would change the curve. Label M_{crit} and M_{DD} on the curves.

A-4.9   An F-15 is flying straight and level (lift equals weight) in 25,000 ft standard day conditions at a velocity of 850 ft/s. Given the following data for the F-15, answer the questions that follow:

parasite drag coefficient $C_{D_0} = 0.018$
wing span $b = 42.8$ ft
wing area $s = 610$ ft^2
weight $W = 50,000$ lb
Oswald efficiency factor $e_0 = 0.915$

(a) What is the wing's aspect ratio R?

(b) What is the lift force L generated by the aircraft?

(c) What is the aircraft's drag coefficient C_D?

(d) What is the drag force D generated by the aircraft?

A-4.10   An aircraft with $C_{D_0} = 0.012$ and $k = 0.18$ is flying with $C_L = 0.26$. The pilot attempts to temporarily reduce the drag on the aircraft by reducing C_L to zero. By what percent will the drag be reduced in this situation? Why would it only be possible to do this temporarily?

A-4.11   Lift-to-drag ratio is a measure of the efficiency of an aircraft. What is the lift-to-drag ratio of an aircraft with $C_{D0} = 0.018$ and $k = 0.13$ operating at $C_L = 0.4$?

A-4.12   A straight-tapered flying-wing aircraft has the following characteristics: span, 20 ft; area, 100 ft^2; root chord, 6 ft; tip chord, 4 ft; leading-edge sweep angle, 40 deg; sweep angle of line of maximum thickness, 24 deg; and Airfoil NACA 0004.

Determine the wing's lift-curve slope and induced drag k value. Assume the flight Mach number is 0.25 in standard sea-level conditions.

A-4.13   An F-104 is 54 ft long and has a maximum cross-sectional area of 21 ft^2. Its total wetted area is 1200 ft^2. Its symmetrical-airfoil wing has a reference planform area of 196 ft^2, and a leading-edge sweep angle of 28 deg. Estimate its C_{D_0} at $M = 0.2$ and 2.0.

5
Performance

5.1 Design Motivation

Aircraft performance analysis is the science of predicting what an aircraft can do, how fast and high it can fly, how quickly it can turn, how much payload it can carry, how far it can go, and how short a runway it can safely use for take-off and landing. Most of the design requirements that a customer specifies for an aircraft are performance capabilities. Therefore, in most cases it is performance analysis that answers the question, "Will this aircraft meet the customer's needs?"

5.2 Equations of Motion

Figure 5.1 shows the forces and geometry for an aircraft in a climb. The *flight-path angle* γ is the angle between the horizon and the aircraft's velocity vector (opposite the relative wind). The angle of attack α is defined between the velocity vector and an aircraft reference line, which is often chosen as the central axis of the fuselage rather than the wing chord line. The choice of the aircraft reference line is arbitrary. The designer is free to choose whatever reference is most convenient, provided care is taken to clearly specify this choice to all users of the aircraft performance data. The *thrust angle* α_T is the angle between the thrust vector and the velocity vector. This will not, in general, be the same as α, because the thrust vector will not generally be aligned with the aircraft reference line.

The equations of motion for the aircraft in Fig. 5.1 are derived by summing the forces on the aircraft in two directions, one parallel to the aircraft's velocity vector and one perpendicular to it. These directions are convenient because lift was defined in Chapter 3 as the component of the aerodynamic force that is perpendicular to the velocity vector, and drag was defined as the component parallel to velocity. The summation parallel to the velocity is

$$\sum F_{\parallel} = ma = T \cos \alpha_T - D - W \sin \gamma \qquad (5.1)$$

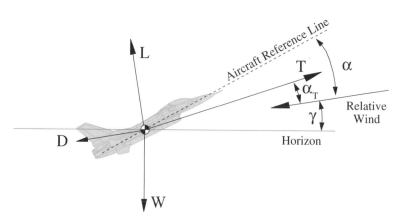

Fig. 5.1   Forces on an aircraft in a climb.

where m is the aircraft's mass and a is its instantaneous acceleration in the direction of the velocity vector. In Eq. (5.1) we have applied Newton's second law of motion, $F = ma$, to equate the sum of the forces on the aircraft parallel to its velocity vector to its mass multiplied by its acceleration parallel to its velocity vector. The acceleration perpendicular to the velocity vector is the centripetal acceleration V^2/r, where r is the radius of turn if the aircraft is turning. The summation perpendicular to velocity is

$$\sum F_\perp = V^2/r = T \sin \alpha_T + L - W \cos \gamma \qquad (5.2)$$

Note that the preceding summation assumes that all forces on the aircraft are in the vertical plane of the drawing. Therefore, if the aircraft is turning, it is turning in the vertical plane (e.g., doing a loop).

Equations (5.1) and (5.2) can be modified to apply to a variety of aircraft maneuvers and flight conditions. For instance, if the aircraft is in a dive ($\gamma < 0$) the same equations still apply. For many aircraft, it is acceptable to assume that the thrust vector is approximately aligned with the velocity vector, so that $\alpha_T = 0$. This simplifies Eqs. (5.1) and (5.2) because $\sin \alpha_T = 0$ and $\cos \alpha_T = 1$. This approximation will be used in the remainder of the performance analyses discussed in this text.

A very simple but extremely useful condition is that of steady, level, unaccelerated flight (SLUF). For SLUF, $\gamma = 0$ and both components of acceleration are zero, so that Eqs. (5.1) and (5.2) simplify to

$$T = D, \qquad L = W \qquad (5.3)$$

Methods for predicting the aerodynamic forces in Eq. (5.3) were discussed in Chapter 4. For the purposes of this chapter, the aircraft weight will be given as the sum of the aircraft empty weight W_e, the weight of the fuel W_f, and the weight of the payload (including pilot and crew) W_p, that is,

$$W = W_e + W_f + W_p \qquad (5.4)$$

Assume for analysis of aircraft that the acceleration of gravity is constant at 9.8 m/s^2 (32.2 ft/s^2) and does not vary significantly with altitude. This leaves thrust as the only quantity in Eq. (5.3) that has not yet been discussed.

5.3 Propulsion

Production of thrust is a topic that could easily occupy an entire chapter or even an entire book. Its treatment here will be limited to the general concepts needed to predict aircraft performance.

5.3.1 Propulsion Choices

The aircraft designer has a wide range of choices for propulsion systems. Each system has characteristics that make it most suitable for particular flight regimes. One of the characteristics of most interest is the ratio of an engine's sea-level output to its own weight T_{SL}/W_{eng}. Another is the engine's thrust-specific fuel consumption (TSFC), which is the ratio of rate of fuel consumption to thrust output:

$$\text{TSFC} \equiv \frac{\dot{W}_f}{T} \qquad (5.5)$$

TSFC is frequently represented by the symbol c_t. If fuel consumption rate has units of pounds per hour and thrust is in pounds, then TSFC has units of reciprocal hours. An engine that is deemed suitable for a particular flight regime would have a relatively high T_{SL}/W_{eng} and a relatively low TSFC in that regime. Figures 5.2 and 5.3 show the variation of T_{SL}/W_{eng} and TSFC with Mach number for several types of engines. Figure 5.4 shows common operating envelopes (ranges of operating altitudes and Mach numbers) of common engine types. Each engine type is described in more detail in the following paragraphs.

5.3.2 Piston Engines

Powerplants of most aircraft from the days of the Wright brothers to the end of World War II were internal combustion piston engines. These devices produce power by mixing air and liquid fuel as they are drawn into variable-volume chambers or cylinders, then compressing and burning the mixture. The explosive increase in the pressure of the burned mixture is converted into power by allowing the gas to push a piston or similar moveable wall of each chamber. The motions of the pistons are converted into rotary motion by linkages called connecting rods that push on a crankshaft in much the same way a bicyclist's foot pushes on a bicycle pedal. Figure 5.5 illustrates the components and action of a simple single-cylinder piston engine. Rotary engines achieve the same effect, although their inner workings are different. They have essentially the same performance characteristics as piston engines.

The power produced by a piston engine varies with the size and number of cylinders, the rate at which the crankshaft rotates, and the density of the air it is using. Engine shaft power (SHP) ratings are normally expressed as horsepower (1 hp = 550 ft lb/s) or kilowatts in standard sea-level conditions at a specified

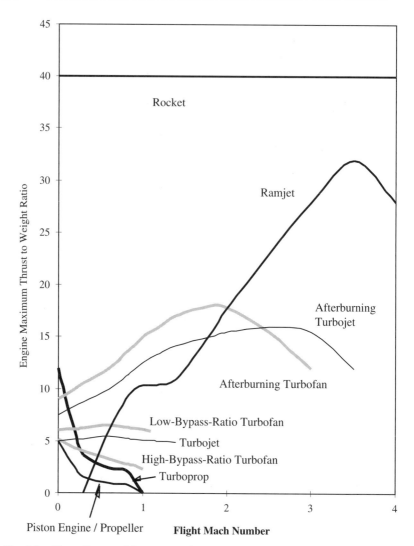

Fig. 5.2   Aircraft propulsion system thrust-to-weight ratios (adapted from Ref. 1).

maximum rotation rate given in revolutions per minute. In general, the shaft power available from a piston engine will be the sea-level rated power adjusted for non-standard density, if the engine is allowed to rotate at its rated rpm (the rpm at which it is designed to operate):

$$\text{SHP}_{\text{avail}} = \text{SHP}_{\text{SL}} \frac{\rho}{\rho_{\text{SL}}} \qquad (5.6)$$

The power produced by a piston engine is converted into thrust by the propeller. This device is composed of two or more blades (really just small wings) attached to a central shaft. As the propeller is rotated by the engine, the blades move through

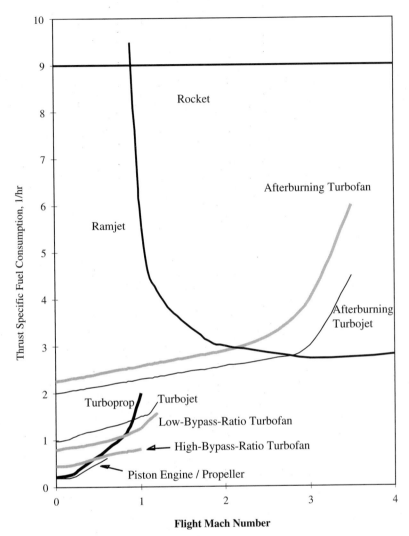

Fig. 5.3   Aircraft propulsion system TSFC (adapted from Ref. 1).

the air like wings and create lift in a direction perpendicular to their motion (parallel to the shaft). The component of this lift created by the propeller that is directed along the propeller shaft is thrust. The concept is illustrated in Fig. 5.6.

The propeller is not 100% efficient at converting engine shaft power into thrust. A portion of the engine's power is used to overcome the aerodynamic drag (form drag, induced drag, and in some cases wave drag) of the propeller blades. A constant speed propeller is designed so that the angle of attack or pitch of the blades can be adjusted to maintain a constant engine rpm. This feature helps keep the propeller's efficiency high over a wider speed range. The variable pitch capability can also be used to allow the engine to turn at its rated rpm, regardless of the aircraft's speed.

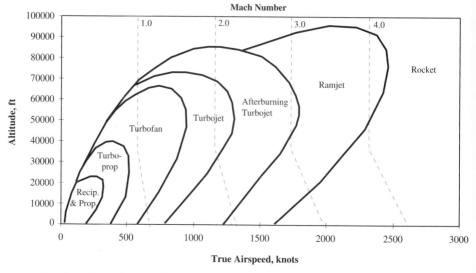

Fig. 5.4   Aircraft propulsion system operating envelopes (adapted from Ref. 2).

The efficiency of the propeller η_P is defined as

$$\eta_P = \text{THP}/\text{SHP} \tag{5.7}$$

where

$$\text{THP} = T \cdot V \tag{5.8}$$

is the thrust power of the engine. Charts that give propeller efficiency and/or thrust as a function of air density, the aircraft's velocity, and the engine RPM are usually available. When this type of data is not available, assume $\eta_P = 0.9$ for a good constant speed propeller at the engine-rated rpm and aircraft speeds below 300 ft/s.

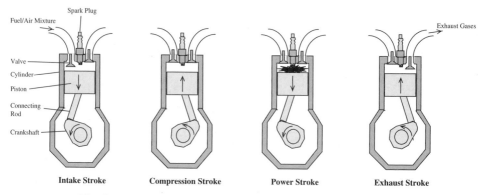

Fig. 5.5   Single-cylinder reciprocating piston-engine power cycle.

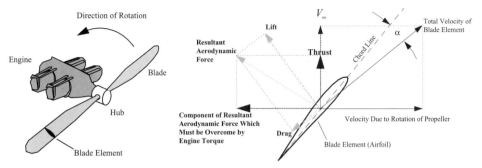

a) Engine and propeller   b) Blade element geometry and aerodynamic forces

Fig. 5.6   Propeller configuration, geometry, and force diagrams.

The available thrust of the engine/propeller combination is then given by

$$T_A = \text{SHP}_{\text{SL}} \frac{\rho}{\rho_{\text{SL}}} \frac{\eta_P}{V} \tag{5.9}$$

5.3.3 Turbojet Engines

Propellers become very inefficient at high subsonic speeds, and no practical supersonic propellers have ever been developed. Propulsion for the high subsonic, transonic, and supersonic flight regimes is usually provided by either turbojet or turbofan engines. These power plants produce thrust without using a propeller. Figure 5.7 illustrates a schematic diagram of a typical turbojet engine. As shown in Fig. 5.7, a turbojet engine takes air in through an inlet diffuser. The diffuser is designed so that its cross-sectional area is greater at its downstream end. This causes the velocity of the air flowing into the inlet to decrease and its static pressure to increase. A compressor then increases the static pressure further as it delivers the air to the combustor or burner. Fuel is mixed with the air and burned in the combustor. The hot gases are exhausted through a turbine, which acts like a windmill to extract

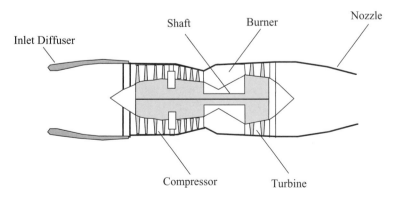

Fig. 5.7   Turbojet engine schematic.

power to turn the compressor through a shaft. Then the gases flow out of the engine through a nozzle, which causes them to accelerate until the static pressure of the exhaust approximately equals the ambient air pressure.

Because of the energy that has been added to the air by the burning fuel, the velocity of the gases exiting the engine is much higher than the velocity of the air entering at the inlet, even though the static pressures are nearly the same. According to Newton's second law, this rate of change in the flow momentum can only occur if the engine is exerting a net force on the air. By Newton's third law, for the action of the engine force on the air, there is an equal and opposite reaction force of the air on the engine. Because the air is accelerating to the rear, the reaction force is toward the front. This reaction force is the thrust generated by a jet engine. It is proportional to the rate at which the momentum of the air flowing through the engine is changing

$$T = \dot{m}(V_e - V_\infty) \tag{5.10}$$

where $\dot{m}$ is the mass flow rate through the engine, V_e is the velocity of the exhaust gases, and V_∞ is the freestream velocity. Note that $\dot{m}$ in Eq. (5.10) includes the fuel that flows through the engine, and its initial velocity is zero relative to the aircraft, not V_∞. Also, if the pressure of the exhaust gases is not the same as the pressure at the front of the engine, pressure thrust or drag is created. These two effects are generally small and will be ignored in this discussion.

Equation (5.8) suggests how the thrust of a turbojet engine varies with altitude and velocity. As altitude increases, density decreases. For the same velocity and engine inlet geometry, $\dot{m} = \rho A V$ [Eq. (3.1)] also decreases. Therefore, maximum thrust varies with air density:

$$T_A = T_{\mathrm{SL}} \left(\frac{\rho}{\rho_{\mathrm{SL}}} \right) \tag{5.11}$$

As V_∞ increases, $\dot{m}$ also increases. Simple turbojet engines with fixed exhaust nozzles are frequently designed so that when the engine is running at full power, the exhaust velocity is the speed of sound a (in the hot exhaust gases). Changing V_∞ does not affect $V_e = a$, so that $(V_e - V_\infty)$ decreases. On the other hand, increasing V_∞ increases $\dot{m} = \rho A V$. The net result of increasing $\dot{m}$ and decreasing $(V_e - V_\infty)$ is that thrust stays approximately constant with velocity.

5.3.4 Afterburners

The amount of energy that can be added to the gases flowing through a normal turbojet engine is limited by the temperature that the gases can safely have when they flow through the turbine. Excessive gas temperatures cause turbine blades to fatigue or deform and fail. However, once the gases have passed through the turbine it is possible to mix more fuel with them and burn it to increase the exhaust velocity. The engine component that does this is called an afterburner. The afterburner is not as efficient in converting heat into kinetic energy as the main engine, and so TSFC increases when afterburner is used. A typical afterburner might increase engine thrust at full throttle by 50%, but increase fuel flow rate by 200%. Full

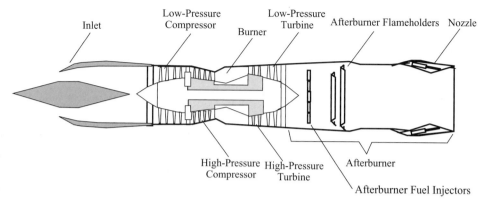

Fig. 5.8   Schematic of a turbojet engine with afterburner.

thrust from an afterburner-equipped engine is called wet or maximum thrust when the afterburner is operating and dry or military thrust when the afterburner is off. Figure 5.8 illustrates a turbojet engine with an afterburner.

The turbojet engine in Fig. 5.8 has two spools. A spool is a compressor and turbine that share a common shaft and therefore rotate at the same speed. The spools rotate independently of each other, so that in a correctly designed engine each rotates at its best rpm. This makes the engine more efficient over a broader range of throttle settings, producing more thrust with a lower TSFC.

Jet engines with afterburners generally have variable-area exhaust nozzles. This is because the nozzle area required to get exhaust gas static pressure approximately equal to P_∞ when the afterburner is not working is significantly different from the exit area required when it is on. These nozzles are also typically shaped so that V_e is greater than the speed of sound, and V_e increases with increasing V_∞ during afterburner operation. As a result, $(V_e - V_\infty)$ stays more nearly constant, and the increasing $\dot{m}$ with increasing V_∞ causes thrust to also increase. The limit to this steady increase with increasing flight speed in afterburning turbojet engine thrust is typically caused by flow separation and/or shock waves that occur at high Mach numbers in front of and inside the engine inlets. These flow disturbances cause total pressure losses that reduce the engine thrust. Internal structural and temperature limits also play a role in limiting the increase in afterburning turbojet thrust with increasing Mach number at low altitudes. Figure 5.9 illustrates the variation of wet and dry thrust with altitude and Mach number for a typical afterburning turbojet. The reduction in the slope of the thrust curve of this engine in afterburner at sea level for $M > 0.75$ is likely caused more by internal structural limits in the engine than inlet shock wave or flow separation effects. A simple approximation for the variation of afterburning turbojet thrust with altitude and Mach number is given by

$$T_A = T_{\text{SL}} \left(\frac{\rho}{\rho_{\text{SL}}} \right) (1 + 0.7 M_\infty) \qquad (5.12)$$

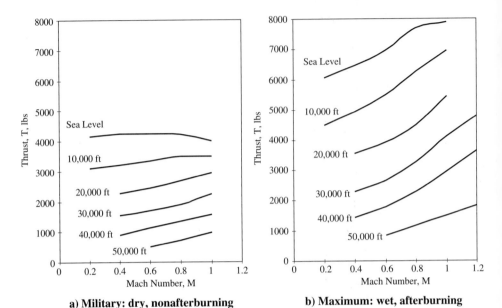

a) Military: dry, nonafterburning   b) Maximum: wet, afterburning

Fig. 5.9   Variation of afterburning turbojet thrust with altitude and Mach number (adapted from Ref. 3).

5.3.5 Turbofan Engines

To reduce the TSFC of a turbojet engine, one of the engine's spools can be connected so that it drives a larger compressor or fan at the front of the engine. Some of the air drawn in and accelerated by this fan does not flow through the engine *core* (compressor, combustor, and turbine). The air from the fan that does not flow through the core is called bypass air. The ratio of the bypass mass flow rate to the mass flow rate of the air flowing through the core is called the *bypass ratio*. A turbofan is more efficient and therefore has a lower TSFC because it accelerates more air (bypass air in addition to core air) for the same amount of fuel burned. Turbofan efficiency increases with increasing bypass ratio, but so does engine size and weight. Figure 5.10 illustrates two types of turbofans. The one on the left has a relatively low bypass ratio, and all of its bypass air flows with the core air into an afterburner. The turbofan on the right has a much higher bypass ratio and no afterburner.

Variation with altitude of the thrust of turbofans generally follows Eq. (5.11). The variation with Mach number depends in part on the bypass ratio. Low-bypass-ratio turbofans behave much like turbojets. High-bypass-ratio turbofans, on the other hand, exhibit a rapid decrease in maximum thrust output with increasing velocity at low altitudes. Figure 5.11 compares thrust curves for an afterburning turbofan with a bypass ratio of 0.7 with those for a nonafterburning turbofan with a bypass ratio of 5.

Note that the thrust curves in Fig. 5.11a reach maximum values at about $M = 1.6$ and then begin to decrease. This is caused by shock waves that form inside and

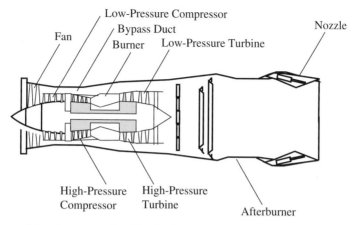

a) Low bypass ratio with afterburner: bypass ratio = 0.2 - 1.0,
T_{SL}/W_{eng} = 6 − 10, TSFC$_{dry}$ = 0.8 - 1.3, TSFC$_{wet}$ = 2.2 - 2.7

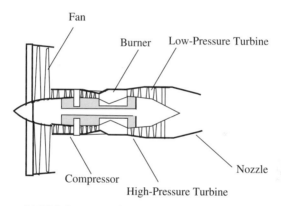

b) High bypass ratio: bypass ratio = 2.0 - 8.0,
T_{SL}/W_{eng} = 4 − 6, TSFC = 0.5 - 0.7

Fig. 5.10   Schematics and characteristics of two typical turbofan engines.

in front of the inlet, causing flow separation and loss of total pressure. Inlets can be designed to operate efficiently at a particular Mach number in spite of the shock waves. At this design Mach number, the shock waves interact with the shape of the inlet to achieve the best possible conservation of total pressure as the flow slows down in the inlet. Some inlets have the capability to change their shape so that they operate efficiently over a much wider range of Mach numbers. This feature makes the inlets much more expensive, but it is essential for aircraft that must fly efficiently above $M = 2.0$ or so. The inlet for the engine of Fig. 5.11a has a fixed geometry with a design Mach number of about 1.5. Thrust curves for the same engine with a variable geometry inlet would extend to much higher Mach numbers before bending over. Nonafterburning and afterburning

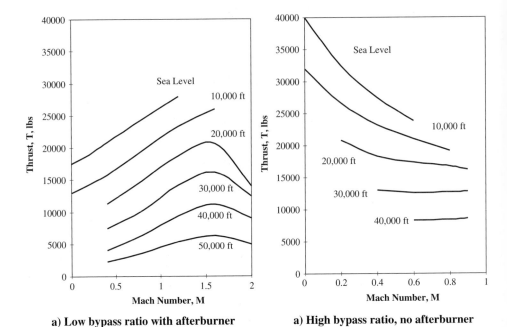

a) Low bypass ratio with afterburner   a) High bypass ratio, no afterburner

Fig. 5.11   Thrust curves for two types of turbofans (adapted from Refs. 3 and 4).

low-bypass-ratio turbofan thrust variation can be modeled with Eqs. (5.11) and (5.12), respectively. The variation of high-bypass-ratio turbofan thrust is approximated by

$$T_A = \left(\frac{0.1}{M_\infty}\right) T_{SL} \left(\frac{\rho}{\rho_{SL}}\right) \qquad (5.13)$$

5.3.6 Turboprops

Turboprop propulsion systems replace the fans of high-bypass-ratio turbofans with propellers. Their operating characteristics are similar to high-bypass-ratio turbofans, and they typically have lower TSFC. However, turboprops lose thrust at high speeds more like piston engine/propeller powerplants. Thrust-to-weight ratio for turboprops is usually higher than for piston engines, but TSFC is generally higher also. Turboprops are usually designed so that the high-energy air from the burners expands almost completely to ambient static pressure in the turbines, so that almost all of the energy is converted into shaft power. Therefore, the sea-level maximum power ratings of turboprops are given as SHP rather than thrust. Any additional thrust produced by a turboprop engine's exhaust is included in the sea-level power rating at the rate of 8 N of thrust per kW (2.5 lb of thrust per horsepower). A turboprop power rating corrected in this way is called an effective shaft power (ESHP). Turboprop thrust available may be approximated as:

$$T_A = \text{ESHP}_{SL} \left(\frac{\rho}{\rho_{SL}}\right) \frac{\eta_P}{V_\infty} \qquad (5.14)$$

5.3.7 Ramjets

At very high Mach numbers, the air that enters a jet engine inlet is slowed and compressed so much that the turbomachinery (compressor and turbine) is not really needed and can be eliminated. The resulting engine is little more than an afterburner connected to the inlet. This device is called a ramjet because the air is compressed by ram effect. Ram effect is the increased static pressure that results when the air is slowed by the inlet. A ramjet cannot function at low speeds because the compression of air in its inlet is not sufficient. This requires ramjet-powered aircraft to be accelerated to operating speed by some other propulsion system. At Mach numbers above about 3.0, however, ramjets are more efficient than afterburning turbojets, and they have much higher thrust-to-weight ratios.

5.3.8 Rockets

For extremely high speeds and for space flight, rocket engines are used. These have the advantage of carrying their own oxidizer with them, so that they do not have to take in air at all. At such high speeds, slowing the air enough to add fuel and burn it would result in impractically high pressures and temperatures. On the other hand, the requirement to carry oxidizer adds significantly to the size and weight of rocket-powered vehicles. Rocket engines can be either solid fueled or liquid fueled. Solid-fueled rockets are very simple, quite like fireworks rockets. The solid fuel contains its own oxidizer. It is placed in a container with a nozzle at one end. The fuel is ignited at the nozzle end and burns inside the container. The hot gases are forced by their high pressure to flow out the nozzle at very high speeds. The acceleration of the fuel/oxidizer is an action of the engine for which the thrust force is the reaction. This force is largely independent of the vehicle's flight velocity.

Liquid-fueled rockets are also quite simple, just a combustion chamber and nozzle, with pumps and lines to supply the fuel and oxidizer from storage tanks. This makes the thrust-to-weight ratio for rocket engines quite high. TSFC is also quite high (about 9/h for liquid-fueled and 16/h for solid-fueled rockets, including both fuel and oxidizer). The high temperatures of rocket-engine exhausts make it difficult to design variable nozzles for them. With a fixed nozzle, rocket engines must be designed to operate best at a particular altitude and Mach number. Performance at other than the design conditions is often poor. These limitations make rocket engines practical for use only in spacecraft and extremely high-speed aircraft.

5.3.9 Thrust Model Summary

Table 5.1 summarizes the equations that will be used in all performance calculations as models for the variation of thrust with density and Mach number or velocity. More detailed thrust, TSFC, and engine cycle models can be found in Ref. 5.

5.3.10 TSFC Models

The operating characteristics and limitations of propulsion systems that determine TSFC are very complex. However, the TSFC curves in Fig. 5.3 show that

Table 5.1   Thrust models for several propulsion concepts

Type	Thrust model	Equation
Piston engine/propeller	$T_A = \text{SHP}_{\text{SL}} \dfrac{\rho}{\rho_{\text{SL}}} \dfrac{\eta_P}{V_\infty}$	$(5.9)^{\text{a}}$
Turboprop	$T_A = \text{ESHP}_{\text{SL}} \left(\dfrac{\rho}{\rho_{\text{SL}}} \right) \dfrac{\eta_P}{V_\infty}$	(5.14)
High bypass-ratio turbofan	$T_A = \left(\dfrac{0.1}{M_\infty} \right) T_{\text{SL}} \left(\dfrac{\rho}{\rho_{\text{SL}}} \right)$	(5.13)
(Use $M = 0.1$ thrust for all $M < 0.1$)		
Turbojet and low-bypass-ratio		
Turbofan		
Dry (no afterburner)	$T_A = T_{\text{SL}} \left(\dfrac{\rho}{\rho_{\text{SL}}} \right)$	$(5.11)^{\text{b}}$
Wet (afterburner operating)	$T_A = T_{\text{SL}} \left(\dfrac{\rho}{\rho_{\text{SL}}} \right)(1 + 0.7M_\infty)$	$(5.12)^{\text{a}}$

[a]Assume $\eta_P = 0.9$. SHP and ESHP in feet pounds per second or watts. Use $V_\infty = 1$ for $V_\infty = 0$.
[b]Valid only for $M_\infty < 0.9$.

for the majority of a turbojet or turbofan engine's operating envelope (as shown in Fig. 5.4), TSFC varies only mildly with Mach number. Piston and turboprop engine TSFCs vary with Mach number, but power specific fuel consumption, the fuel flow required for a given power output, remains relatively constant with Mach number and with variations in air temperature. Power specific fuel consumption is usually called BSFC (for brake specific fuel consumption) because it is measured as the brake power output for a given fuel flow. Brake power is measured by connecting the engine to a brake (this also can be done on a dynamometer) that absorbs power and measures the engine's torque and rpm. Because the propeller is not involved in this measurement, propeller efficiency must be included to determine fuel consumption for a given thrust power output.

Small variations in TSFC and BSFC with Mach number and air temperature will be ignored in this text. Because of internal temperature and material strength limitations, a much more significant variation of TSFC with air temperature occurs for turbine engines. TSFC values for turbine engines generally vary according to the following relationship:

$$c_t = c_{t_{\text{SL}}} \sqrt{\dfrac{T}{T_{\text{SL}}}} \qquad (5.15)$$

Note that the ratio of the square roots of the absolute temperatures in Eq. (5.15) can also be expressed, using Eq. (3.35) as the ratio of the speed of sound in ambient conditions to the standard sea-level speed of sound:

$$c_t = c_{t_{\text{SL}}} \left(\dfrac{a}{a_{\text{SL}}} \right) \qquad (5.16)$$

5.3.11 Installed Thrust and TSFC

For a variety of reasons, the thrust produced by an engine is frequently less when it is installed in an aircraft than when it is tested uninstalled. Some of the sources of this thrust loss include viscous losses in the inlets, loss of momentum of cooling air, power and compressed air bleed requirements to run engine accessories, etc. Whenever possible, use installed sea-level thrust and TSFC ratings supplied by manufacturers as the reference values for thrust and TSFC models. However, if only uninstalled ratings are available decrease thrust and increase TSFC by 20% to approximate the installed values. This correction only applies to turbine engines. All of the thrust and TSFC values shown in Figs. 5.2, 5.3, 5.9, and 5.11 are installed values.

Example 5.1

A new afterburning low-bypass-ratio turbofan engine produces 15,000 lb of thrust in military power and 22,000 lb of thrust in afterburner in static, sea-level conditions. Its TSFC for these conditions is 0.8/hr in military power and 2.2/h in afterburner. What are its military and afterburner thrust and TSFC at $h = 20,000$ ft and $M = 0.8$?

Solution: Because only an altitude is specified, standard atmosphere conditions will be assumed. From the standard atmosphere table, for an altitude of 20,000 ft, $\rho = 0.001267$ slug/ft^3, and $a = 1036.9$ ft/s. For military power, the thrust at 20,000 ft is given by Eq. (5.11):

$$T_A = T_{SL}\left(\frac{\rho}{\rho_{SL}}\right) = 15{,}000\text{ lb}\left(\frac{0.001267\text{ slug/ft}^3}{0.002377\text{ slug/ft}^3}\right)$$

$$= 7995\text{ lb in military power}$$

The military power TSFC is given by Eq. (5.16):

$$c_t = c_{t_{SL}}\left(\frac{a}{a_{SL}}\right) = (0.8/\text{h})\left(\frac{1036.9\text{ ft/s}}{1116.2\text{ ft/s}}\right) = 0.74/\text{h in military power}$$

Similarly, the thrust in afterburner is predicted by Eq. (5.12):

$$T_A = T_{SL}\left(\frac{\rho}{\rho_{SL}}\right)(1 + 0.7M_\infty) = 22{,}000\text{ lb}\left(\frac{0.001267\text{ slug/ft}^3}{0.002377\text{ slug/ft}^3}\right)[1 + 0.7(0.8)]$$

$$= 18{,}293\text{ lb}$$

and the afterburning TSFC is

$$c_t = c_{t_{SL}}\left(\frac{a}{a_{SL}}\right) = (2.2/\text{h})\left(\frac{1036.9\text{ ft/s}}{1116.2\text{ ft/s}}\right) = 2.04/\text{h in afterburner}$$

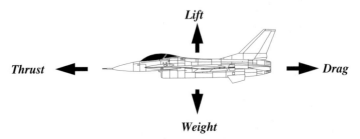

Fig. 5.12   Steady, level, unaccelerated flight.

5.4   Drag Curves

Consider again the case of steady, level, unaccelerated flight with the thrust vector aligned with the velocity. Figure 5.12 illustrates this situation. If the aircraft is in steady flight, not accelerating, forces must be in balance. Therefore, lift must equal weight, and thrust must equal drag. The lift requirement can be used to determine the required lift coefficient at any freestream velocity or Mach number:

$$L = W = C_L q S \tag{5.17}$$

$$C_L = \frac{W}{q S} \tag{5.18}$$

Once C_L is known, the aircraft's drag polar can be used to determine C_D and then D at that velocity:

$$C_D = C_{D_0} + k C_L^2, \quad D = C_D q S \tag{5.19}$$

If the calculation of drag is performed for a range of velocities, and for a fixed aircraft weight and altitude, a drag curve is generated. The drag is also called thrust required T_R because it is the thrust required from the engine to sustain steady, level flight for the given conditions. Figure 5.13 shows a drag curve for a typical subsonic aircraft. The drag is shown as the sum of parasite drag and induced drag. Note that parasite and induced drag are equal, each making up half the total drag, at the point on the curve where drag is a minimum. A thrust available model for an appropriately sized nonafterburning low-bypass-ratio turbofan engine [Eq. (5.11)] is also shown on Fig. 5.13. The thrust and drag curves are not drawn for velocities much faster than the speed where thrust available equals thrust required because the aircraft does not have enough thrust to sustain steady, level flight at those speeds. The speed where thrust available equals thrust required is the maximum level flight speed V_{max} for the aircraft for these conditions.

Thrust and drag curves are not drawn in Fig. 5.13 for $V < 70$ kn. This is because, for this particular aircraft, values of C_L that would be required to maintain level flight at speeds below 70 kn exceed the aircraft's $C_{L_{max}}$. The speed where the value of C_L required in order to maintain level flight is just equal to $C_{L_{max}}$ is called the

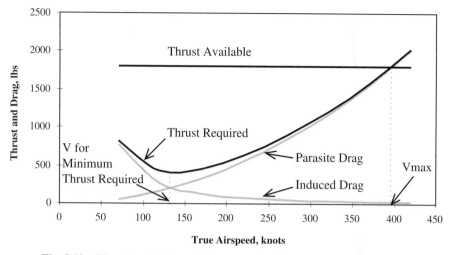

Fig. 5.13 Thrust available and thrust required for a subsonic jet aircraft.

aircraft's *stall speed* V_{stall}. An expression for V_{stall} can be derived by substituting $C_{L_{max}}$ for C_L in Eq. (5.18):

$$V_{stall} = \sqrt{\frac{2W}{\rho S C_{L_{max}}}} \qquad (5.20)$$

If, as in Fig. 5.13, the aircraft has sufficient thrust to maintain level flight at low speed, then the plane's minimum level-flight speed $V_{min} = V_{stall}$. However, for many aircraft, $V_{min} > V_{stall}$ because thrust required exceeds thrust available at low speeds. Figure 5.14 illustrates both situations.

Example 5.2

An aircraft with $C_{D_0} = 0.020$, $k_1 = 0.12$, and $k_2 = 0$ is flying at $h = 30,000$ ft and $M_\infty = 0.8$. If the aircraft has a wing area of 375 ft² and it weighs 25,000 lb, what is its drag coefficient, and how much drag is it generating? If the aircraft is in SLUF, how much thrust is its engine producing? If its $C_{L_{max}} = 1.8$, what is its stall speed at that altitude?

Solution: The atmospheric conditions for this situation are obtained from the standard atmosphere table for $h = 30,000$ ft as $\rho = 0.00089$ slug/ft³ and $a = 994.8$ ft/s, so

$$V_\infty = M_\infty a = (0.8)(994.8 \text{ ft/s}) = 795.8 \text{ ft/s}$$

and

$$q = \tfrac{1}{2}\rho V^2 = \tfrac{1}{2}(0.00089 \text{ slug/ft}^3)(795.8)^2 = 281.8 \text{ lb/ft}^2$$

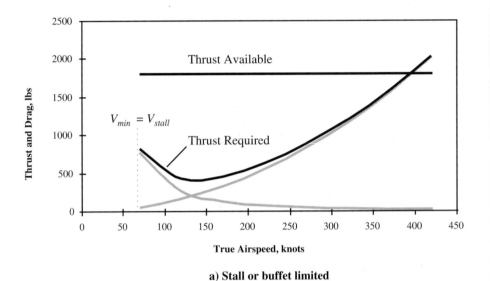

a) Stall or buffet limited

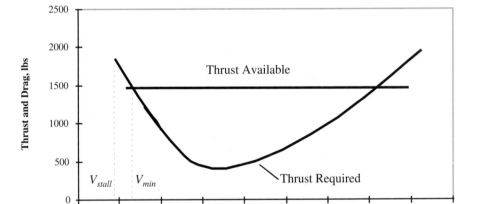

b) Thrust limited

Fig. 5.14   Two possible ways that V_{min} can be limited.

so, using Eq. (5.18)

$$C_L = \frac{W}{qS} = \frac{25{,}000 \text{ lb}}{(281.8 \text{ lb/ft}^2)(375 \text{ ft}^2)} = 0.2366$$

and, using Eq. (5.19)

$$C_D = C_{D_0} + kC_L^2 = 0.020 + 0.12(0.2366)^2 = 0.027$$

$$D = C_D qS = 0.027(281.8 \text{ lb/ft}^2)(375 \text{ ft}^2) = 2823 \text{ lb}$$

If the aircraft is in SLUF, then

$$T = T_R = D = 2823 \text{ lb}$$

Finally, its stall speed in SLUF is given by Eq. (5.20) as

$$V_{\text{stall}} = \sqrt{\frac{2W}{\rho S C_{L_{\max}}}} = \sqrt{\frac{2\,(25{,}000 \text{ lb})}{(0.00089 \text{ slug/ft}^3)(375 \text{ ft}^2)(1.8)}} = 288.5 \text{ ft/s}$$

5.5 Lift-to-Drag Ratio

The algebraic form of the aircraft drag polar we have assumed leads to some simple and helpful relationships. These mostly pertain to the parameter $(L/D)_{\max}$, which describes the maximum efficiency of the aircraft. Because lift is the aerodynamic force that allows aircraft to fly and carry payloads and passengers, it represents the usefulness or economic value of the aircraft. Drag, on the other hand, must be opposed by thrust, which is typically created by burning fuel. Drag therefore represents the cost of providing the economic value, mostly the cost of fuel. L/D, therefore, represents the aircraft's benefit-to-cost ratio, and $(L/D)_{\max}$ describes the maximum achievable value of this ratio.

To find where this ratio is a maximum on thrust and drag curves like Fig. 5.13, we need only remember that these curves are created by assuming lift equals weight and that weight is held constant. Therefore, if lift is constant $(L/D)_{\max}$ is achieved where drag is minimum. This point is identified on Fig. 5.13 as "V for minimum thrust required."

An interesting phenomenon occurs at the speed for $(L/D)_{\max}$, which is once more a consequence of the assumed algebraic form of the drag polar. Figure 5.13 was created by carefully calculating parasite and induced drag values for a fixed assumed aircraft weight and drag polar, then summing the two to get total drag. Note that at the speed for $(L/D)_{\max}$, parasite and induced drag are exactly equal. This is always true when we assume a drag polar of the form of Eq. (5.19).

We can use the fact that parasite drag equals induced drag at the speed for $(L/D)_{\max}$ to develop some additional algebraic relationships that will be helpful in solving performance problems. First, if we divide the equality by qS on both sides

$$C_{D_0} qS = qSkC_L^2$$

$$C_{D_0} = kC_L^2$$

Therefore, the C_L for $(L/D)_{max}$ is

$$C_L = \sqrt{\frac{C_{D_0}}{k}} \qquad (5.21)$$

Also at $(L/D)_{max}$

$$C_D = C_{D_0} + kC_L^2 = 2C_{D_0} = 2kC_L^2 \qquad (5.22)$$

and

$$\frac{L}{D} = \frac{C_L q S}{C_D q S} = \frac{C_L}{C_D} = \frac{C_L}{C_{D_0} + kC_L^2}$$

Combining Eqs. (5.21) and (5.22) for $(L/D)_{max}$,

$$\left(\frac{L}{D}\right)_{max} = \left(\frac{C_L}{C_D}\right)_{max} = \left(\frac{\sqrt{C_{D_0}/k}}{2C_{D_0}}\right)_{max} = \frac{1}{2\sqrt{kC_{D_0}}} \qquad (5.23)$$

Note that these relationships are only strictly valid for aircraft with drag polars like Eq. (5.19), and that this assumed simple drag polar form is only an approximation to the variation of drag on real aircraft. However, for most aircraft the approximation is a good one, and Eqs. (5.21–5.23) are useful tools in performance analysis.

5.6   Power Curves

For propeller-driven aircraft, engine performance is specified in terms of power. A chart that is analogous to Fig. 5.13 is easily developed, but with drag expressed as power required P_R using the relationship

$$P_R = T_R \cdot V_\infty = D \cdot V_\infty \qquad (5.24)$$

Figure 5.15 illustrates a power-required curve for a typical propeller-driven aircraft. The power-available model for an appropriately sized reciprocating engine/propeller combination, obtained by multiplying Eq. (5.9) by V_∞, is also plotted on the figure. The airspeed where power available equals power required is the aircraft's V_{max} for that altitude and aircraft weight. For a propeller-driven aircraft, the airspeed where power required is a minimum is, among other things, the speed at which the aircraft can maintain level flight at that altitude and weight with the minimum engine throttle setting.

Power curves are also useful in predicting the performance of turbojet- and turbofan-driven aircraft. Figure 5.16 illustrates power-available and power-required curves for the same aircraft whose thrust-available and thrust-required curves are shown in Figure 5.13. The curves are obtained by multiplying thrust and drag at each point by the freestream velocity. Note that minimum power required occurs at a lower velocity than minimum thrust required. For this flight condition, induced drag is three times as great as parasite drag, that is, $3C_{D_0} = kC_L^2$.

Example 5.3

What is the power required for the situation in Example 5.2?

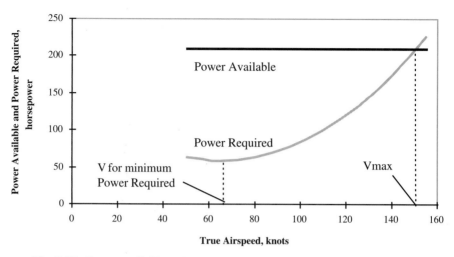

Fig. 5.15   Power available and power required for a propeller-driven aircraft.

Solution: Power required is given by Eq. (5.24):

$$P_R = T_R \cdot V_\infty = D \cdot V_\infty = (2823 \text{ lb})(795.8 \text{ ft/s}) = 2,247,000 \text{ ft lb/s} = 4084 \text{ hp!}$$

5.7   Curve Shifts

Changes in aircraft weight and configuration change the power required and thrust required curves. Figure 5.17 illustrates these changes for thrust required.

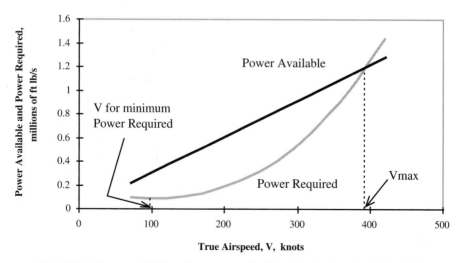

Fig. 5.16   Power available and power required for the jet aircraft of Fig. 5.13.

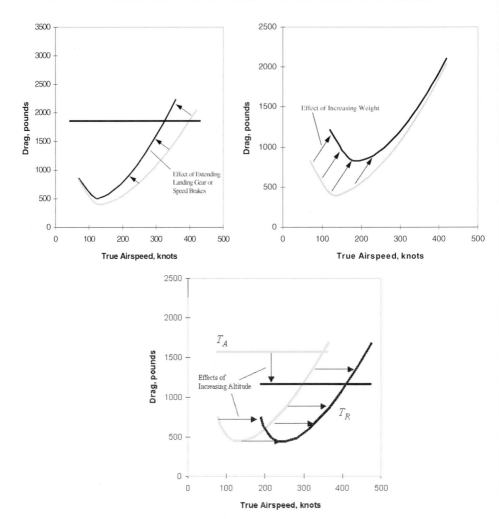

Fig. 5.17   Changes to thrust and drag curves with configuration, weight, and altitude changes.

Generally, changing aircraft configuration involves extending landing gear, speed brakes, or high-lift devices, all of which increase C_{D_0} without changing k significantly. High-lift devices usually have the largest effect on C_{D_0}, but they also increase $C_{L_{max}}$, so their effect on the curve is more complex. Figure 5.17 shows the effect of deploying speed brakes, spoilers, or landing gear, which only increase C_{D_0}. Note that parasite drag is increased at all speeds, but induced drag is unchanged. Because parasite drag is largest at high speeds, the net effect is to shift the drag curve up and to the left. Increasing weight changes the induced drag without changing parasite drag. Because induced drag is greatest at low speeds, the net effect is to shift the curve up and to the right.

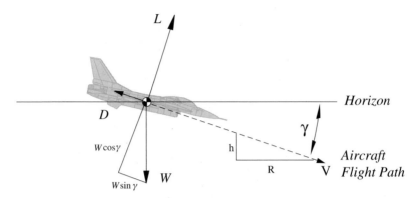

Fig. 5.18   Aircraft in a power-off glide.

Figure 5.17 also illustrates the effect of changes in altitude on the thrust and drag vs velocity curves. Because the true airspeed for a given dynamic pressure increases as density decreases, the effect of increasing altitude is to shift the curves to the right, without changing their shape. This is true as long as true airspeeds do not increase above the aircraft's critical Mach number. Above M_{crit}, curve shapes change as a result of increased wave and shock-induced separation drag. The effect of increasing altitude and decreasing density in reducing thrust available is also shown on Fig. 5.17. The effects of weight, configuration, and altitude changes on power-required curves are very similar to the curve shifts just described for thrust required.

5.8   Glides

Figure 5.18 shows an aircraft in a power-off glide. The aircraft's flight-path angle γ is taken as positive downward, and the thrust is zero. With these changes Eqs. (5.1) and (5.2) simplify to

$$\sum F_{\parallel} = ma = 0 = -D + W \sin \gamma$$

$$D = W \sin \gamma \tag{5.25}$$

$$\sum F_{\perp} = mV^2/r = 0 = L - W \cos \gamma$$

$$L = W \cos \gamma \tag{5.26}$$

5.8.1   Maximum Glide Range

To determine the aircraft speed that will produce the maximum glide range, first note in Fig. 5.18 that the aircraft's distance traveled through the air has two components, the vertical altitude lost in the glide h and the horizontal distance or range traveled R. For a fixed initial altitude, the range is maximized when the magnitude of the flight-path angle is as small as possible. The limit to how small γ

can get, while still sustaining steady flight, is set by the force balance in Eq. (5.25). Combining Eqs. (5.25) and (5.26),

$$\frac{D}{L} = \frac{1}{L/D} = \frac{W \sin \gamma}{W \cos \gamma} = \tan \gamma = \frac{h}{R} \tag{5.27}$$

The message in Eq. (5.27) is that the aircraft will achieve its flattest glide angle and its longest glide range when the aircraft is flown at the speed for $(L/D)_{max}$. Another useful result is

$$\frac{L}{D} = \frac{R}{h} \tag{5.28}$$

It is significant that weight is not a variable in Eq. (5.28). Because L/D is a function of C_L and $(L/D)_{max}$ is achieved for a specific value of C_L [Eq. (5.21)], the velocity for maximum glide range increases with weight, but the glide ratio R/h does not change.

Example 5.4

An aircraft with a drag polar $C_D = 0.012 + 0.2C_L^2$ has a wing area of 698 ft^2 and weighs 40,000 lb. What is its $(L/D)_{max}$ and at what speed in feet/second and knots is this achieved in standard sea level conditions? What if the aircraft weight decreased to 30,000 lb?

Solution: $(L/D)_{max}$ is given by Eq. (5.23):

$$\left(\frac{L}{D}\right)_{max} = \frac{1}{2\sqrt{kC_{D_0}}} = \frac{1}{2\sqrt{(0.2)(0.012)}} = 10.2$$

Then we use Eq. (5.21) to find the C_L for $(L/D)_{max}$:

$$C_L = \sqrt{\frac{C_{D_0}}{k}} = \sqrt{\frac{0.012}{0.2}} = 0.245$$

And we use lift-equal-to-weight to solve for the speed for $(L/D)_{max}$.

$$C_L = \frac{W}{qS} = \frac{W}{\frac{1}{2}\rho V^2 S}$$

$$V = \sqrt{\frac{W}{\rho S C_L}} = \sqrt{\frac{40,000 \text{ lb}}{(0.002377 \text{ slug/ft}^3)(698 \text{ ft}^2)(0.245)}} = 313.7 \text{ ft/s}$$

And converting to knots,

$$313.7 \text{ ft/s} \left(\frac{1 \text{ kn}}{1.69 \text{ ft/s}}\right) = 185.6 \text{ kn}$$

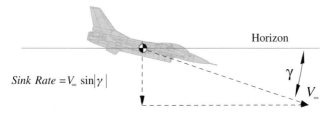

$$\text{Sink Rate} = V_\infty \sin|\gamma|$$

Fig. 5.19 Velocity components for an aircraft in a glide.

If the aircraft's gross weight changes to 30,000 lb, only the speed for $(L/D)_{max}$ will change, because weight was not a factor in calculating the magnitude of $(L/D)_{max}$.

$$V = \sqrt{\frac{W}{\rho S C_L}} = \sqrt{\frac{30{,}000 \text{ lb}}{(0.002377 \text{ slug/ft}^3)\,(698 \text{ ft}^2)\,(0.245)}} = 271.7 \text{ ft/s}$$

And converting to knots,

$$271.7 \text{ ft/s} \left(\frac{1 \text{ kn}}{1.69 \text{ ft/s}}\right) = 160.7 \text{ kn}$$

5.8.2 Minimum Sink Rate

Of particular interest to those who design, build, and/or fly sailplanes is the speed for minimum sink rate. Sailplanes are unpowered aircraft that must be towed into the air, but that use vertical air currents to stay aloft for hours or even days. They are able to do this because they are designed so that their minimum sink rate (minimum downward vertical velocity in steady flight) is less than the upward vertical velocity of the air currents. Therefore, although the aircraft is descending through the air mass the air is rising faster than the plane is descending through it, so that the plane's altitude increases. Figure 5.19 illustrates the components of a glider's velocity relative to the air mass. Note that sink rate is $V_\infty \sin \gamma$, which, from Eq. (5.22), is

$$\text{sink rate} = V_\infty \sin \gamma = \frac{V_\infty D}{W} = \frac{P_R}{W} \tag{5.29}$$

The significant conclusion from Eq. (5.29) is that the speed for minimum sink rate is the speed for minimum power required, as defined in Sec. 5.6.

Example 5.5

A sailplane's drag polar is $C_D = 0.01 + 0.02 C_L^2$. It has a mass of 500 kg and a wing area of 20 m^2. What is its maximum glide ratio and minimum sink rate at sea level, and at what speeds are these achieved?

Solution: The aircraft's maximum glide ratio is equal to its $(L/D)_{max}$:

$$\left(\frac{R}{h}\right)_{max} = \left(\frac{L}{D}\right)_{max} = \frac{1}{2\sqrt{k C_{D_0}}} = \frac{1}{2\sqrt{(0.01)\,(0.02)}} = 35.4$$

The speed for $(L/D)_{max}$ is determined by first finding the C_L for $(L/D)_{max}$:

$$C_L = \sqrt{\frac{C_{D_0}}{k}} = \sqrt{\frac{0.01}{0.02}} = 0.707$$

$$L = W = C_L q S, \quad q = \frac{W}{C_L S} = \frac{mg}{C_L S} = \frac{(500 \text{ kg}) (9.8 \text{ m/s}^2)}{0.707(20 \text{ m}^2)} = 346.5 \text{ N/m}^2$$

At sea level, from the standard atmosphere chart $\rho = 1.225 \text{ kg/m}^3$, and so

$$V_\infty = \sqrt{\frac{2q}{\rho}} = \sqrt{\frac{2(346.5 \text{ N/m}^2)}{1.225 \text{ kg/m}^3}} = 23.8 \text{ m/s}$$

for best glide range. The velocity for minimum sink rate is the velocity for minimum power required, where

$$3C_{D_0} = kC_L^2, \quad C_L = \sqrt{\frac{3C_{D_0}}{k}} = \sqrt{\frac{3(0.01)}{0.02}} = 1.225$$

$$L = W = C_L q S, \quad q = \frac{W}{C_L S} = \frac{mg}{C_L S} = \frac{(500 \text{ kg}) (9.8 \text{ m/s}^2)}{1.225(20 \text{ m}^2)} = 200 \text{ N/m}^2$$

At sea level, from the standard atmosphere chart $\rho = 1.225 \text{ kg/m}^3$, and so

$$V_\infty = \sqrt{\frac{2q}{\rho}} = \sqrt{\frac{2(200 \text{ N/m}^2)}{1.225 \text{ kg/m}^3}} = 18.1 \text{ m/s}$$

for minimum sink rate. Because $3C_{D_0} = kC_L^2$ for this condition, $C_D = 4C_{D_0}$, and the drag at this speed is

$$D = C_D q S = 0.04 (200 \text{ N/m}^2)(20 \text{ m}^2) = 160 \text{ N}$$

Then the minimum rate of sink is given by Eq. (5.29):

$$\text{minimum sink rate} = \frac{V_\infty D}{W} = \frac{V_\infty D}{mg} = \frac{18.1 \text{ m/s} (160 \text{ N})}{500 \text{ kg} (9.8 \text{ m/s}^2)} = 0.59 \text{ m/s}$$

5.9 Climbs

Figure 5.1 depicts an aircraft in a climb. Assuming thrust is approximately aligned with the flight-path vector and that the maneuver is a steady climb, the situation simplifies to that shown in Fig. 5.20. Equations (5.1) and (5.2) simplify to

$$\sum F_\parallel = ma = 0 = T - D - W \sin \gamma$$

$$\sin \gamma = \frac{T - D}{W} \tag{5.30}$$

$$\sum F_\perp = mV^2/r = 0 = L - W \cos \gamma$$

$$L = W \cos \gamma \tag{5.31}$$

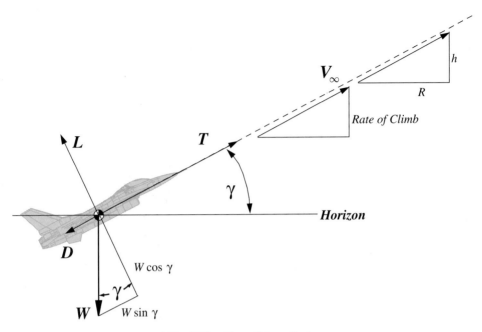

Fig. 5.20   Aircraft in a climb.

5.9.1   Maximum Climb Angle

The requirement to climb at maximum angle (maximum height gained for minimum ground distance traveled) is normally the result of some obstacle (e.g., trees, buildings, mountains, etc., or an altitude restriction imposed by a regulatory agency) in the flight path, which must be cleared. Equation (5.30) suggests that the maximum sustainable climb angle will be achieved for conditions that produce the maximum $T - D$ and minimum aircraft weight. For nonafterburning turbojets and low-bypass-ratio turbofans [thrust model Eq. (5.10)], maximum $T - D$ will occur at the velocity for D_{min} and $(L/D)_{max}$ because thrust is constant with velocity. Maximum $T - D$ for aircraft with other types of propulsion systems can be found graphically by comparing thrust and drag curves.

5.9.2   Maximum Rate of Climb

The requirement to climb at maximum rate normally stems from a need to quickly, and with minimum fuel expenditure, get to higher altitudes where the aircraft's maximum range and best cruise airspeeds are higher (and on a hot day in Texas, where the air is cooler). As shown on Fig. 5.20, the rate of climb is the vertical component of the aircraft's velocity:

$$\text{rate of climb} = R/C = V_\infty \sin \gamma = \frac{V_\infty(T - D)}{W} = \frac{P_{avail} - P_{req}}{W} \qquad (5.32)$$

From Eq. (5.32) it is clear that maximum sustained rate of climb for propeller-driven aircraft [thrust model Eq. (5.9) or (5.14)] is achieved for the lowest possible aircraft weight and at the airspeed for minimum power required. For aircraft powered by other types of propulsion systems, the airspeed for maximum rate of climb can be found by comparing power-available and power-required curves. The speed where excess power $(P_{avail} - P_{req})$ is greatest is the speed for maximum rate of climb.

Example 5.6

What are the maximum angle of climb and maximum rate of climb at sea level for the aircraft described in Figs. 5.13 and 5.16 and the speeds at which these occur? Assume the aircraft weighs 6000 lb.

Solution: Maximum angle of climb occurs at the speed where $T - D$ is a maximum. On Fig. 5.13 this occurs at the speed for D_{min}, approximately 130 kn, because thrust available does not vary with velocity. The drag at this speed is approximately 400 lb, as shown on Fig. 5.21. The thrust is 1800 lb, and so

$$\sin \gamma = \frac{T-D}{W}, \quad \gamma = \sin^{-1}\left(\frac{T-D}{W}\right) = \sin^{-1}\left(\frac{1800 \text{ lb} - 400 \text{ lb}}{6000 \text{ lb}}\right) = 13.5 \text{ deg}$$

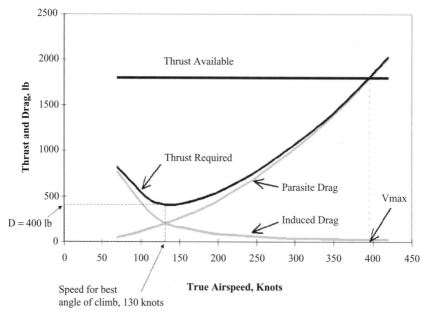

Fig. 5.21   Graphical determination of speed for best angle of climb and drag at that speed for the aircraft of Figs. 5.13 and 5.16.

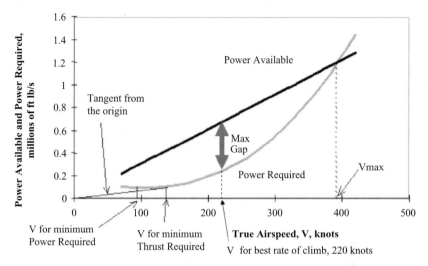

Fig. 5.22 Graphical determination of speed for best rate of climb for the aircraft of Figs. 5.13 and 5.16.

Maximum rate of climb occurs at the velocity where $P_{avail} - P_{req}$ is a maximum. On Fig. 5.16 this occurs at approximately 220 kn, as shown on Fig. 5.22. The power required at this point is approximately 240,000 ft lb/s, and the power available is 680,000 ft lb/s, so that

$$\text{maximum rate of climb} = \frac{P_{avail} - P_{req}}{W}$$

$$= \frac{680,000 \text{ ft lb/s} - 240,000 \text{ ft lb/s}}{6000 \text{ lb}} = 73.3 \text{ ft/s} = 4400 \text{ ft/min}$$

5.9.3 Ceilings

Design performance requirements for an aircraft can be specified in terms of a *ceiling* or maximum attainable altitude. In Fig. 5.17 it is apparent that $T_{avail} - T_{req}$ decreases with increasing altitude. At some altitude, thrust available decreases to the point that it just equals the minimum drag. Maximum angle of climb and maximum rate of climb are zero, and in fact the aircraft can only sustain this altitude by flying at the minimum drag airspeed. This altitude is referred to as the aircraft's absolute ceiling. It is not a very practical altitude because in theory it would take an infinite amount of time for the aircraft to climb that high. Ceilings that are more commonly specified in design requirements are the service ceiling, the altitude where a 100 ft/min rate of climb can be sustained, and the combat ceiling, the altitude where 500 ft/min rate of climb can be sustained.

5.10 Range and Endurance

For many aircraft, the ability to fly for long distances and/or long periods of time is among the most important design requirements. It is hard to imagine an airline buying a transport aircraft that has to land every 100 miles to refuel, a resources agency buying a pollution-monitoring aircraft that can only stay on station for an hour at a time, or an air force buying a fighter that requires multiple refuelings from tanker aircraft to complete its mission. It is common to see airliners fly non-stop from Chicago to Frankfurt, Moscow, or Tokyo, but these capabilities had to be specifically designed into the aircraft. The range and endurance that an aircraft can achieve depend on its aerodynamics (primarily, its drag polar), the characteristics of its propulsion system, the amount of fuel the aircraft can carry, and the way it is operated.

5.10.1 Turbojet and Turbofan Aircraft Endurance

Because TSFC is modeled as constant with Mach number for turbojets and turbofans, the drag (thrust required) curve of Fig. 5.13 can be viewed as a fuel-flow-required curve. This is because multiplying the drag values everywhere by a constant value of c_t would change the scale but not the shape of the curve. For a given thrust required = drag and a specified ΔW_f (the weight of fuel available to be burned), the endurance is given by

$$E = \frac{\Delta W_f}{\dot{W}_f} = \frac{\Delta W_f}{c_t D} \tag{5.33}$$

Equation (5.33) makes it clear that maximum endurance for a turbojet or turbofan aircraft is achieved for maximum fuel weight and minimum TSFC, when the aircraft flies at the speed for minimum drag or thrust required. Recall that because the drag curve of Fig. 5.13 was computed assuming a constant weight and lift equal to weight, the minimum drag condition is also the condition for maximum lift-to-drag ratio $(L/D)_{\max}$. Equation (5.33) is not very useful in its current form for predicting endurance because the aircraft's weight, and therefore its drag, will change as it burns fuel. An approximate endurance estimate can be made by using the average aircraft weight for the endurance problem to calculate an average drag:

$$E = \frac{\Delta W_f}{c_t D_{\text{avg}}} \tag{5.34}$$

Equation (5.34) is known as the average value method for predicting endurance, and the accuracy of its results is often quite good. For a more accurate prediction of endurance, it is necessary to write Eq. (5.32) in differential form and then integrate it with respect to the weight change:

$$dt = \frac{-dW}{c_t D}$$

or

$$E = -\int_{W_1}^{W_2} \frac{dW}{c_t D} \tag{5.35}$$

Note that the negative sign on dW is required because the burning of a positive amount of fuel in Eq. (5.34) results in a negative change in the aircraft's weight.

At this point, it is difficult to integrate Eq. (5.35) because a relationship between weight and drag has not yet been established. If it is assumed that the endurance task is flown at a constant aircraft angle of attack, hence a constant C_L and L/D, then using the fact that lift equals weight,

$$E = \int_{W_2}^{W_1} \frac{1}{c_t} \frac{L}{D} \frac{dW}{W} = \frac{1}{c_t} \frac{C_L}{C_D} \int_{W_2}^{W_1} \frac{dW}{W}$$

$$E = \frac{1}{c_t} \frac{C_L}{C_D} \ln\left(\frac{W_1}{W_2}\right) \tag{5.36}$$

Equation (5.36) reaffirms the fact that maximum endurance is achieved for maximum fuel, minimum TSFC, and maximum L/D. Note that maintaining a constant value of C_L at a constant altitude throughout the endurance task will require the aircraft to fly slower as its weight decreases.

5.10.2 Turbojet and Turbofan Aircraft Range

The range of an aircraft is its endurance multiplied by its velocity. For the average value method, Eq. (5.34) multiplied by velocity is

$$R = E \cdot V_\infty = \frac{\Delta W_f}{c_t D_{avg}} \cdot V_\infty = \frac{\Delta W_f}{c_t(D_{avg}/V_\infty)} \tag{5.37}$$

The quantity $1/c_t(D_{avg}/V_\infty)$ in Eq. (5.37) has units of distance per pound of fuel, much like the miles per gallon rating used for automobiles. Indeed, air nautical miles per pound of fuel is a parameter commonly used by pilots in planning flights. Maximizing this parameter will maximize the aircraft's range. The ratio D_{avg}/V_∞ is the slope of a line drawn on a thrust and drag vs velocity plot (Fig. 5.13) from the origin to any point on the drag curve. Because D_{avg}/V_∞ must be minimized to maximize range, the line from the origin that has the lowest possible slope but still touches the drag curve (a line from the origin tangent to the drag curve) is used to identify the best range velocity. Figure 5.23 illustrates such a line and the maximum range airspeed it identifies.

The average value method is an approximation, albeit often a good one. As with endurance, a more accurate expression for range can be obtained by assuming angle of attack and L/D do not change, then writing Eq. (5.37) in differential form and integrating with respect to weight,

$$dx = V_\infty \cdot dt = \frac{-V_\infty dW}{c_t D}$$

$$R = -\int_{W_1}^{W_2} \frac{V_\infty}{c_t} \frac{L}{D} \frac{dW}{W} = \frac{1}{c_t} \frac{C_L}{C_D} \int_{W_2}^{W_1} \frac{V_\infty dW}{W}$$

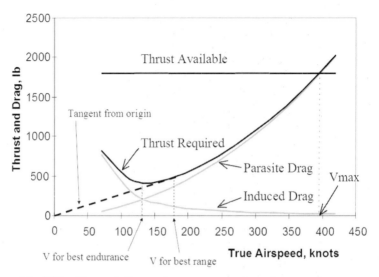

Fig. 5.23   Airspeeds for maximum range and maximum endurance.

But to maintain a constant C_L at a constant altitude, V_∞ must change with changing weight, $V_\infty = \sqrt{2W/\rho S C_L}$, so that

$$R = \int_{W_2}^{W_1} \sqrt{\frac{2W}{\rho S C_L}} \frac{1}{c_t} \frac{C_L}{C_D} \frac{dW}{W} = \sqrt{\frac{2}{\rho S} \frac{1}{c_t} \frac{C_L^{1/2}}{C_D}} \int_{W_2}^{W_1} \frac{dW}{W^{1/2}}$$

$$R = \sqrt{\frac{2}{\rho S} \frac{2}{c_t} \frac{C_L^{1/2}}{C_D}} \left(W_1^{1/2} - W_2^{1/2} \right) \tag{5.38}$$

Equation (5.38) asserts that range is maximized when density is low (high altitude), TSFC is low (high altitude up to the tropopause), the weight of fuel available is high, and when $C_L^{1/2}/C_D$ is a maximum or the reciprocal $C_D/C_L^{1/2}$ is a minimum. For the assumed drag polar, the condition for minimizing $C_D/C_L^{1/2}$ can be found by expressing the ratio in terms of the drag polar, taking the derivative, and setting it equal to zero:

$$C_D/C_L^{1/2} = \frac{C_{D_0} + k C_L^2}{C_L^{1/2}} = \frac{C_{D_0}}{C_L^{1/2}} + k C_L^{3/2}$$

$$\frac{d\left(C_D/C_L^{1/2}\right)}{dC_L} = 0 = -\frac{1}{2}\left(\frac{C_{D_0}}{C_L^{3/2}}\right) + \frac{3}{2}\left(k C_L^{1/2}\right)$$

$$0 = -C_{D_0} + 3k C_L^2 \quad \text{or} \quad C_{D_0} = 3k C_L^2 \tag{5.39}$$

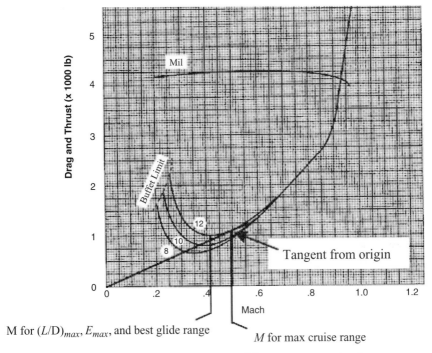

Fig. 5.24   Portion of T-38 thrust and drag curve at sea level.

The validity of this conclusion can be confirmed by comparing Fig. 5.23 with Fig. 5.13. The point on Fig. 5.23 where range is maximized is exactly the same point on Fig. 5.13 where parasite drag is three times as great as induced drag.

Example 5.7

You are flying a T-38 in standard sea-level conditions (over Death Valley in the winter). Your aircraft weighs 12,000 lb. Thrust and drag charts for the T-38 are available in Appendix B at the end of this book. Answer the following questions about this situation:

1) At what Mach number should you fly for maximum endurance at this altitude?

Solution: On the T-38 sea-level thrust and drag vs Mach-number curve, a portion of which is shown in Fig. 5.24, maximum endurance occurs at the Mach number for minimum drag, which for $W = 12,000$ lb is $M = 0.41$.

2) At what Mach number should you fly for maximum range at this altitude?

Solution: On Fig. 5.24, maximum range occurs at the Mach number where a line drawn from the origin is tangent to the drag curve, which for $W = 12,000$ lb is $M = 0.495$.

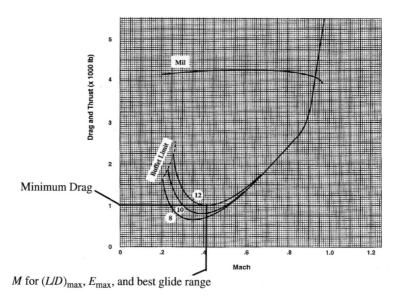

M for $(L/D)_{max}$, E_{max}, and best glide range

Fig. 5.25   Portion of T-38 sea-level thrust and drag curves showing minimum drag value.

3) At what Mach number should you fly for maximum power-off glide range at this altitude?

Solution: On Fig. 5.24, maximum glide range occurs at the Mach number for minimum drag, which for $W = 12,000$ lb is $M = 0.41$.

4) What is your maximum glide ratio for the situation in question 3?

Solution: As shown in Fig. 5.25, maximum glide ratio can be determined by reading off the minimum drag value and then assuming lift = weight. The maximum glide ratio is equal to the maximum lift-to-drag ratio, and so $(L/D_{max}) = W/D_{min} = 12,000\,\text{lb}/1000\,\text{lb} = 12$.

5) At what Mach number should you fly for minimum power-off sink rate at this altitude?

Solution: Minimum sink rate occurs at the speed for minimum power required, which on the T-38 power required vs Mach curve occurs at $M = 0.36$. Figure 5.26 shows this.

6) What is your minimum sink rate for the situation in Question 5?

Solution: The sink rate is equal to the power required divided by the weight. From the sea-level power-required curve, the minimum power required for $W = 12,000$ lb

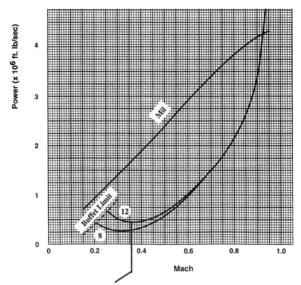

M for minimum power required and minimum sink rate

Fig. 5.26 T-38 sea-level power curves showing Mach for minimum power required.

is 450,000 ft lb/s, and so the minimum sink rate is 450,000 ft lb/s/12,000 lb = 37.5 ft/s = 2250 ft/min.

7) At what Mach number should you fly for maximum angle of climb with military thrust at this altitude?

Solution: The climb angle can be found from sin $\gamma = (T - D)/W$. Figure 5.27 shows a portion of the T-38 thrust and drag vs Mach-number curves found in the Appendix. On Fig. 5.27, maximum angle of climb in military thrust occurs at the Mach number for minimum drag because military thrust does not vary significantly with Mach number. For $W = 12,000$ lb, minimum drag is at $M = 0.41$.

8) What is your maximum angle of climb with military thrust for the situation in Question 7?

Solution: The climb angle can be found from sin $\gamma = (T - D)/W$. On the T-38 thrust and drag vs Mach-number curve, minimum drag is 1000 lb at $M = 0.41$, and thrust at this Mach number is 4250 lb. For $W = 12,000$ lb, the maximum climb angle is $\gamma = \sin^{-1}[(T-D)/W] = \sin^{-1}[(4250\,\text{lb}-1000\,\text{lb})/12,000\,\text{lb}] = 15.7\,\text{deg}$.

9) At what Mach number should you fly for maximum rate of climb with military thrust at this altitude?

Solution: The climb rate $= (P_A - P_R)/W$. Figure 5.28 is a portion of the T-38 sea-level power vs Mach-number curves from Appendix B. On Fig. 5.28,

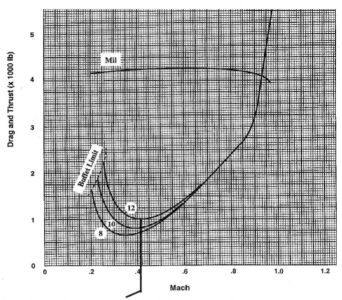

M for $(L/D)_{max}$, E_{max}, best glide range, and best mil thrust angle of climb

Fig. 5.27   Portion of the T-38 sea-level thrust and drag vs Mach-number curves show-ing the Mach number for best glide and best military thrust angle of climb.

maximum excess power occurs at the Mach number where the maximum excess power, the maximum power available minus power required, occurs. One tech-nique for quickly finding this point is to draw the parallel tangent, a line parallel to the power-available curve tangent to the power-required curve. The point of tangency identifies where $P_A - P_R$ is a maximum. On Fig. 5.28, this occurs at $M = 0.67$.

10) What is your maximum rate of climb for the situation in Question 9?

Solution: The climb rate $= (P_A - P_R)/W$. On Fig. 5.28, maximum excess power for $W = 12,000$ lb occurs at $M = 0.67$, where $P_A = 3,250,000$ ft lb/s and $P_R = 1,350,000$ ft lb/s. Then maximum climb rate $= (P_A - P_R)/W = 158.3$ ft/s $= 9500$ ft/min.

You climb your T-38 to 20,000 ft and level off. Conditions match a standard day at this altitude. Your aircraft now weighs 11,000 lb, and you have 2000 lb of useable fuel. Answer the following questions about this situation:

11) What is your maximum range at this altitude if you use all of your usable fuel to cruise?

Solution: The speed for maximum range is found as in Question 2, but this time the weight and altitude are different. Using the T-38 thrust and drag curves for $h = 20,000$ ft (Fig. 5.29 shows the applicable portion) and an average cruise

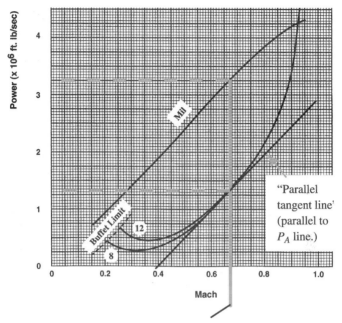

M for maximum rate of climb and maximum excess power

Fig. 5.28   Portion of T-38 sea-level power vs Mach-number curves showing best rate of climb speed and excess power.

weight of 10,000 lb (the starting weight minus half the fuel weight), the average Mach number for best cruise is 0.61, and the average drag for cruise (the drag at the average weight) is 950 lb. The average speed for cruise is the average Mach number times the speed of sound at 20,000 ft, $V = Ma = 0.61(1036.9 \text{ ft/s}) = 632.5 \text{ ft/s} = 632.5 \text{ ft/s}/(1.69 \text{ ft/s per kn}) = 374.3 \text{ kn}$. The T-38 normal power TSFC at sea level is given in Appendix B as 1.09/h, and so its TSFC at 20,000 ft is

$$c_t = c_{t_{\text{sea level}}} \left(\frac{a}{a_{\text{sea level}}} \right) = 1.09/\text{h} \left(\frac{1036.9 \text{ ft/s}}{1116.4 \text{ ft/s}} \right) = 1.01/\text{h}$$

Then, using the average value method

$$R = E \cdot V_\infty = \frac{\Delta W_f}{c_t D_{\text{avg}}} \cdot V_\infty = \frac{2000 \text{ lb}}{(1.01/\text{h})(950 \text{ lb})} (374.3 \text{ nm/h}) = 780.2 \text{ n mile}$$

12) What is your maximum endurance at this altitude if you use all of your usable fuel to loiter?

Solution: Using the TSFC calculated for Question 11 and the minimum drag = 880 lb on Fig. 5.29 and $W = 10,000$ lb, the average value method yields

$$E = \frac{\Delta W_f}{c_t D_{\text{avg}}} = \frac{2000 \text{ lb}}{(1.01/\text{h})(880 \text{ lb})} = 2.25 \text{ h}$$

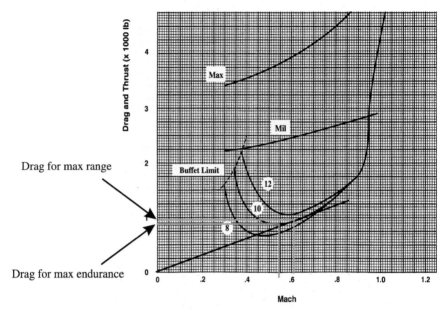

Fig. 5.29   Portion of T-39 thrust and drag vs Mach number curves for 20,000 ft.

Example 5.8

A turbojet-powered trainer aircraft weighs 5000 lb and is flying at $h = 25,000$ ft with 1000 lb of fuel onboard. Its drag polar is $C_D = 0.018 + 0.095\,C_L^2$, its wing area is 180 ft^2, and the TSFC of its engines is 1.0/h at sea level. What is its maximum range and endurance to tanks dry at this altitude, and at what speed should the pilot initially fly to achieve each?

Solution: Maximum endurance is achieved at the speed for $(L/D)_{\max}$. This speed can be determined by first calculating the required value of C_L, then solving for the speed required to achieve $L = W$ at that C_L:

$$C_L = \sqrt{\frac{C_{D_0}}{k}} = \sqrt{\frac{0.018}{0.095}} = 0.435$$

$$L = W = C_L q S, \quad q = \frac{W}{C_L S} = \frac{5000\ \text{lb}}{0.435\,(180\,\text{ft}^2)} = 63.86\ \text{lb/ft}^2$$

Using $\rho = 0.001066$ at $h = 25,000$ ft obtained from the standard atmosphere table and the definition of q,

$$V_\infty = \sqrt{\frac{2q}{\rho}} = \sqrt{\frac{2\,(63.86\ \text{lb/ft}^2)}{0.001066\ \text{slug/ft}^3}} = 346.1\ \text{ft/s}$$

for maximum endurance. Note that this is only the initial velocity for maximum endurance, and that as fuel is burned the velocity for best endurance will decrease. To calculate the maximum endurance time, it is first necessary to determine the magnitude of $(L/D)_{max}$ using Eq. (5.23):

$$\left(\frac{L}{D}\right)_{max} = \left(\frac{C_L}{C_D}\right)_{max} = \frac{1}{2\sqrt{kC_{D_0}}} = \frac{1}{2\sqrt{0.095\,(0.018)}} = 12.1$$

The TSFC is also predicted using Eq. (5.16) with $a = 1016.1$ at $h = 25{,}000$ ft and $a_{\text{sea level}} = 1116.1$ ft/s obtained from the standard atmosphere table

$$c_t = c_{t_{\text{sea level}}} \left(\frac{a}{a_{\text{sea level}}}\right) = 1.0/\text{h} \left(\frac{1016.1\ \text{ft/s}}{1116.2\ \text{ft/s}}\right) = 0.91/\text{h}$$

Then the endurance is calculated using Eq. (5.36) with $W_1 = 5000$ lb and

$$W_2 = W_1 - W_f = 5000\,\text{lb} - 1000\,\text{lb} = 4000\,\text{lb}$$

$$E = \frac{1}{c_t}\frac{C_L}{C_D}\ell n\left(\frac{W_1}{W_2}\right) = \frac{1}{0.91/\text{h}}\,(12.1)\,\ell n\left(\frac{5000\,\text{lb}}{4000\,\text{lb}}\right) = 2.97\,\text{h}$$

Similarly, the velocity for maximum range is obtained by solving Eq. (5.39) for C_L and Eq. (4.3) for q:

$$C_{D_0} = 3kC_L^2, \quad C_L = \sqrt{\frac{C_{D_0}}{3k}} = \sqrt{\frac{0.018}{3(0.095)}} = 0.251$$

$$q = \frac{W}{C_L S} = \frac{5000\,\text{lb}}{0.251\,(180\,\text{ft}^2)} = 110.7\ \text{lb/ft}^2$$

$$V_\infty = \sqrt{\frac{2q}{\rho}} = \sqrt{\frac{2(110.7\ \text{lb/ft}^2)}{0.001066\ \text{slug/ft}^3}} = 455.7\ \text{ft/s}$$

for maximum range. As with the velocity for maximum endurance, the velocity for best range will decrease as fuel is burned. The value calculated for C_L is now used to calculate C_D, after which the maximum range is predicted using Eq. (5.38):

$$C_D = 0.018 + 0.095C_L^2 = 0.018 + 0.095(0.251)^2 = 0.024$$

$$R = \sqrt{\frac{2}{\rho S}\frac{2}{c_t}\frac{C_L^{1/2}}{C_D}\left(W_1^{1/2} - W_2^{1/2}\right)}$$

$$= \sqrt{\frac{2}{(0.001066\ \text{slug/ft}^3)(180\,\text{ft}^2)}\frac{2}{0.91/\text{h}}\frac{(0.251)^{1/2}}{0.024}\left[(5000\ \text{lb})^{1/2}-(4000\ \text{lb})^{1/2}\right]}$$

$$= \sqrt{10.4\frac{\text{ft}^2}{\text{lb s}^2}\,(2.197\,\text{h})(20.87)\,(7.465\ \text{lb}^{1/2})}\bigg/\left(1.69\frac{\text{ft/s}}{\text{kn}}\right) = 653.9\ \text{n mile}$$

Note that the first term in Eq. (5.38) produces units of feet per second per square-root pounds [ft/s · (lb)$^{0.5}$], whereas the second term has units of hours, and the desired answer is in nautical miles. It is necessary to divide by the factor 1.69 ft/s/kn to resolve this.

5.10.3 Propeller-Driven Aircraft Endurance

Because fuel consumption for piston engines and turboprops is proportional to power output, the power curves are the best tools for determining a propeller-driven aircraft's endurance and range. The average value method prediction for the endurance of propeller-driven aircraft is

$$E = \frac{\Delta W_f}{\dot{W}_f} = \frac{\Delta W_f}{(c/\eta_{\text{prop}}) P_{R_{(\text{avg})}}} = \frac{\Delta W_f}{(c/\eta_{\text{prop}}) D_{\text{avg}} V_{\infty}} \tag{5.40}$$

where c with no subscript is a commonly used symbol for BSFC and η_{prop} is the propeller efficiency factor. The speed for maximum endurance of a propeller-driven aircraft is easily chosen from a power required curve such as Fig. 5.14 as the speed for minimum power required.

The more accurate form of Eq. (5.40) is

$$E = \int_{W_2}^{W_1} \frac{\eta_{\text{prop}}}{c} \frac{dW}{D V_{\infty}} = \int_{W_2}^{W_1} \frac{\eta_{\text{prop}}}{c} \frac{L}{D V_{\infty}} \frac{dW}{W}$$

Using the assumption of constant C_L, so that $V_{\infty} = \sqrt{2W/\rho S C_L}$,

$$E = \frac{\eta_{\text{prop}}}{c} \frac{C_L^{3/2}}{C_D} \sqrt{\frac{\rho S}{2}} \int_{W_2}^{W_1} \frac{dW}{W^{3/2}} = \frac{\eta_{\text{prop}}}{c} \frac{C_L^{3/2}}{C_D} \sqrt{\frac{\rho S}{2}} (-2) \left(W_1^{-1/2} - W_2^{-1/2} \right)$$

$$E = \frac{\eta_{\text{prop}}}{c} \frac{C_L^{3/2}}{C_D} \sqrt{2\rho S} \left(W_2^{-1/2} - W_1^{-1/2} \right) \tag{5.41}$$

Equation (5.41) is known as the Breguet endurance equation, named after a famous French aviation pioneer and aircraft builder to whom the original derivation of the equation is often attributed. Equations (5.36) and (5.38) are also often referred to as Breguet equations because they are derived in a similar fashion. Equation (5.41) asserts that maximum endurance is achieved for conditions of high propeller efficiency, low BSFC, high density (low altitude and temperature), high weight of fuel available, and a maximum value of the ratio $C_L^{3/2}/C_D$. That the maximum value of this ratio is obtained for conditions of minimum power required is easily shown using the expression for power required:

$$P_R = V_{\infty} \cdot D = V_{\infty} \cdot \frac{W}{C_L/C_D}$$

but

$$V_{\infty} = \sqrt{2W/\rho S C_L}$$

so that

$$P_R = \frac{W}{C_L/C_D}\sqrt{\frac{2W}{\rho S C_L}} = \sqrt{\frac{2W^3}{\rho S}}\frac{1}{C_L^{3/2}/C_D} = \frac{\text{constant}}{C_L^{3/2}/C_D} \qquad (5.42)$$

Clearly from Eq. (5.42), P_R is minimized when $C_L^{3/2}/C_D$ is a maximum, or the reciprocal $C_D/C_L^{3/2}$ is minimized. For the assumed drag polar, the condition for minimizing $C_D/C_L^{3/2}$ can be found by expressing the ratio in terms of the drag polar, taking the derivative, and setting it equal to zero:

$$C_D/C_L^{3/2} = \frac{C_{D_0} + kC_L^2}{C_L^{3/2}} = \frac{C_{D_0}}{C_L^{3/2}} + kC_L^{1/2}$$

$$\frac{d(C_D/C_L^{3/2})}{dC_L} = 0 = -\frac{3}{2}\left(\frac{C_{D_0}}{C_L^{5/2}}\right) + \frac{1}{2}\left(\frac{k}{C_L^{1/2}}\right)$$

$$0 = -3C_{D_0} + kC_L^2 \qquad \text{or} \qquad 3C_{D_0} = kC_L^2 \qquad (5.43)$$

This result can be confirmed by comparing thrust curves of Fig. 5.13 with Fig. 5.15, the power curves for the same aircraft. The velocity for minimum power required is the velocity on Fig. 5.15, where induced drag is three times as much as parasite drag. Note that this speed is significantly slower than the speed for $(L/D)_{\max}$.

5.10.4 Propeller-Driven Aircraft Range

To complete the discussion of range and endurance, the average value method expression for propeller-driven aircraft range is obtained by multiplying Eq. (5.30) by V_∞:

$$R = V_\infty \cdot E = V_\infty \frac{\Delta W_f}{\dot{W}_f} = V_\infty\left[\frac{\Delta W_f}{(c/\eta_{\text{prop}})P_{\text{req (avg)}}}\right] = V_\infty\left[\frac{\Delta W_f}{(c/\eta_{\text{prop}})D_{\text{avg}}V_\infty}\right]$$

$$R = \frac{\Delta W_f}{(c/\eta_{\text{prop}})D_{\text{avg}}} \qquad (5.44)$$

Because the form of Eq. (5.44) is identical to that of Eq. (5.34), further analysis will produce essentially the same results for propeller-driven aircraft range as were found for jet aircraft endurance. The Breguet range equation for propeller-driven aircraft is

$$R = \frac{\eta_{\text{prop}}}{c}\frac{C_L}{C_D}\ln\left(\frac{W_1}{W_2}\right) \qquad (5.45)$$

Equation (5.45) suggests that propeller-driven aircraft range is not influenced by air density (altitude), except to the degree that air density and temperature influences BSFC. Propeller-driven aircraft range is maximized by flying in conditions that are characterized by maximum propeller efficiency, minimum BFSC, maximum weight of fuel available, and minimum drag (maximum L/D or C_L/C_D). Recall

214 INTRODUCTION TO AERONAUTICS: A DESIGN PERSPECTIVE

that maximum L/D occurs at the speed where parasite drag equals induced drag $(C_{D_0} = kC_L^2)$.

Example 5.9

An aircraft is being designed to fly on Mars (where the acceleration of gravity is 3.72 m/s^2) at an altitude where $\rho = 0.01$ kg/m^3. The aircraft will be powered by a piston engine driving a propeller. The engine has, when tested, burned 50 kg of fuel and 400 kg of oxidizer in 1 h while producing 104 kW of shaft power. The propeller efficiency has been measured in Mars-like conditions at 0.85. The aircraft's drag polar is $C_D = 0.03 + 0.07C_L^2$, and its wing area is 50 m^2. What will be the aircraft's maximum range and endurance at this altitude on 500 kg of propellants if its mass with propellants is 1500 kg?

Solution: The atmosphere of Mars is composed almost entirely of carbon dioxide, so that the term *propellant* refers in this case to both fuel and oxidizer, which the aircraft must carry and consume in order for the engine to operate. The BFSC for this engine therefore must be based on total propellant consumption,

$$\text{BSFC} = c = \frac{\dot{W}_f}{\text{SHP}} = \frac{(450 \text{ kg/h})(3.72 \text{ m/s}^2)}{104 \text{ kW}} = 16.1 \text{ N/(kW h)}$$

The C_L for maximum endurance is obtained by solving Eq. (5.43),

$$3C_{D_0} = kC_L^2, \quad C_L = \sqrt{\frac{3C_{D_0}}{k}} = \sqrt{\frac{3(0.03)}{0.07}} = 1.13$$

then

$$C_D = C_{D_0} + kC_L^2 = C_{D_0} + 3C_{D_0} = 4C_{D_0} = 0.12$$

Using 1500 kg (3.72 m/s^2) = 5580 N as the initial weight and 3720 N as the final weight, the maximum endurance is

$$E = \frac{\eta_{\text{prop}}}{c} \frac{C_L^{3/2}}{C_D} \sqrt{2\rho S} \left(W_2^{-1/2} - W_1^{1/2} \right)$$

$$E = \frac{0.85}{16.1} \frac{\text{kw h}}{\text{N}} \frac{(1.13)^{3/2}}{0.12} \sqrt{2(0.01 \text{ kg/m}^3)(50 \text{ m}^2)} \left[(3720 \text{ N})^{-1/2} - (5580 \text{ N})^{-1/2} \right]$$

$$= 0.528 \frac{\text{kw h}}{\text{N}} \sqrt{N \frac{\text{s}^2}{\text{m}^2}} \left(\frac{0.003}{\text{N}^{1/2}} \right) = 0.00158 \frac{\text{kw h}}{\text{w}} = 1.58 \text{ h}$$

Maximum range for piston engine/propeller-driven aircraft is achieved at the speed for $(L/D)_{\text{max}}$, and the magnitude of $(L/D)_{\text{max}}$ is

$$\left(\frac{L}{D} \right)_{\text{max}} = \left(\frac{C_L}{C_D} \right)_{\text{max}} = \frac{1}{2\sqrt{kC_{D_0}}} = \frac{1}{2\sqrt{0.07(0.03)}} = 10.9$$

The maximum range is then given by Eq. (5.45):

$$R = \frac{\eta_{\text{prop}}}{c}\frac{C_L}{C_D}\ell n\left(\frac{W_1}{W_2}\right) = \frac{0.85}{16.1}\frac{\text{kw h}}{\text{N}}(10.9)\,\ell n\left(\frac{5580\ \text{N}}{3720\ \text{N}}\right)$$

$$= 0.677\frac{(1000\ \text{N m/s})\,(\text{h})}{\text{N}}\left(\frac{3600\ \text{s}}{1\ \text{h}}\right)(0.405)$$

$$= 987{,}000\ \text{m} = 987\ \text{km}$$

5.10.5 Altitude Variations

The choice of an appropriate cruising and/or loitering (maximum enduring) altitude is an important consideration for aircraft designers, as well as for pilots and crewmembers planning their flights. The cruising and loitering altitude choices can be influenced by weather conditions, winds, traffic congestion (the Lear Jet and other business jets are designed to cruise above the heavy jet airliner traffic at or near the tropopause), navigation and terrain constraints, training requirements, enemy threat system lethal envelopes and warning system coverage (for military aircraft), and the cruise speed that can be achieved at a particular altitude. High speeds and short travel times are among the most important advantages of aircraft over surface transportation. High cruise speeds allow airliners and military aircraft to make more flights in the same time period and therefore generate more revenue or more combat effectiveness. The effect of altitude on cruise speed for maximum range can be seen in Fig. 5.30, which shows the shift in drag and power required curves with changes in altitude. As altitude increases, the decreasing air density causes V_∞ for maximum range to increase. This benefit is limited by the ability of the engine(s) to generate sufficient thrust, increasing wind velocities with increasing altitude, the time and fuel required to climb to higher altitudes, and Mach number effects.

5.10.6 Best Cruise Mach/Best Cruise Altitude (BCM/BCA)

It was shown in Chapter 4 that aircraft which are capable of flying at speeds near and above their critical Mach number experience significant changes in their drag polars at high speeds. The rapid drag rise at the drag-divergence Mach number, just above M_{crit}, in many cases reduces the speed for maximum range from that which would be predicted by Eq. (5.38). Figure 5.31 shows thrust available and thrust required curves for an afterburning turbofan-powered supersonic fighter aircraft. The afterburning and nonafterburning thrust available curves were generated using Eqs. (5.11) and (5.12). Curves were plotted for sea level and for 45,000 ft mean sea level (MSL). The second altitude was chosen because it is the altitude where the airspeed for $(L/D)_{\text{max}}$ equals the airspeed corresponding to M_{crit}. If the aircraft has sufficient thrust to fly at this altitude without using afterburner, the altitude where this condition is satisfied results in the absolute maximum range for that aircraft. The altitude and speed for this optimum cruise condition is referred to

a) Turbojet-powered trainer aircraft

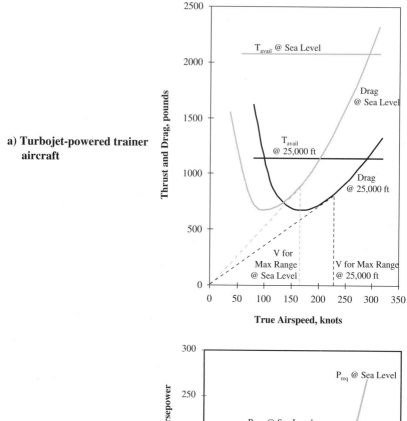

b) Piston-powered trainer aircraft

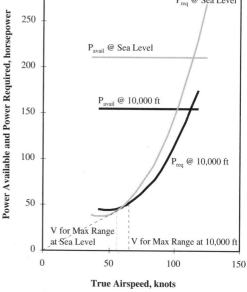

Fig. 5.30   Altitude effects on cruise speed for maximum range.

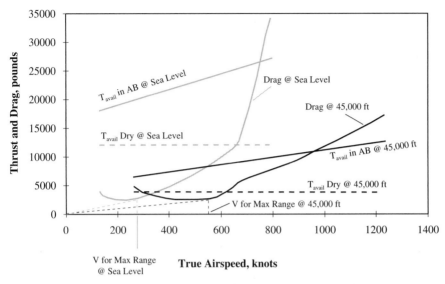

Fig. 5.31 Afterburning turbofan-powered fighter aircraft thrust and drag curves at sea level and at 45,000 ft.

as the best cruise Mach/best cruise altitude (BCM/BCA). Note that for the case shown maximum range cruise airspeed at 45,000 ft is about twice the maximum range cruise airspeed at sea level, but the drag is less. Because TSFC is lower at 45,000 ft, the aircraft's range is more than doubled at the higher altitude. Note also that because the speed for $(L/D)_{max}$ varies with aircraft weight the altitude and Mach number for BCM/BCA will change as the aircraft burns fuel. A more complete discussion of BCM/BCA is contained in Ref. 5.

5.10.7   Thrust and Power Curve Summary

Figure 5.32 compares the thrust and power curves of Figs. 5.13 and 5.16 and marks on them the airspeeds for various types of maximum performance. It is left as an exercise for the reader to construct a similar summary chart for propeller-driven aircraft.

5.11   Takeoff and Landing

Regardless of an aircraft's design mission, it must takeoff and land to start and finish its flight. Almost every conventional takeoff and landing (CTOL, as opposed to vertical takeoff and/or vertical landing using vectored thrust, etc.) aircraft was designed to meet specified maximum takeoff and landing distances. The ability to use shorter runways allows airliners to serve smaller cities or fly from a small runway near the business district of a large city, a light aircraft to land in a farmer's field and park next to his house, and military aircraft to operate from improvised runways close to the front lines or from established airfields whose runways have been damaged.

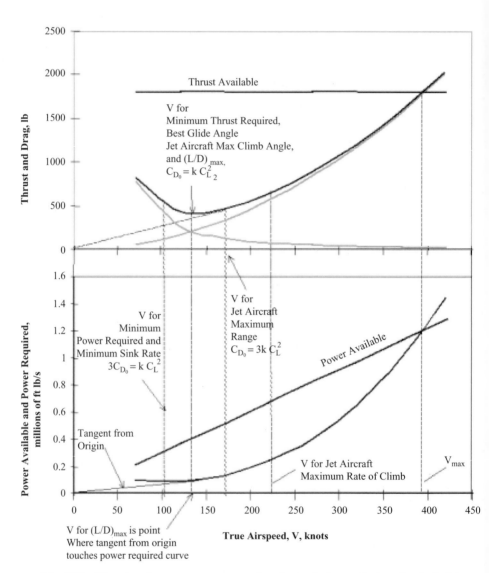

Fig. 5.32   Thrust and power curve comparison for typical nonafterburning turbojet and low-bypass-ratio turbofan-powered aircraft.

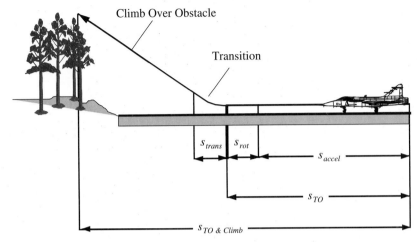

Fig. 5.33   Takeoff.

5.11.1 Takeoff Distance

For an aircraft to take off, it uses excess thrust to accelerate to a safe flying speed. Normally airspeed 1.2 times the aircraft's stalling speed at its takeoff weight and configuration is considered safe to become airborne. This safe flying airspeed is called *takeoff speed* V_{TO}. At or just prior to reaching takeoff speed, the pilot raises the aircraft's nose to establish a pitch attitude and angle of attack called the *takeoff attitude*. Once the takeoff attitude is established and the aircraft has sufficient speed, it generates enough lift to begin flying. Takeoff distance, then, is the distance required for the aircraft to accelerate to takeoff speed and rotate. Some aircraft design requirements specify a rotation time, usually around 3 s, which must be allowed, and the distance covered by the aircraft added to the takeoff distance after it has reached takeoff speed. Design requirements can also specify required takeoff performance in terms of the distance required to accelerate, rotate, transition to a climb, and climb over an obstacle with a specified height. Figure 5.33 illustrates these steps or phases in a takeoff.

The distance labeled s_{accel} in Fig. 5.33 is the distance required for the aircraft to accelerate to takeoff speed. If the pilot initiates takeoff rotation so as to reach the takeoff attitude at the point s_{accel}, then the aircraft will lift off as it reaches V_{TO}. Keep in mind that some design specifications will force the designer to add 3 more seconds of takeoff acceleration to the takeoff distance calculation beyond s_{accel}. This situation is shown in Fig. 5.33 as $s_{TO} = s_{accel} + s_{rot}$ where s_{rot} is the distance covered during the 3-s rotation allowance. If the design requirement does not specify a rotation allowance, then $s_{TO} = s_{accel}$. The analysis that follows will consider this simpler case.

Figure 5.34 illustrates the forces on an aircraft during its takeoff acceleration. The rolling friction of the wheels on the runway is modeled as a rolling-friction coefficient μ multiplied by the normal force N, exerted by the aircraft on the runway and as a reaction by the runway on the aircraft. Typical values for the

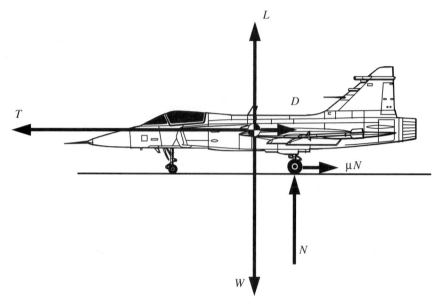

Fig. 5.34   Forces during takeoff acceleration.

rolling-friction coefficient are 0.02–0.04 for paved runways and 0.08–0.1 for turf. Assuming the thrust vector is parallel to the surface of the runway and that the runway is level, summing the forces in the vertical direction yields

$$\sum F_\perp = mV^2/r = 0 = N + L - W, \quad N = W - L \qquad (5.46)$$

and in the horizontal direction

$$\sum F_\| = ma = T - D - \mu N \qquad (5.47)$$

Combining Eqs. (5.46) and (5.47) yields

$$ma = \left(\frac{W_{TO}}{g}\right) a = T - D - \mu(W_{TO} - L)$$

$$a = \frac{dV}{dt} = \frac{g[T - D - \mu(W_{TO} - L)]}{W_{TO}} \qquad (5.48)$$

where W_{TO} is the takeoff weight. The velocity V_1 at any time t_1 during the takeoff acceleration is obtained by integrating Eq. (5.48) with respect to time:

$$V_1 = \int_0^{t_1} \frac{dV}{dt} \, dt \qquad (5.49)$$

If the initial velocity at the start of the takeoff is zero and the acceleration can be approximated as being constant during the takeoff, then Eq. (5.49) simplifies to

$$V_1 = \int_0^{t_1} \frac{dV}{dt}\,dt = \frac{dV}{dt}\int_0^{t_1} dt = a(t_1 - 0) = a\,t_1 \qquad (5.50)$$

The time to complete the takeoff t_{TO} is obtained by substituting the takeoff speed (the speed at which the airplane leaves the runway) V_{TO} for V_1 in Eq. (5.49) and solving for t_{TO}.

$$t_{TO} = V_{TO}/a$$

Now, $V_{TO} = 1.2\, V_{stall}$, and so

$$V_{TO} = 1.2\sqrt{\frac{2W_{TO}}{\rho S C_{L_{max}}}} \qquad (5.51)$$

$$t_{TO} = 1.2\sqrt{\frac{2W}{\rho S C_{L_{max}}}}\bigg/ g\frac{[T - D - \mu(W - L)]}{W}$$

$$= \frac{1.2W}{g[T - D - \mu(W - L)]}\sqrt{\frac{2W}{\rho S C_{L_{max}}}} \qquad (5.52)$$

For the same assumptions takeoff distance s_{TO} is obtained by integrating Eq. (5.50) with respect to time:

$$s_{TO} = \int_0^{t_{TO}} a\,t\,dt = a\int_0^{TO} t\,dt = \tfrac{1}{2}a\,t_{TO}^2 - 0 = \tfrac{1}{2}a\,t_{TO}^2 \qquad (5.53)$$

Substituting Eqs. (5.48) and (5.52) into (5.53) yields

$$s_{TO} = \frac{1}{2}\frac{g[T - D - \mu(W - L)]}{W}\frac{1.44W^2}{\{g[T - D - \mu(W - L)]\}^2}\frac{2W}{\rho S C_{L_{max}}}$$

$$s_{TO} = \frac{1.44W_{TO}^2}{\rho S C_{L_{max}} g[T - D - \mu(W_{TO} - L)]} \qquad (5.54)$$

In practice, the force terms in Eq. (5.54) can vary significantly during the takeoff acceleration. Reasonable results can be obtained by using an average acceleration, however. The average acceleration is obtained by evaluating the acceleration forces at $0.7V_{TO}$, so that Eq. (5.54) becomes

$$s_{TO} = \frac{1.44W_{TO}^2}{\rho S C_{L_{max}} g[T - D - \mu(W_{TO} - L)]_{0.7V_{TO}}} \qquad (5.55)$$

Equation (5.55) makes it clear that short takeoff distances can be achieved for high thrust, low weight, high $C_{L_{max}}$ with low drag, large wing area, low rolling-friction coefficient (good tires and a smooth runway), and high density (low altitude and cold temperatures). A further simplification can be used for aircraft with very high thrust available, nearly equal to their takeoff weight. For these aircraft, the thrust is so great that the retarding forces are negligible by comparison, and Eq. (5.53) simplifies to

$$s_{TO} = \frac{1.44 W_{TO}^2}{\rho S C_{L_{max}} g T} \qquad (5.56)$$

Caution: Equation (5.56) is valid only for very high-performance aircraft with thrust nearly equal to or even greater than their weight. Applying this equation to lower-performance aircraft such as light planes or airliners can dangerously underpredict takeoff distance. Use Eq. (5.55) instead for such aircraft.

Example 5.10

The nonafterburning turbojet engines of a Cessna T-37 jet trainer produce approximately 1700 lb of installed thrust for takeoff at sea level. Its takeoff weight is 6575 lb. Its $C_{L_{max}} = 1.6$ for takeoff and its drag polar in its takeoff configuration is $C_{D_0} = 0.03 + 0.057 C_L^2$. Its reference planform area is 184 ft². Normal takeoff procedure requires the pilot to rotate the aircraft to the takeoff attitude just prior to reaching takeoff velocity, and so for the majority of the takeoff roll the aircraft's $C_L = 0.8$. What will be the aircraft's takeoff distance at sea level with no wind?

Solution: Thrust for this aircraft is very much less than its weight, and so it is probably not reasonable to ignore the drag on takeoff. The takeoff speed is

$$V_{TO} = 1.2 \sqrt{\frac{2 W_{TO}}{\rho S C_{L_{max}}}} = 1.2 \sqrt{\frac{2(6575 \text{ lb})}{(0.002377 \text{ slug/ft}^3)(184 \text{ ft}^2)(1.6)}} = 164.5 \text{ ft/s}$$

but $C_L = 0.8$ during the takeoff roll, and so at $V = 0.7 V_{TO} = 0.7(137.1 \text{ ft/s}) = 115.15 \text{ ft/s}$.

$$q = \tfrac{1}{2}\rho V^2 = \tfrac{1}{2}(0.002377 \text{ slug/ft}^3)(115.15 \text{ ft/s})^2 = 15.76 \text{ lb/ft}^2$$

$$L = C_L q S = 0.8(15.76 \text{ lb/ft}^2)(184 \text{ ft}^2) = 2320 \text{ lb}$$

$$C_D = C_{D_0} + k C_L^2 = 0.03 + 0.057(0.8)^2 = 0.0665$$

$$D = C_D q S = 0.0665(15.76 \text{ lb/ft}^2)(184 \text{ ft}^2) = 192.8 \text{ lb}$$

then

$$s_{TO} = \frac{1.44 W_{TO}^2}{\rho S C_{L_{max}} g [T - D - \mu(W_{TO} - L)]}$$

$$= \frac{1.44(6575 \text{ lb})^2}{(0.002377 \text{ slug/ft}^3)(184 \text{ ft}^2)(1.6)(32.2 \text{ ft/s}^2)[1700 \text{ lb} - 192.8 \text{ lb} - 0.03(6575 \text{ lb} - 2,320 \text{ lb})]}$$

$$s_{TO} = 2003 \text{ ft}$$

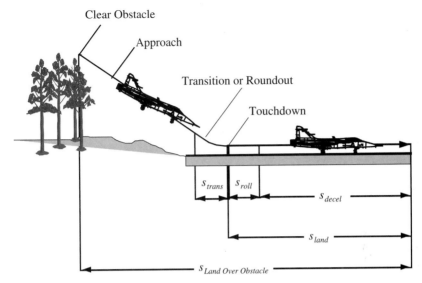

Fig. 5.35   Landing.

5.11.2 Landing Distance

As Fig. 5.35 illustrates, the landing maneuver is broken up into approximately the same steps as takeoff. As with takeoff, the details of the design requirements for landing distance vary. The landing speed V_L is usually specified as 1.3 V_{stall}. The approach or descent to landing is also generally flown at V_L, or slightly faster. Some customers and/or regulatory agencies might specify landing distances over a fixed obstacle. Others might specify that the aircraft pass over the end or threshold of the landing runway at a specified height or that it touch down a specified distance down the runway. The design specifications might require the landing analysis to include three or more seconds of free roll (deceleration only caused by normal rolling friction and air drag) after touchdown before the brakes are applied. A landing analysis can also include the effects of reverse thrust or a drag parachute that is deployed at or slightly before touchdown. The simple case of no free roll allowance, so that $s_L = s_{decel}$, will be considered here.

Figure 5.36 shows the forces on an aircraft during a landing deceleration. The forces are similar to those for the takeoff, except that thrust is zero and μ, which is now called braking coefficient, has a much higher value because brakes are applied. Braking coefficient values are 0.4–0.6 for dry concrete, 0.2–0.3 for wet concrete, and 0.05–0.1 for an icy runway.

The same analysis steps as for takeoff yield

$$s_L = \frac{1.69 W_L^2}{\rho S C_{L_{max}} g [D + \mu (W_L - L)]_{0.7 V_L}} \tag{5.57}$$

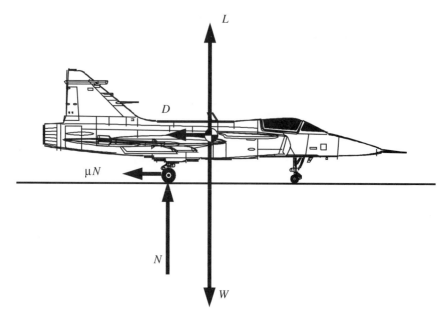

Fig. 5.36   Forces during landing deceleration.

Note that the factor 1.69 in Eq. (5.57) instead of 1.44 as in Eq. (5.46) is because $V_L = 1.3\,V_{\text{stall}}$, while $V_{\text{TO}} = 1.2\,V_{\text{stall}}$. As with the average force values for takeoff, the average deceleration forces are evaluated at $0.7V_L$.

5.12   Turns

Turning performance is important to military fighter aircraft, pylon racers, crop dusters, and to a lesser degree to any aircraft that must maneuver in tight quarters, for instance to takeoff and land at an airfield in a canyon or among skyscrapers in the center of a city. The most important characteristics of turning performance that are frequently specified as design requirements are turn rate and turn radius. This performance can be specified either as an instantaneous or a sustained capability. As the names imply, a sustained turn rate or radius is performance the aircraft can maintain for a long period of time—minutes or even hours. An instantaneous turn rate or radius is a capability the aircraft can achieve momentarily, but then the maximum performance might begin to decrease immediately.

5.12.1   Level Turns

The most commonly performed turning maneuver for an aircraft is the level turn. In this maneuver, the aircraft maintains a constant altitude (and in a sustained turn, a constant airspeed). Its velocity vector changes directions but stays in a horizontal plane. Figure 5.37 shows front and top views of an aircraft in a level

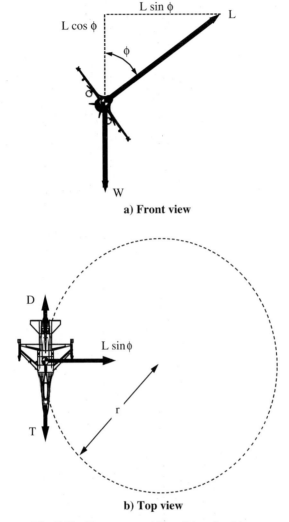

a) Front view

b) Top view

Fig. 5.37   Forces on an aircraft in a level turn.

turn. Summing forces in the vertical direction yields

$$\sum F_{\text{vert}} = 0 = L \cos \phi - W$$

$$W = L \cos \phi \qquad (5.58)$$

$$1/\cos \phi = L/W \equiv n \qquad (5.59)$$

where the parameter n defined in Eq. (5.59) is known as the *load factor*. This is also known as the g that the aircraft is pulling, but the symbol g is already being used to denote the acceleration of gravity. Equation (5.59) states that there is a one-to-one correspondence between bank angle and load factor in a level turn, regardless of the aircraft type. Note that in deriving Eq. (5.59) the acceleration in the vertical direction is set to zero, because the aircraft's motion is assumed to remain in a horizontal plane. Summing forces perpendicular to the velocity vector in the horizontal plane gives

$$\sum F_{\text{horz}} = mV^2/r = \frac{W}{g}\frac{V^2}{r} = L\sin\phi = \sqrt{L^2 - W^2} = W\sqrt{n^2 - 1}$$

$$\frac{V^2}{gr} = \sqrt{n^2 - 1}$$

which can be solved for the turn radius r

$$r = \frac{V^2}{g\sqrt{n^2 - 1}} \tag{5.60}$$

Once again, this result is independent of aircraft type, and so a Boeing 747 and an F-16 at the same bank angle and airspeed will have the same turn radius. The rate of turn $\omega = V/r$, and so

$$\omega = \frac{V}{V^2/g\sqrt{n^2 - 1}} = \frac{g\sqrt{n^2 - 1}}{V} \tag{5.61}$$

which is also independent of aircraft type. This analysis assumes no component of thrust perpendicular to the velocity vector. Aircraft with thrust vectoring capability can significantly exceed the turn capability predicted by Eqs. (5.59–5.61).

No information about whether the turn is sustained or just instantaneous is available or needed in Fig. 5.37a, and so Eqs. (5.59–5.61) apply to both types of turns. Referring to Fig. 5.28b, summing forces parallel to the aircraft's velocity vector yields

$$\sum F_{\|} = ma = \frac{W}{g}\frac{dV}{dt} = T - D$$

which for a sustained turn simplifies to

$$T = D$$

This seems a simple result, but drag for a turning aircraft is not the same as for one in SLUF. Most of the additional drag in a turn results from additional induced drag caused by the additional lift required to turn. Recall from Eq. (5.59) that $L = nW$, and so

$$C_L = \frac{nW}{qS}$$

so

$$D = C_{D_0}qS + qSk\left(\frac{nW}{qS}\right)^2 = C_{D_0}qS + \frac{k(nW)^2}{qS} \tag{5.62}$$

The significant result from Eq. (5.62) is that induced drag in a turn increases with n^2. This can make it very difficult for many aircraft to sustain turns at high load factors.

Example 5.11

Two aircraft, one on Mars and one on Earth, are performing level turns at identical true airspeeds of 100 m/s and identical bank angles of 60 deg. How do their load factors, turn radii, and rates of turn compare?

Solution: In each case, the load factor depends only on the bank angle, so that it will be the same on either planet:

$$n = 1/\cos \phi = 1/\cos 60 \text{ deg} = 2.0$$

The turn radius on each planet is calculated using Eq. (5.60):

$$r = \frac{V^2}{g\sqrt{n^2 - 1}} = \frac{(100 \text{ m/s})^2}{3.72 \text{ m/s}^2\sqrt{2^2 - 1}} = 1552 \text{ m}$$

on Mars and

$$r = \frac{V^2}{g\sqrt{n^2 - 1}} = \frac{(100 \text{ m/s})^2}{9.8 \text{ m/s}^2\sqrt{2^2 - 1}} = 589 \text{ m}$$

on Earth. The turn rates are calculated using Eq. (5.61):

$$\omega = \frac{g\sqrt{n^2 - 1}}{V} = \frac{3.72 \text{ m/s}^2\sqrt{2^2 - 1}}{100 \text{ m/s}} = 0.0644 \text{ radian/s} = 3.69 \text{ deg/s}$$

on Mars and

$$\omega = \frac{g\sqrt{n^2 - 1}}{V} = \frac{9.8 \text{ m/s}^2\sqrt{2^2 - 1}}{100 \text{ m/s}} = 0.1697 \text{ radians/s} = 9.72 \text{ deg/s}$$

on Earth.

5.12.2 Vertical Turns

Many aircraft perform turns that are not limited to a horizontal plane. The simplest of these is a turn made purely in a vertical plane. Such a maneuver completed through 360 deg of turn to the original flight conditions is called a loop. Figure 5.38 shows the forces on an aircraft at three points in a loop. The maneuver is started with a pull-up, a vertical turn from initial straight and level conditions. At the top of the loop, the aircraft is performing a pull-down from the inverted flight condition. A pull-down can also be entered from straight-and-level flight by rolling (changing the bank angle ϕ) the aircraft until it is inverted. A vertical turn initiated by rolling inverted and then pulling down, completing the turn to level flight headed in the opposite direction is known as a split-S.

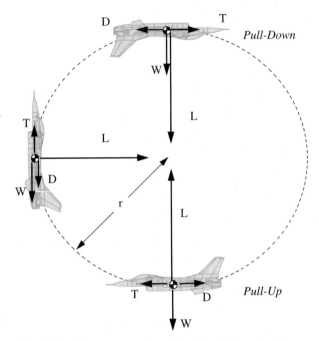

Fig. 5.38   Forces on an aircraft at three points in a loop.

For the pull-up, summing the forces perpendicular to the velocity vector yields

$$\sum F_\perp = mV^2/r = \frac{WV^2}{gr} = L - W$$

$$\frac{V^2}{gr} = L/W - 1 = n - 1$$

$$r = \frac{V^2}{g(n-1)} \tag{5.63}$$

$$\omega = \frac{g(n-1)}{V} \tag{5.64}$$

Likewise, for the pull-down

$$\sum F_\perp = mV^2/r = \frac{WV^2}{gr} = L + W$$

$$\frac{V^2}{gr} = L/W + 1 = n + 1$$

$$r = \frac{V^2}{g(n+1)} \tag{5.65}$$

$$\omega = \frac{g(n+1)}{V} \tag{5.66}$$

Finally, for the case where the aircraft's velocity vector is vertical

$$\sum F_\perp = mV^2/r = \frac{WV^2}{gr} = L$$

$$\frac{V^2}{gr} = L/W = n$$

$$r = \frac{V^2}{gn} \tag{5.67}$$

$$\omega = \frac{gn}{V} \tag{5.68}$$

Note that Eqs. (5.67) and (5.68) are approximately true for all of the turns and for most other, more complex turn geometries, especially when n is large. For modern fighter aircraft that routinely use load factors of nine, Eqs. (5.67) and (5.68) are reasonably good approximations for all turning situations.

Example 5.12

The procedure for performing a loop in the T-37 begins with diving the aircraft to reach the entry airspeed of 250 KIAS, then pulling straight up (check wings level) at $n = 4$. As the nose comes up, the speed begins to bleed off, and at about 200 KIAS or a little less you start to feel a tickle in the stick. This tickle is the first signs of stall, and it occurs before the drag due to lift (wing-tip vortices and flow separation) gets too large. If you pull into heavy buffet at this point, the excessive drag will cause the aircraft to slow down too much, and you will not make it over the top. But, if you pull any less than "on the tickle" your load factor will not be enough, and you also will not make it over the top. So, you pull on the tickle, relaxing backpressure on the stick as the aircraft slows down. As the pitch attitude exceeds 90 deg, you throw your head back and look to find the horizon, and then pull down to it being careful to keep the wings level. If you do everything correctly, the speed over the top is about 100 kn, and the load factor is about 1.5 or so. Then, as you continue to pull down inverted the speed builds up, as does the backpressure and load factor required to stay on the tickle. When the load factor reaches 4, you hold that until you complete the maneuver.

From the cockpit, a T-37 loop flown as just described looks like a circle, as shown in Fig. 5.39. But what would it look like from the ground or from another aircraft. Use the information on Fig. 5.39 to calculate the turn radius at the three points shown, and then estimate the actual shape of the T-37's flight path as it performs a loop.

Solution: Equation (5.63) applies at the bottom of the loop, where $V = 250$ KIAS. If this is at 15,000 ft and position error is negligible, then 250 KIAS = 250 KCAS and

$$V_e = fV_c = 0.987\,(250\ \text{KCAS}) = 246.8\ \text{KEAS}$$

$$V = V_e\sqrt{\frac{\rho_{SL}}{\rho}} = 245.8\ \text{KEAS}\sqrt{\frac{0.002377\ \text{slugs/ft}^3}{0.001496\ \text{slugs/ft}^3}} = 310\ \text{KTAS}\frac{1.69\ \text{ft/s}}{\text{KTAS}}$$

$$= 523.6\ \text{ft/s}$$

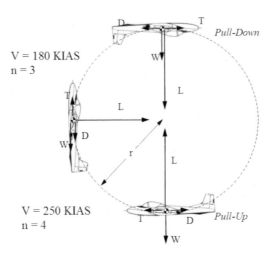

$$V = 100 \text{ KIAS}$$
$$n = 1.5$$

Pull-Down

$$V = 180 \text{ KIAS}$$
$$n = 3$$

$$V = 250 \text{ KIAS}$$
$$n = 4$$

Pull-Up

Fig. 5.39   T-37 loop.

and then

$$r = \frac{V^2}{g(n-1)} = \frac{(523.6 \text{ ft/s})^2}{32.2 \text{ ft/s}^2 (4-1)} = 2838 \text{ ft}$$

at the bottom of the loop. Now, with a nearly 3000-ft radius the aircraft will climb nearly 3000 ft as it executes the first quarter of the loop, so that its altitude when it is going straight vertical will be close to 18,000 ft. So, for this case, again assuming negligible position error so 180 KIAS = 180 KCAS,

$$V_e = f V_c = 0.990(180 \text{ KCAS}) = 178.2 \text{ KEAS}$$

$$V = V_e \sqrt{\frac{\rho_{\text{SL}}}{\rho}} = 178.2 \text{ KEAS} \sqrt{\frac{0.002377 \text{ slug/ft}^3}{0.001355 \text{ slug/ft}^3}} = 236 \text{ KTAS} \frac{1.69 \text{ ft/s}}{\text{KTAS}}$$

$$= 398.9 \text{ ft/s}$$

and then

$$r = \frac{V^2}{g(n)} = \frac{(398.9 \text{ ft/s})^2}{32.2 \text{ ft/s}^2 (3)} = 1647 \text{ ft}$$

at the side of the loop. Now, the radius decreased so much in this first part of the loop that we predict the aircraft will only climb another 1000 ft or less as it executes the second quarter of the loop, so that its altitude when it is going over the top will be close to 19,000 ft. So, for this case, again assuming negligible position

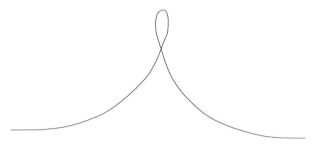

Fig. 5.40   T-37 flight path during a loop.

error so 100 KIAS = 100 KCAS,

$$V_e = fV_c = 0.997(100 \text{ KCAS}) = 99.7 \text{ KEAS}$$

$$V = V_e\sqrt{\frac{\rho_{\text{SL}}}{\rho}} = 99.7 \text{ KEAS}\sqrt{\frac{0.002377 \text{ slugs/ft}^3}{0.001310 \text{ slugs/ft}^3}} = 134.3 \text{ KTAS}\frac{1.69 \text{ ft/s}}{\text{KTAS}}$$

$$= 227 \text{ ft/s}$$

and then

$$r = \frac{V^2}{g(n+1)} = \frac{(227 \text{ ft/s})^2}{32.2 \text{ ft/s}^2(1.5+1)} = 640 \text{ ft}$$

at the top of the loop! So, we predict that a T-37 loop would look like Fig. 5.40, with a large radius at the bottom and a small one at the top.

5.13   V–n Diagrams

The turn analysis up to this point has said nothing of the limitations the aircraft might have on its ability to generate the lift or sustain the structural loading needed to perform a specified turn. These limitations are often summarized on a chart known as a V–n diagram. Figure 5.41 is a V–n diagram for a subsonic jet trainer. The maximum positive and negative load factors that the aircraft structure can sustain are shown as horizontal lines on the chart because for this particular aircraft these structural limits are not functions of velocity. At low speeds, the maximum load factor is limited by the maximum lift the aircraft can generate because

$$L_{\max} = n_{\max}W = C_{L_{\max}}qS = C_{L_{\max}}1/2\rho V^2 S$$

so

$$n_{\max} = \frac{C_{L_{\max}}\rho S}{2W}V^2 \tag{5.69}$$

The maximum lift boundary is also known as the stall boundary. Equation (5.69) also leads to a more general form of the stall speed equation:

$$V_{\text{stall}} = \sqrt{\frac{2nW}{\rho SC_{L_{\max}}}} \tag{5.70}$$

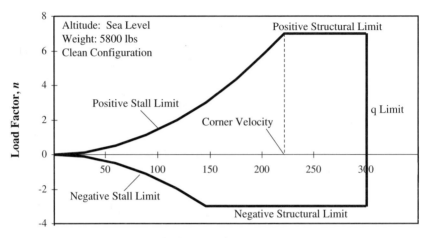

Calibrated Airspeed, V_c, knots

Fig. 5.41   V–n diagram for a subsonic jet trainer.

The vertical line on Fig. 5.41 that is labeled "q limit" indicates the maximum structural airspeed of the aircraft for these conditions. In this case, the maximum structural airspeed is not a function of load factor. On many aircraft, maximum structural speed decreases at high positive and negative load factors. The feature of the aircraft that sets the maximum speed varies. For the aircraft of Fig. 5.39, the maximum speed limit is actually set by the aircraft's critical Mach number. Flight above this speed is prohibited because shock-induced separation causes control difficulties. For other aircraft, the limit is set by the structural strength required by the wings, windscreen, etc. to resist the high dynamic pressures and high stagnation point pressures at these speeds, hence the name q limit. For many high-speed aircraft the maximum speed is actually a temperature limit because at faster speeds skin friction and shock waves generate so much heat that the aircraft skin will melt! For the F-104, the q limit is set by maximum engine inlet temperature.

5.13.1   Corner Velocity

Figure 5.41 and Eqs. (5.69) and (5.70) can be used to determine the airspeed at which an aircraft can make its quickest, tightest turn. From Eq. (5.69) it is clear that the lowest speed at which the maximum load factor can be generated will produce the smallest turn radius. Equation (5.70) dictates the same condition to produce the highest turn rate. The velocity labeled *corner velocity* on Fig. 5.39 is the velocity at which the stall limit and the structural limit make a "corner" on the graph. This velocity satisfies the conditions for quickest, tightest turn because a faster velocity would not see an increase in n as a result of the structural limit, and a slower velocity would see n limited to less than its maximum value by the stall. The term "corner velocity" might also have been chosen to reflect the fact that the aircraft makes its sharpest corner at that speed. An expression for corner velocity

V^* is obtained by substituting the aircraft's maximum load factor into Eq. (5.67):

$$V^* = \sqrt{\frac{2n_{\max} W}{\rho S C_{L_{\max}}}} \qquad (5.71)$$

Example 5.13

An aircraft with a wing loading W/S of 70 lb/ft^2 and $C_{L_{\max}} = 1.5$ has a maximum structural load limit of 9. What is its corner velocity at sea level?

Solution: Corner velocity is calculated using Eq. (5.71):

$$V^* = \sqrt{\frac{2n_{\max} W}{\rho S C_{L_{\max}}}} = \sqrt{\frac{2n_{\max}}{\rho C_{L_{\max}}} \frac{W}{S}} = \sqrt{\frac{2(9)(70 \text{ lb/ft}^2)}{0.002377 \text{ slug/ft}^3 (1.5)}} = 594.5 \text{ ft/s}$$

5.14 Energy Height and Specific Excess Power

One of the design requirements for a multirole fighter listed in Table 1.2 was a certain specific excess power achieved for specified conditions. Specific excess power is a measure of an aircraft's ability to increase its specific energy H_e, the sum of its kinetic and potential energy divided by its weight.

$$H_e = \frac{\text{P.E.} + \text{K.E.}}{W} = \frac{mgh + 1/2mV^2}{W} = h + \frac{V^2}{2g} \qquad (5.72)$$

Specific energy is also called energy height because it has units of height. Energy is changed by doing work, raising an object against the pull of gravity to a higher altitude, accelerating an object to a faster velocity, or both. The rate of doing work is power, but only part of an aircraft's power can be used for this. A portion of an aircraft's power available must be used to balance the power required to overcome the aircraft's drag. The work done by this portion of the aircraft's power is converted by air viscosity into heat and air turbulence. The aircraft's power available that is in excess of its power required is its excess power. It is this portion of its power that can be used to increase its potential and/or kinetic energy.

$$P_{\text{avail}} - P_{\text{required}} = V(T - D) = \frac{d(\text{P.E.} + \text{K.E.})}{dt}$$

Dividing both sides of the equation by weight gives an expression for specific excess power P_s:

$$P_s \equiv \frac{P_{\text{avail}} - P_{\text{required}}}{W} = \frac{V(T-D)}{W} = \frac{d}{dt}\left(\frac{\text{P.E.} + \text{K.E.}}{W}\right) = \frac{dH_e}{dt} = \frac{d}{dt}\left(h + \frac{V^2}{2g}\right)$$

$$P_s = \frac{V(T - D)}{W} = \frac{dh}{dt} + \frac{V}{g}\frac{dV}{dt} \qquad (5.73)$$

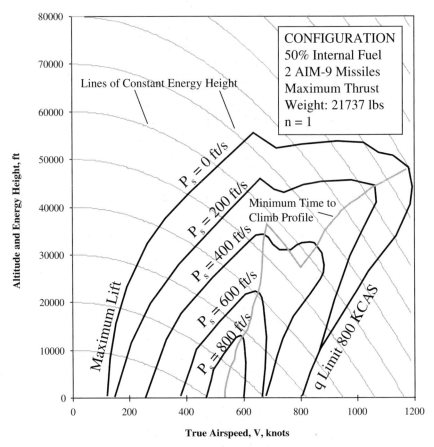

Fig. 5.42 P_s diagram for a multirole fighter aircraft at $n = 1$.

5.14.1 P_s Diagrams

Equation (5.73) is a powerful tool for evaluating and comparing aircraft performance. It can be used to verify that a given aircraft design meets a specific P_s requirement, such as the one listed in Table 1.2. It is more common for P_s to be calculated and plotted for an aircraft for a range of altitudes and Mach numbers to create a P_s diagram. Figure 5.42 is a typical P_s diagram for a multirole fighter aircraft. Note that the P_s values are indicated by contour lines, lines connecting points with equal values of P_s. Lines of constant energy height are also plotted on the diagram. Figure 5.42 is only valid for one aircraft weight, configuration, and load factor.

In essence, the P_s diagram functions as a three-dimensional (with altitude as the third dimension) power-available/power-required curve. The $P_s = 0$ contour is the aircraft's operating envelope. For all combinations of altitude and Mach number inside this envelope, the aircraft has sufficient thrust to sustain level flight. Where $P_s > 0$, the aircraft can climb and/or accelerate. The aircraft's absolute ceiling is the highest point on the $P_s = 0$ contour. Likewise, its service ceiling would be the highest point on its $P_s = 100$ ft/min (not feet/second as on Fig. 5.42) contour. The aircraft's absolute maximum speed in level flight occurs at the altitude and Mach number where the $P_s = 0$ contour reaches furthest to the right.

5.14.2 Zoom Climbs

An aircraft can operate briefly outside its level-flight envelope. This can be done either by diving (so that, as in a glide, a component of weight acts opposite the drag) to reach airspeeds above its maximum level-flight speed or by performing a *zoom* or *zoom climb* (as in "zooming to meet our thunder" in the U.S. Air Force Song.) A zoom climb occurs when an aircraft climbs so as to convert airspeed into altitude. If an aircraft is flown to the edge of its operating envelope (so that $T = D$) and then forced to climb, it will move along a constant energy height line on the P_s diagram. It will decelerate as it climbs, but (at least initially, because $T = D$) its total energy will remain constant. As the aircraft slows down, it might deviate from the H_e = constant line as drag changes and no longer equals thrust. If the aircraft whose P_s diagram is shown in Fig. 5.40 were flown to its absolute ceiling, $h = 57,000$ ft and $V = 630$ kn, this would correspond to an energy height of 74,000 ft. If it entered a zoom climb from this condition and thrust remained equal to drag, it would move along the $H_e = 74,000$-ft line decelerating until it reached zero velocity at an altitude of 74,000 ft.

5.14.3 Minimum Time to Climb

The P_s diagram can be used to determine a strategy for climbing to a given altitude in absolute minimum time, as when the F-15A *Streak Eagle* set minimum-time-to-climb records in 1975. The maximum rate of climb at any given altitude is achieved at the speed where P_s is maximum. However, because the aircraft can be zoomed at the end of its climb to get to a particular altitude faster, the minimum time to climb is achieved by changing energy height, not just height, as fast as possible. The aircraft increases energy height the fastest when it moves perpendicular to H_e = constant lines at the point where P_s is maximum on each line. A trajectory is shown on Fig. 5.40 that satisfies this requirement to cross each energy height line where P_s is maximum. At one point along the trajectory, the aircraft descends and accelerates following an H_e = constant line, then continues on the climb profile. The constant-energy-height descent/acceleration moves the aircraft quickly through the transonic regime to an altitude and supersonic speed, where P_s is maximum along the higher H_e = constant lines.

5.14.4 Maneuvering P_s

Figure 5.43 is a P_s diagram for the same aircraft at a load factor, $n = 5$. Note that P_s has decreased everywhere on the diagram. This is because more induced drag and therefore more power required result from the five times greater lift required to generate a load factor $n = 5$.

Example 5.14

For the aircraft whose P_s diagrams are depicted in Figs. 5.42 and 5.43, determine the following:

1) What is the aircraft's maximum 1-g level-flight, speed and at what altitude does it occur?

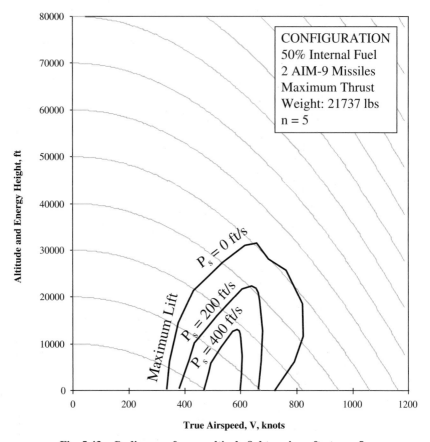

Fig. 5.43 P_s diagram for a multirole fighter aircraft at $n = 5$.

2) What is the aircraft's maximum zoom altitude?

3) What is the aircraft's best rate of climb at sea level, and at what velocity does it occur?

4) What is this aircraft's minimum level flight speed at sea level? What causes this limit?

5) What is this aircraft's maximum level flight speed at sea level? What causes this limit?

6) What is the maximum altitude at which this aircraft can sustain a 5-g turn, and at what speed does it occur?

7) What is the maximum speed at which this aircraft can sustain a 5-g turn, and at what altitude does it occur?

8) What is the minimum speed at which this aircraft can sustain a 5-g turn, at what altitude and airspeed does it occur, and what causes this limit?

Solution: All of the answers to these questions can be determined by looking at the P_s diagrams.

1) The maximum speed occurs at the point where the $P_s = 0$ contour goes the farthest to the right. On Fig. 5.42 this is $V = 1200$ kn at $h = 45,000$ ft.

2) The maximum zoom altitude is the maximum energy height line touched by the $P_s = 0$ contour. On Fig. 5.42 this is $H_e = h_{zoom} = 110,000$ ft.

3) The best rate of climb at sea level occurs where P_s is maximum. On Fig. 5.42 this is at approximately 530 kn at sea level, and the maximum rate of climb is greater than 800 ft/s or 48,000 ft/min.

4) The minimum level-flight speed at sea level is depicted on Fig. 5.42 as the aerodynamic limit line, which means the minimum speed is limited by stall, buffet, or maximum usable angle of attack. The speed depicted on Fig. 5.42 is approximately 120 kn.

5) The maximum level-flight speed at sea level is depicted on Fig. 5.42 as the q limit line, which means the maximum speed is limited by the maximum dynamic pressure that the aircraft structure can sustain. The speed depicted on Fig. 5.42 is 800 kn. In reality, this limit could be caused by engine inlet limitations, aircraft skin temperature limits, or even just the fact that the aircraft has not been flight tested beyond this limit. When test pilots "push the edge of the envelope," they are demonstrating that the aircraft is safe to fly in areas of the flight envelope that are achievable but have not been demonstrated yet. Note that the q limit in Fig. 5.42 is identified by a maximum allowable calibrated airspeed, a performance measurement easily monitored by the pilot.

6) The maximum altitude at which this aircraft can sustain $n = 5$ is the highest point on the $P_s = 0$ contour on Fig. 5.43. This point is at $h = 32,000$ ft and $V = 670$ kn.

7) The maximum speed at which this aircraft can sustain $n = 5$ is the farthest right point on the $P_s = 0$ contour on Fig. 5.43. This point is at $h = 17,000$ ft and $V = 820$ kn. Insufficient thrust prevents sustaining 5 g at a higher speed.

8) The minimum speed at which this aircraft can sustain $n = 5$ is the farthest left point on the $P_s = 0$ contour on Fig. 5.43. This point is at $h = 0$ ft and $V = 330$ kn. This speed is limited by stall, buffet, or maximum usable angle of attack.

5.14.5 Aircraft P_s Comparisons

Figure 5.44 illustrates one of the most important uses of P_s diagrams. It was created by calculating the differences between the P_s values of two different fighter aircraft, aircraft A and aircraft B, at each point on a P_s diagram. Regions of the diagram with similar values of P_s differences are shaded the same. The level flight envelope for aircraft A is shown in black and for aircraft B in gray. In the center of the diagram, differences in P_s between the two aircraft are less than 100 ft/s, so that there is no clear advantage for either plane. On the right side of the diagram, higher Mach numbers and lower altitudes, aircraft A has a P_s advantage greater than 100 ft/s over aircraft B. At very high Mach numbers, aircraft A has exclusive use of a range of velocities and altitudes that are outside aircraft B's level flight envelope. Likewise, at low speeds aircraft B has an advantage. Aircraft B has exclusive use of a range of very low speeds and high altitudes.

Comparative P_s diagrams such as this are very useful to fighter pilots as they plan how to conduct an aerial battle against an adversary aircraft of a particular type. In the case shown in Fig. 5.42, the pilot of aircraft A would attempt to bring

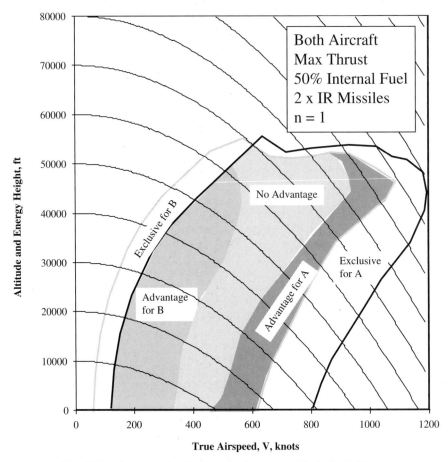

Fig. 5.44   Comparative P_s diagram for two multirole jet fighters.

the fight to lower altitude and stay at high speed. At the same time, the pilot of aircraft B would attempt to keep the fight high and slow to lower speeds, where that aircraft has the advantage. Similar diagrams made for higher load factors are also used because most extended aerial fights involve a great deal of turning.

5.15   Chapter Summary

Performance is a large, complex topic, and so any summary of the subject can easily get too long to be useful. The following notes hit the very "high points" of performance. As you review them, if the meanings of the symbols and short comments do not immediately come to your mind, go to the appropriate section in this chapter, and read more details.

5.15.1   SLUF

$$L = W, \quad T = D$$

5.15.2 Propulsion

$$T_A = T_{SL}\left(\frac{\rho}{\rho_{SL}}\right)$$

$$c_t = c_{t_{SL}}\left(\frac{a}{a_{SL}}\right)$$

5.15.3 Thrust and Power Curves

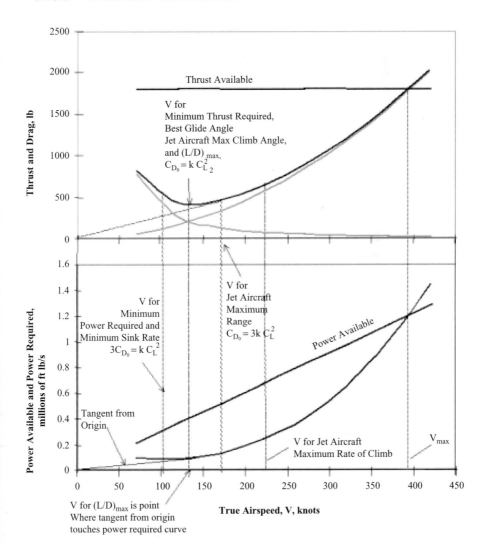

5.15.4 Curve Shifts

Configuration change (increased prasite drag):

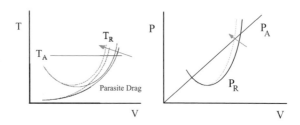

Weight increase:

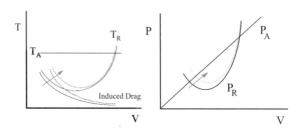

Altitude increase:

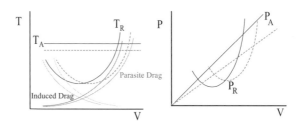

5.15.5 Drag Polar Relationships

1) Minimum power required

$$\left(C_L^{3/2}/C_D\right)_{\text{max}} \quad 3C_{D_0} = kC_L^2$$

a) Maximum glide endurance (minimum sink)

2) Minimum thrust required

$$(C_L/C_D)_{\text{max}} = (L/D)_{\text{max}} \quad C_{D_0} = kC_L^2$$

a) Maximum glide range
b) Maximum jet endurance
c) Maximum climb angle

3) Minimum (D/V)

$$\left(C_L^{1/2}/C_D\right)_{\text{max}} \quad C_{D_0} = 3kC_L^2$$

a) Maximum range jet

5.15.6 (L/D)max

$$C_L = \sqrt{\frac{C_{D_0}}{k}}, \quad C_{D_0} = kC_L^2, \quad C_D = 2C_{D_0} = 2kC_L^2, \quad \left(\frac{L}{D}\right)_{\text{max}} = \frac{1}{2\sqrt{kC_{D_0}}}$$

5.15.7 Glides

1) Maximum glide range
   a) Occurs at $T_{R\text{min}}$ or $(L/D)_{\text{max}}$
   b)

$$C_{D_0} = kC_L^2$$

   c)

$$\frac{L}{D} = \frac{R}{h}$$

2) Maximum glide endurance
   a) Occurs at $P_{R\text{min}}$
   b)

$$3C_{D_0} = kC_L^2$$

   c) Sink rate is equal to

$$V_\infty \sin\gamma = \frac{V_\infty D}{W} = \frac{P_R}{W}$$

5.15.8 Climbs

1) Maximum climb angle
   a) For jets with $T = \text{const}$, occurs at $T_{R\text{min}}$ or $(L/D)_{\text{max}}$
   b)

$$C_{D_0} = kC_L^2$$

   c)

$$\sin\gamma = \frac{T - D}{W}$$

2) Maximum rate of climb
   a) Occurs at $(P_{\text{Avail}} - P_R)_{\text{max}}$
   b) Use parallel tangent technique

c) Rate of climb is equal to

$$\frac{V_\infty (T - D)}{W} = \frac{P_{\text{avail}} - P_{\text{req}}}{W}$$

3) Ceilings
   a) Absolute—rate of climb $= 0$
   b) Service—100 ft/min rate of climb can be sustained
   c) Combat ceiling—500 ft/min rate of climb can be sustained

5.15.9   Range and Endurance

1) Jets
   a) Maximum endurance
      i) Occurs at $T_{R\text{min}}$ or $(L/D)_{\text{max}}$
      ii)
$$C_{D_0} = kC_L^2$$
      iii) Average value method
$$E = \frac{\Delta W_f}{c_t D_{\text{avg}}}$$
      iv) Breguet equation
$$E = \frac{1}{c_t} \frac{C_L}{C_D} \ln \left(\frac{W_1}{W_2} \right)$$

   b) Maximum range
      i) Occurs at $(D/V)_{\text{min}}$ or $(C_L^{1/2}/C_D)_{\text{max}}$
      ii)
$$C_{D_0} = 3kC_L^2$$
      iii) Average value method
$$R = \frac{\Delta W_f}{c_t \left(D_{\text{avg}}/V_\infty \right)}$$
      iv) Breguet equation
$$R = \sqrt{\frac{2}{\rho S}} \frac{2}{c_t} \frac{C_L^{1/2}}{C_D} \left(W_1^{1/2} - W_2^{1/2} \right)$$

2) Propeller-driven aircraft
   a) Maximum endurance
      i) Occurs at $P_{R\text{min}}$
      ii)
$$3C_{D_0} = kC_L^2$$
      iii) Average value method
$$E = \frac{\Delta W_f}{(c/\eta_{\text{prop}}) D_{\text{avg}} V_\infty}$$

iv) Breguet equation

$$E = \frac{\eta_{\text{prop}}}{c} \frac{C_L^{3/2}}{C_D} \sqrt{2\rho S} \left(W_2^{-1/2} - W_1^{-1/2}\right)$$

b) Maximum range
   i) Occurs at $T_{R\text{min}}$ or $(L/D)_{\text{max}}$
   ii)

$$C_{D_0} = kC_L^2$$

iii) Average value method

$$R = \frac{\Delta W_f}{(c/\eta_{\text{prop}})D_{\text{avg}}}$$

iv) Breguet equation

$$R = \frac{\eta_{\text{prop}}}{c} \frac{C_L}{C_D} \ln\left(\frac{W_1}{W_2}\right)$$

5.15.10 Takeoff and Landing

1) Takeoff

$$s_{\text{TO}} = \frac{1.44 W_{\text{TO}}^2}{\rho S C_{L_{\text{max}}} g \left[T - D - \mu\left(W_{\text{TO}} - L\right)\right]_{0.7 V_{\text{TO}}}}$$

2) Landing

$$s_L = \frac{1.69 W_L^2}{\rho S C_{L_{\text{max}}} g \left[D + \mu\left(W_L - L\right)\right]_{0.7 V_L}}$$

5.15.11 Turns

1) Level turns
   a) Radius

$$r = \frac{V^2}{g\sqrt{n^2 - 1}}$$

   b) Rate

$$\omega = \frac{g\sqrt{n^2 - 1}}{V} \text{ radians/s}$$

2) Pull-up
   a) Radius

$$r = \frac{V^2}{g(n - 1)}$$

   b) Rate

$$\omega = \frac{g(n - 1)}{V} \text{ radians/s}$$

3) Pull-down
   a) Radius

$$r = \frac{V^2}{g(n + 1)}$$

b) Rate

$$\omega = \frac{g(n+1)}{V} \text{ radians/s}$$

4) Vertical and average
  a) Radius

$$r = \frac{V^2}{g(n)}$$

  b) Rate

$$\omega = \frac{g(n)}{V} \text{ radians/s}$$

5.15.12   V–n Diagrams

1) Load factor
  a)

$$L = nW, \quad n = L/W$$

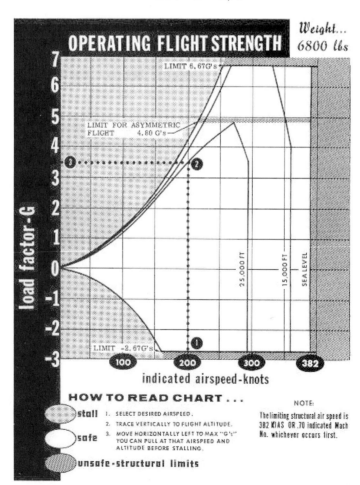

2) Stall speed
   a) Solve $L = W = C_L q S$ for V.
   b) Substitute $C_{L_{max}}$ for C_L and V_{stall} for V.
   c)

$$V_{stall} = \sqrt{\frac{2nW}{\rho S C_{L_{max}}}}$$

3) Corner velocity
   a) Fastest, tightest turn
   b)

$$V^* = \sqrt{\frac{2n_{max}W}{\rho S C_{L_{max}}}}$$

5.15.13 P_s Diagrams

1) Energy height

$$H_e = h + \frac{V^2}{2g}$$

2) Specific excess power

$$P_s = \frac{V(T-D)}{W} = \frac{dh}{dt} + \frac{V}{g}\frac{dV}{dt}$$

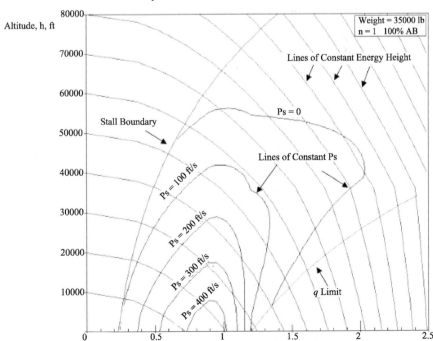

Specific Excess Power

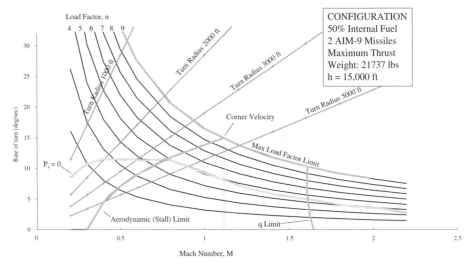

Fig. 5.45   Maneuverability diagram.

5.16   More Details—Maneuverability Diagrams and Constraint Analysis

5.16.1   Maneuverability Diagrams

Another very useful performance diagram combines the P_s and V–n diagrams, but plots them in such a way that turn rate and radius can also be read from them. Figure 5.45 is an example of this chart, which is referred to as a maneuverability diagram. The diagram is made for a fixed aircraft weight, configuration, and altitude, and so load factor is a variable. The major axes of the chart are airspeed or Mach number and turn rate. Contour lines of constant load factor and turn radius are added, and then the aerodynamic limits and $P_s = 0$ curves are plotted.

The resulting diagram is extremely useful for planning air combat because it displays maximum instantaneous turn performance as well as sustainable ($P_s = 0$) turn capability. Maneuverability diagrams for different aircraft are often compared in much the same way P_s diagrams are compared. An interesting result clearly shown on Fig. 5.43 is the fact that the absolute minimum turn radius is not achieved at corner velocity, but at a slower speed. This is because of a degradation in $C_{L\max}$ for this particular aircraft at speeds above its critical Mach number. The difference in radius is usually small, however, and the absolute maximum turn rate is at V^*, and so corner velocity is usually the velocity of choice for turning. For aircraft with corner velocities well below their critical Mach number, so that $C_{L\max}$ is constant all along the stall boundary, minimum turn radius occurs at corner velocity along with maximum turn rate.

Example 5.15

What are the maximum instantaneous turn rate, minimum instantaneous turn radius, maximum sustained turn rate, and minimum sustained turn radius at $h = 15,000$ ft for the aircraft whose maneuverability diagram is depicted in Fig. 5.45?

Solution: The maximum instantaneous turn rate at $h = 15,000$ ft occurs at corner velocity where the stall limit line and maximum load factor line meet. On Fig. 5.45, this rate is 15 deg/s at $M = 1.15$. The minimum instantaneous turn radius actually occurs at a much lower velocity, although the approximation made by saying it also occurs at corner velocity is quite good. The actual minimum instantaneous turn radius occurs where the aerodynamic limit line extends farthest toward the upper-left corner of the diagram and crosses the lowest-valued constant turn radius line. On Fig. 5.45 this occurs at $M = 0.6$ and $r = 4000$ ft. Because this point falls inside the $P_s = 0$ contour, this is also the minimum sustained turn radius at this altitude. Finally, the maximum sustained turn rate is the point where the $P_s = 0$ contour intersects the aerodynamic limit line, $\omega = 11.5$ deg/s at $M = 0.78$.

5.16.2   Constraint Analysis: Designing to a Requirement

It was stated at the beginning of this chapter that performance analysis in most cases answers the question of whether a particular aircraft design will meet a customer's needs. The methods discussed up to this point in this text enable the engineer to take an existing aircraft design, estimate its aerodynamics and thrust characteristics, and predict its performance capabilities. The challenge for aircraft designers is to turn this process around and use the analysis methods to design an aircraft that will have the desired performance capabilities.

Table 1.2 lists specific performance requirements for a multirole fighter. But how does an aircraft designer know how to design an aircraft to meet those requirements. There are so many interrelated variables to control and choices to make that aircraft designers use an analysis method called *constraint analysis* to narrow down the choices and help them focus on the most promising concepts. Constraint analysis calculates ranges of values for an aircraft concept's takeoff wing loading W_{TO}/S and takeoff thrust loading or takeoff thrust-to-weight ratio T_{SL}/W_{TO}, which will allow the design to meet specific performance requirements. In many cases, constraint analysis will eliminate some aircraft concepts from further consideration. In other instances, constraint analysis will identify two conflicting design requirements that no single aircraft configuration can satisfy.

5.16.3   Constraint Analysis Master Equation

The methodology of constraint analysis is based on a modification of the equation for specific excess power:

$$P_s = \frac{V(T - D)}{W} = \frac{dh}{dt} + \frac{V}{g}\frac{dV}{dt} \tag{5.73}$$

$$\frac{T}{W} - \frac{D}{W} = \frac{1}{V}\frac{dh}{dt} + \frac{1}{g}\frac{dV}{dt} \tag{5.74}$$

Substitute the following relations into Eq. (5.74):
1) $T = \alpha T_{SL}$, where α, the thrust lapse ratio depends on ρ/ρ_{SL} and M.
2) $W = \beta W_{TO}$, where $\beta =$ the weight fraction for a given constraint.
3)

$$D = C_D q S = \left(C_{D_0} + k_1 C_L^2\right) q S$$

4)

$$C_L = \frac{L}{qS} = \frac{nW}{qS}$$

This produces the "master equation" for constraint analysis:

$$\frac{T_{SL}}{W_{TO}} = \frac{\beta}{\alpha} \left\{ \frac{q}{\beta} \left[\frac{C_{D_0}}{(W_{TO}/S)} + k_1 \left(\frac{n\beta}{q} \right)^2 \left(\frac{W_{TO}}{S} \right) \right] + \frac{1}{V} \frac{dh}{dt} + \frac{1}{g} \frac{dV}{dt} \right\} \quad (5.75)$$

Equation (5.75) is written in a form that expresses T_{SL}/W_{TO} as a function of W_{TO}/S. All other variables in Eq. (5.75) are specified by each design requirement. For instance, one of the design requirements in Table 1.2 is a maximum sustained level turn load factor of $n = 4$ at a Mach number of 1.2 at 20,000 ft MSL. For this constraint, the climb and acceleration terms in Eq. (5.75) are zero, and the thrust lapse is determined from the specified flight conditions (obtain the value of density from the standard atmosphere model for 20,000 ft) and the appropriate thrust model from Table 5.1.

Another requirement in Table 1.2 is a P_s of 800 ft/s at $n = 1$, $M = 0.9$, $h = 5000$ ft, and maneuvering weight. In this case, the two right-hand terms of Eq. (5.75) together must equal 800 ft/s, and as with the preceding example, all other variables are specified by the design requirement. For each requirement, Eq. (5.75) is used to calculate the T_{SL}/W_{TO} values required to meet that requirement for a range of W_{TO}/S values. When the results are plotted, the line is called a constraint line because all values of T_{SL}/W_{TO} below the line will not meet the design requirement. When several constraint lines are plotted on a single set of axes, a constraint diagram like Fig. 5.46 is formed. The portion of the constraint diagram that is above all of the constraint lines is called the *solution space* because all combinations of T_{SL}/W_{TO} and W_{TO}/S within that portion of the diagram will satisfy all of the design requirements.

Performing a constraint analysis allows an aircraft designer to make much more intelligent choices about aircraft configuration, engine size, etc. These choices involve choosing a *design point*, specific values of T_{SL}/W_{TO} and W_{TO}/S from within the solution space which the aircraft concept will be designed to achieve. If

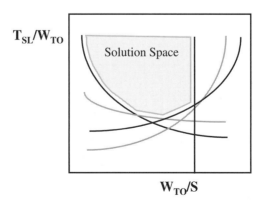

Fig. 5.46   Sample constraint diagram.

the analysis is reasonably accurate, then a design that achieves the specified thrust and wing loading values will meet the design requirements.

Constraint analysis is always an approximation because it depends so heavily on accurate predictions of the aerodynamic characteristics of an aircraft that is not yet built! The wise designer will choose a design point that is a small amount above or away from all of the constraint lines, so that the final product will still meet all the requirements even if its aerodynamics differ from the original predictions. In Fig. 5.46 note that the solution space does not lie "above" all of the constraint lines because one of them is a vertical line. The solution space can be more correctly described as lying "inside" all of the boundaries set by the various constraint lines.

Constraint analysis occasionally reveals two design requirements that conflict so completely with each other that their constraint lines do not permit a solution space, or they only have a solution for unreasonably high values of T_{SL}/W_{TO}. When this happens, it is time to talk to the customer and determine which constraint can be relaxed or what kind of compromise can be made to allow a solution.

5.16.4   Takeoff and Landing Constraints

Equation (5.75) models most constraints that deal with in-flight performance, but for takeoff and landing constraints different equations must be developed. For the takeoff constraint, the takeoff distance equation, Eq. (5.55), is rewritten in terms of T_{SL}/W_{TO} and W_{TO}/S using the very conservative and commonly true assumption that lift is approximately zero prior to rotation.

$$s_{TO} = \frac{1.44\beta^2 W_{TO}^2}{\rho S C_{L_{max}} g \, [T - D - \mu(\beta W_{TO} - L)]_{0.7 V_{TO}}}$$

$$\frac{[T - D]_{0.7 V_{TO}} - \mu(\beta W_{TO})}{\beta W_{TO}} = \frac{1.44\beta}{\rho C_{L_{max}} g s_{TO}} \left(\frac{W_{TO}}{S}\right)$$

$$\frac{\alpha T_{SL}}{\beta W_{TO}} = \frac{1.44\beta}{\rho C_{L_{max}} g s_{TO}} \left(\frac{W_{TO}}{S}\right) + \frac{[D]_{0.7 V_{TO}}}{\beta W_{TO}} + \mu$$

$$\frac{\alpha T_{SL}}{\beta W_{TO}} = \frac{1.44\beta}{\rho C_{L_{max}} g s_{TO}} \left(\frac{W_{TO}}{S}\right) + \frac{[C_{D_0} q S]_{0.7 V_{TO}}}{\beta W_{TO}} + \mu$$

$$\frac{T_{SL}}{W_{TO}} = \frac{1.44\beta^2}{\alpha \rho C_{L_{max}} g s_{TO}} \left(\frac{W_{TO}}{S}\right) + \frac{[C_{D_0} q]_{0.7 V_{TO}}}{\beta (W_{TO}/S)} + \mu \qquad (5.76)$$

As with the in-flight constraints, values of T_{SL}/W_{TO} greater than the constraint line will allow the aircraft to meet the performance requirement. Thrust lapse and weight fraction are included in Eq. (5.76) to allow for takeoff requirements that specify other than sea-level conditions and maximum gross weight. The discussion of takeoff performance in Sec. 5.11 recommended calculating drag at $0.7 V_{TO}$. Therefore, the thrust lapse, parasite drag coefficient, and the value of q in Eq. (5.76) should be evaluated for this speed.

$$V_{TO} = 1.2 \sqrt{\frac{2 W_{TO}}{\rho S C_{L_{max}}}} = 1.2 \sqrt{\frac{W_{TO}}{S}} \sqrt{\frac{2}{\rho C_{L_{max}}}}$$

q at $0.7V_{\text{TO}}$ is given by

$$q = \frac{1}{2}\rho\,(0.7V_{\text{TO}})^2 = \frac{1}{2}\rho\left[(0.7)(1.2)\sqrt{\frac{W_{\text{TO}}}{S}}\sqrt{\frac{2}{\rho C_{L_{\max}}}}\right]^2 = \left(\frac{W_{\text{TO}}}{S}\right)\frac{0.7}{C_{L_{\max}}}$$

So Eq. (5.76) becomes

$$\frac{T_{\text{SL}}}{W_{\text{TO}}} = \frac{1.44\beta^2}{\alpha\rho C_{L_{\max}}g s_{\text{TO}}}\left(\frac{W_{\text{TO}}}{S}\right) + \frac{0.7C_{D_0}}{\beta C_{L_{\max}}} + \mu \tag{5.77}$$

Equation (5.77) is the equation of a straight line because all terms except $T_{\text{SL}}/W_{\text{TO}}$ and W_{TO}/S are constants for a given constraint. The assumption of zero lift prior to rotation is a conservative one, especially on soft or grassy fields. The actual takeoff time and distance can be slightly reduced to less than what is assumed in Eq. (5.77) by causing the aircraft to produce a small fraction of its maximum lift coefficient during the takeoff roll.

The landing constraint is slightly different because T_{SL} is not present in Eq. (5.57), so the constraint equation is written only in terms of W_{TO}/S:

$$\frac{W_L}{S} = \beta\frac{W_{\text{TO}}}{S} = \frac{s_L\rho C_{L_{\max}}g\,[D + \mu\,(W_L - L)]_{0.7\,V_L}}{1.69\,W_L}$$

$$\beta\frac{W_{\text{TO}}}{S} = \frac{s_L\rho C_{L_{\max}}g\,\left[q S\,(C_{D_0} + kC_{L_b}^2 - \mu C_{L_b}) + \mu\beta W_{\text{TO}}\right]_{0.7\,V_L}}{1.69\,\beta W_{\text{TO}}} \tag{5.78}$$

where C_{L_b} is the lift coefficient maintained during braking. Minimum stopping distance for most aircraft is achieved by reducing lift to a minimum to put maximum weight on the wheels, then using maximum braking. Lift usually can be reduced to nearly zero by retracting flaps, lowering the nosewheel to the runway, and/or deploying spoilers. As with takeoff, the effect of drag on landing can be reasonably approximated by calculating it using the dynamic pressure and drag polar at $0.7V_L$.

$$V_L = 1.3\sqrt{\frac{2W_L}{\rho S C_{L_{\max}}}} = 1.3\sqrt{\frac{W_{\text{TO}}}{S}}\sqrt{\frac{2\beta}{\rho C_{L_{\max}}}}$$

q at $0.7\,V_L$ is given by

$$q = \frac{1}{2}\rho\,(0.7V_L)^2 = \frac{1}{2}\rho\left[(0.7)(1.3)\sqrt{\frac{W_{\text{TO}}}{S}}\sqrt{\frac{2\beta}{\rho C_{L_{\max}}}}\right]^2 = \left(\frac{W_{\text{TO}}}{S}\right)\frac{0.83}{C_{L_{\max}}}$$

Therefore, for zero lift during braking and drag evaluated at $0.7V_L$,

$$\frac{W_{\text{TO}}}{S} = \frac{s_L\rho C_{L_{\max}}g\left[\left(\dfrac{0.083}{C_{L_{\max}}}\right)\left(\dfrac{W_{\text{TO}}}{S}\right)\left(\dfrac{S}{W_{\text{TO}}}\right)(C_{D_0}) + \mu\beta\right]}{1.69\beta^2}$$

$$\frac{W_{\text{TO}}}{S} = \frac{s_L\rho g(\mu\beta C_{L_{\max}} + 0.083C_{D_0})}{1.69\beta^2} \tag{5.79}$$

With typical values of $\mu = 0.5$ for braking on dry pavement and typical values of $C_{D_0} = 0.05$ unless a deceleration parachute is used, aerodynamic drag in this condition is typically much less than the deceleration force available from the wheel brakes, especially at low speeds. For this common situation and conservative assumption, (5.79) simplifies to

$$\frac{W_{TO}}{S} = \frac{s_L \rho C_{L_{max}} g \mu}{1.69 \beta} \tag{5.80}$$

The assumption of zero lift and negligible drag is quite realistic and conservative. Most new aircraft designs have spoilers, computer-controlled flaps, etc. that cut the lift to near zero for maximum braking after landing. This capability can certainly be built into new designs. With maximum weight on wheels and good brakes, the drag on the aircraft is almost negligible, unless a drag chute is used. In the absence of a drag chute, ignoring drag makes the landing distance constraint line slightly more restrictive than reality. That means it gives the constraint analysis a slight margin of safety. For a more general formulation for the constraint line equations, see Ref. 6.

5.16.5 Constraint Analysis Example

Consider the multirole fighter design requirements from Table 1.2 and the aerodynamic model for the F-16 developed in Chapter 4. Table 1.2 specifies the following performance requirements:
1) Combat turn (maximum AB): 9.0-g sustained at 5000 ft/M = 0.9.
2) Combat turn (maximum AB): 4.0-g sustained at 20,000 ft/M = 1.8.
3) Takeoff and braking distance: $s_{TO} = s_L = 2000$ ft.
Calculate and plot the constraint lines for these requirements.

Solution: We first calculate the subsonic combat turn. We are given the following information: $h = 5000$ ft, $M = 0.9$, $a = 1096.9$ ft/s, $\alpha = 1.4$, $n = 9$, $V = 987.2$ ft/s, $\rho = 0.002048$ slug/ft^3, $C_{D_0} = 0.0243$, $k_1 = 0.121$, $q = 997.9$ lb/ft^2, and $\beta = 0.8$.
Substituting these values into the master equation yields

$$\frac{T_{SL}}{W_{TO}} = \frac{\beta}{\alpha} \left\{ \frac{q}{\beta} \left[\frac{C_{D_0}}{(W_{TO}/S)} + k_1 \left(\frac{n\beta}{q} \right)^2 \left(\frac{W_{TO}}{S} \right) \right] \right\}$$

$$= \frac{q}{\alpha} \frac{C_{D_0}}{(W_{TO}/S)} + \frac{k_1}{\alpha} \frac{(n\beta)^2}{q} \left(\frac{W_{TO}}{S} \right)$$

$$\frac{T_{SL}}{W_{TO}} = \frac{997.9 \text{ lb/ft}^2}{1.4} \frac{0.0243}{(W_{TO}/S)} + \frac{0.121}{1.4} \frac{[9(0.8)]^2}{997.9 \text{ lb/ft}^2} \left(\frac{W_{TO}}{S} \right)$$

$$= \frac{17.32 \text{ lb/ft}^2}{(W_{TO}/S)} + \frac{4.45 \times 10^{-3}}{(\text{lb/ft}^2)} \left(\frac{W_{TO}}{S} \right)$$

T_{SL}/W_{TO} results for this constraint for several W_{TO}/S values are listed in Table 5.2.

Table 5.2   F-16 subsonic combat
turn constraint

W_{TO}/S, psf	T_{SL}/W_{TO}
40	0.61
50	0.57
60	0.56
70	0.56
80	0.57

Next, we calculate points for the supersonic combat turn constraint for the following conditions: $h = 20{,}000$ ft, $M = 1.2$, $\alpha = 1036.9$ ft/s, $\alpha = 0.98$, $n = 4$, $V = 1244$ ft/s, $\rho = 0.001267$ slug/ft^3, $C_{D_0} = 0.0412$, $k_1 = 0.169$, $q = 980.8$ lb/ft^2, and $\beta = 0.8$.

Substituting these values into the master equation yields

$$\frac{T_{SL}}{W_{TO}} = \frac{980.8 \text{ lb/ft}^2}{0.98} \frac{0.0412}{(W_{TO}/S)} + \frac{0.196}{0.98} \frac{[4(0.8)]^2}{980.8 \text{ lb/ft}^2} \left(\frac{W_{TO}}{S}\right)$$

$$= \frac{41.23 \text{ lb/ft}^2}{(W_{TO}/S)} + \frac{1.8 \times 10^{-3}}{(\text{lb/ft}^2)} \left(\frac{W_{TO}}{S}\right)$$

T_{SL}/W_{TO} results for this constraint for several W_{TO}/S values are listed in Table 5.3.

Next, we calculate points for the takeoff. We are given the following information: $h = 0$ ft, $C_{L_{max\,TO}} = 1.21$, $C_{D_0} = 0.0519$, $\alpha = 1.105$, $\beta = 1.0$, and $\mu = 0.03$.

$$\frac{T_{SL}}{W_{TO}} = \frac{1.44\beta^2}{\alpha \rho C_{L_{max}} g s_{TO}} \left(\frac{W_{TO}}{S}\right) + \frac{0.7 C_{D_0}}{\beta C_{L_{max}}} + \mu$$

$$= \frac{1.44}{(1.105)(0.002377 \text{ slug/ft}^3)(1.21)\left(32.2 \text{ ft/s}^2\right)(2000 \text{ ft})} \left(\frac{W_{TO}}{S}\right)$$

$$+ \frac{0.7(0.05)}{1.21} + 0.03$$

$$= \frac{0.0067}{\text{lb/ft}^2} \frac{W_{TO}}{S} + 0.0589$$

Table 5.3   F-16 supersonic combat
turn constraint

W_{TO}/S, psf	T_{SL}/W_{TO}
40	1.10
50	0.91
60	0.79
70	0.71
80	0.66

Table 5.4   F-16 takeoff constraint

W_{TO}/S, psf	T_{SL}/W_{TO}
40	0.32
50	0.38
60	0.45
70	0.51
80	0.58

T_{SL}/W_{TO} results for this constraint for several W_{TO}/S values are listed in Table 5.4.

Finally, we calculate the wing loading defining the landing constraint, which we will plot on the constraint diagram as a vertical line.

$$h = 0 \text{ ft} \quad C_{L_{\max \text{Lnd}}} = 1.37 \quad \mu = 0.5 \quad \alpha = 1.0 \quad \beta = 1.0$$

Assume lift is reduced to zero and drag is negligible compared to braking (no drag parachute used)

$$\frac{W_{TO}}{S} = \frac{s_L \rho C_{L_{\max}} g \mu}{1.69\beta} = \frac{(2000 \text{ ft})(0.002377 \text{ slug/ft}^3)(1.37)(32.2 \text{ ft/s}^2)(0.5)}{1.69(1.0)}$$

$$= 62 \text{ lb/ft}^2$$

The resulting constraint diagram is shown in Fig. 5.47.

Allowing some margin for error and growth, an initial design point of $W_{TO}/S = 60$ lb/ft^2 and $T_{SL}/W_{TO} = 0.85$ would be selected. However, this is a relatively low wing loading compared to typical modern fighter aircraft, as shown in Fig. 5.48. If the landing constraint could be relaxed, a much higher wing loading and lower thrust-to-weight ratio would allow the aircraft to meet the other design requirements.

The results of this initial constraint analysis would probably prompt discussions between the designer and the customer to determine how important the landing distance constraint is to the success of the design. If the constraint is not relaxed, then the designer will either proceed with design development using this design point or develop design features such as more aggressive high-lift devices, drag chutes, or reverse thrust capability to allow the aircraft to meet the requirement at a more typical design point. The decision on which solution to apply will undoubtedly be based on cost. In either case, constraint analysis has identified the design drivers, the supersonic turn requirement, and the landing distance constraint, which, because they are the most restrictive, will have the greatest influence on the shape of the final design.

5.17 Performance Analysis Example

An example will make the methods just discussed easier to understand. Consider the aerodynamic model for the F-16 that was generated in the aerodynamic analysis example, Sec. 4.9. That analysis calculated $C_{L_{\max}} = 1.21$ for takeoff and $C_{L_{\max}} = 1.37$ for landing. Drag polar results are summarized in Table 5.5.

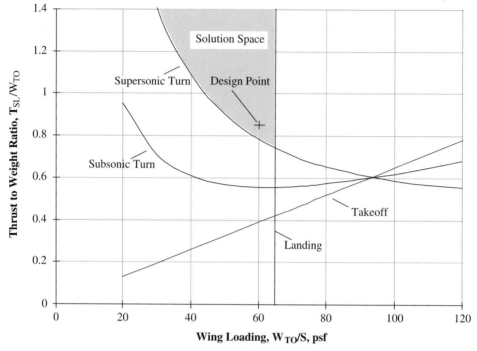

Fig. 5.47   Constraint diagram for F-16 example.

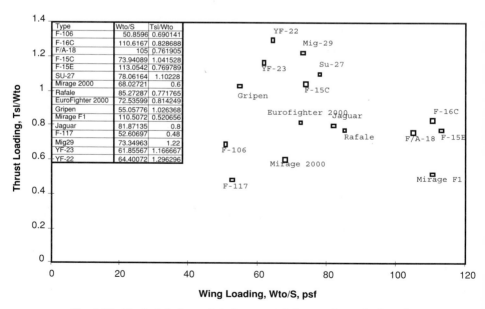

Type	Wto/S	Tsl/Wto
F-106	50.8596	0.690141
F-16C	110.6167	0.828688
F/A-18	105	0.761905
F-15C	73.94089	1.041528
F-15E	113.0542	0.769789
SU-27	78.06164	1.10228
Mirage 2000	68.02721	0.6
Rafale	85.27287	0.771765
EuroFighter 2000	72.53599	0.814249
Gripen	55.05776	1.026368
Mirage F1	110.5072	0.520656
Jaguar	81.87135	0.8
F-117	52.60697	0.48
Mig29	73.34963	1.22
YF-23	61.85567	1.166667
YF-22	64.40072	1.296296

Fig. 5.48   Typical design points for recent fighter and attack aircraft.

Table 5.5 Drag polar calculated for the F-16

Mach number	C_{D_0}	k_1
0.1	0.0169	0.117
0.86	0.0169	0.117
1.05	0.0430	0.128
1.5	0.0382	0.252
2.0	0.0358	0.367

In addition, performance analysis requires a model for engine thrust and TSFC. The F-16C with the Pratt and Whitney F-100-220 engine has the following static $(M = 0)$ sea-level installed thrust and TSFC characteristics: $T_{SL\,dry} = 11,200$ lb with $c_t = 0.8$ /h and $T_{SL\,wet} = 17,500$ lb with $c_t = 2.46$/h. The variation of dry and wet thrust with Mach number and altitude will be modeled with Eqs. (5.11) and (5.12), respectively. The variation of TSFC with altitude will be modeled with Eq. (5.15).

The final information on the aircraft that must be known is the aircraft empty, payload, and fuel weights for which the performance is to be evaluated. Takeoff performance is evaluated for the aircraft's maximum takeoff gross weight W_{TO}, which for the F-16 is 36,800 lb. Table 1.2 specifies that the turning performance and P_s requirements be evaluated at maneuvering weight, which is defined as the basic aircraft with 50% internal fuel and standard air-to-air armament. For the F-16, this configuration includes full 20-mm cannon armament and two AIM-9 missiles for a total weight of 21,737 lb. This weight will be used for the climb and glide performance as well. The F-16 carries 6972 lb of fuel internally, so that its maximum weight in the air-to-air configuration is 25,223 lb and its zero-fuel weight is 18,251 lb. For loiter and cruise problems, a reasonable starting weight is 25,000, and a typical ending weight might be 20,000 lb.

5.17.1 Glide Performance

The simplified performance analysis just presented is strictly valid only for aircraft that develop minimum drag at zero lift and for which $k_2 = 0$. This was not the case for the F-16, but its k_2 is quite small. The error caused by ignoring this effect is also small, though not negligible. The nonzero value of k_2 for the F-16 will be ignored in the following analysis to demonstrate the method. However, it is important to realize that this can cause a significant error in the analysis for some aircraft, particularly airliners and transports that have highly cambered wings. On a positive note, the actual performance of an aircraft with nonzero k_2 is usually better (except at very high speeds) than the performance predicted by assuming $k_2 = 0$. This makes the predictions of the simplified method conservative and, when used as a first approximation, safe because the actual airplane will do better than predicted.

The best glide range is achieved when L/D is maximum, where

$$\left(\frac{L}{D}\right)_{max} = \left(\frac{C_L}{C_D}\right)_{max} = \frac{1}{2\sqrt{kC_{D_0}}} = \frac{1}{2\sqrt{0.117(0.0169)}} = 11.24$$

so that the F-16 will glide 11.24 n miles for every 6076 ft (1 n mile) of altitude lost. Glide speed can be found using the fact that induced drag equals parasite drag for the best glide condition. For $h = 10,000$ ft,

$$C_{D_0} = k_1 C_L^2$$

$$C_L = \sqrt{\frac{C_{D_0}}{k_1}} = \sqrt{\frac{0.0169}{0.117}} = 0.38$$

$$V_{glide} = \sqrt{\frac{2W}{\rho S C_L}} = \sqrt{\frac{2(21,737\text{ lb})}{0.001756\text{ slug/ft}^3(300\text{ ft}^2)\,0.38}} = 466\text{ ft/s}$$

Minimum sink rate is achieved where $3C_{D_0} = k_1 C_L^2$:

$$C_L = \sqrt{\frac{3C_{D_0}}{k_1}} = \sqrt{3\left(\frac{0.0169}{0.117}\right)} = 0.66$$

$$V_{min\ sink} = \sqrt{\frac{2W}{\rho S C_L}} = \sqrt{\frac{2(21,737\text{ lb})}{0.001756\text{ slug/ft}^3(300\text{ ft}^2)0.66}} = 353\text{ ft/s}$$

$$D = C_D q S = 4C_{D_0}^{1/2}\rho V^2 S = 2(0.0169)(0.001756\text{ slug/ft}^3)(353\text{ ft/s})^2(300\text{ ft}^2)$$
$$= 2218\text{ lb}$$

$$\text{min sink rate} = V_\infty \sin\gamma = \frac{V_\infty D}{W} = \frac{353\text{ ft/s}(2218\text{ lb})}{21,737\text{ lb}} = 36\text{ ft/s}$$

5.17.2 Climb Performance

Climbs can be performed in afterburner or military (no afterburner) thrust. The best angle of climb in military thrust is achieved at $(L/D)_{max}$ where, at 10,000 ft,

$$D = C_D q S = 2C_{D_0}^{1/2}\rho V^2 S = (0.0169)(0.001756\text{ slug/ft}^3)(466\text{ ft/s})^2(300\text{ ft}^2)$$
$$= 1933\text{ lb}$$

Military thrust is modeled using Eq. (5.11):

$$T_{avail} = T_{SL}\left(\frac{\rho}{\rho_{SL}}\right) = 11,200\text{ lb}\left(\frac{0.001756}{0.002377}\right) = 8273\text{ lb}$$

and so

$$\sin\gamma = \frac{T-D}{W} = \frac{8273\text{ lb} - 1933\text{ lb}}{21,737\text{ lb}} = 0.292$$
$$\gamma = 17.0\text{ deg}$$

Maximum climb angle in max thrust is achieved at the airspeed where $T - D$ is maximum. This speed will not be the speed for $(L/D)_{max}$ because afterburner thrust is not constant with velocity. Maximum thrust increases with increasing velocity according to Eq. (5.12), and $(T - D)_{max}$ occurs at a higher velocity than the velocity for $(L/D)_{max}$. Maximum rate of climb will occur where $V(T - D)$ is maximum. The rate of climb in military thrust at $h = 10,000$ ft for the conditions for maximum climb angle is

$$\text{rate of climb} = R/C = \frac{V(T-D)}{W} = \frac{466 \text{ ft/s} (8273 \text{ lb} - 1933 \text{ lb})}{21,737 \text{ lb}}$$

$$= 136.1 \text{ ft/s} = 8164 \text{ ft/min}$$

5.17.3 Loiter and Cruise Performance

Best endurance at $h = 10,000$ ft will be achieved at the speed for $(L/D)_{max}$. The engine TSFC at that altitude will be

$$c_t = c_{t(\text{sea level})} \sqrt{\frac{T}{T_{\text{sea level}}}} = 0.8/\text{h} \sqrt{\frac{483.1°\text{R}}{518.69°\text{R}}} = 0.77/\text{h}$$

Using the Breguet endurance equation

$$E = \frac{1}{c_t} \frac{C_L}{C_D} \ell_n \left(\frac{W_1}{W_2}\right) = \frac{1 \text{ h}}{0.77} (11.24) \ell_n \left(\frac{25,000 \text{ lb}}{20,000 \text{ lb}}\right) = 3.26 \text{ h}$$

Maximum range is achieved for the airspeed where $C_{D_0} = 3k_1 C_L^2$:

$$C_L = \sqrt{\frac{C_{D_0}}{3k_1}} = \sqrt{\frac{0.0169}{3(0.117)}} = 0.22$$

$$C_D = C_{D_0} + k_1 C_L^2 = C_{D_0} + 1/3\, C_{D_0} = 4/3\, C_{D_0} = 0.0225$$

$$\frac{C_L^{1/2}}{C_D} = \frac{(0.22)^{1/2}}{0.0225} = 20.8$$

$$R = \sqrt{\frac{2}{\rho S}} \frac{2}{c_t} \frac{C_L^{1/2}}{C_D} \left(W_1^{1/2} - W_2^{1/2}\right)$$

$$= \sqrt{\frac{2}{0.001756 \text{ slug/ft}^3 (300 \text{ ft}^2)}} \frac{2 \text{ h}}{0.77} (20.8)(25,000^{1/2} - 20,000^{1/2}) \text{ lb}^{1/2}$$

$$= 676.5 \text{ (ft/s)(h)}(1 \text{ n miles}/6076 \text{ ft})(3600 \text{ s/h}) = 400.6 \text{ n miles}$$

5.17.4 Specific Excess Power

Table 1.2 requires $P_s = 800$ ft/s at $M = 0.9$ at $h = 5000$ ft. At that Mach number, the value of C_{D_0} is approximated by a straight line between $C_{D_0} = 0.0169$

at $M = 0.86$ and $C_{D_0} = 0.043$ at $M = 1.05$:

$$C_{D_0} = C_{D_0.86} + (0.9 - 0.86)\frac{C_{D_0 1.05} - C_{D_0.86}}{1.05 - 0.86}$$

$$= 0.043 + (0.9 - 0.86)\frac{0.043 - 0.0169}{1.05 - 0.86} = 0.022$$

The same is done for k_1:

$$k_1 = (k_1)_{0.86} + (0.9 - 0.86)\frac{(k_1)_{1.05} - (k_1)_{0.86}}{1.05 - 0.86} = 0.119$$

Then at $M = 0.9$ and $h = 5000$ ft, $V_\infty = Ma = 0.9(1097.1 \text{ ft/s}) = 987.4$ ft/s:

$$C_L = \frac{W}{qS} = \frac{21,737}{1/2(0.002048 \text{ slug/ft}^3)(987.4 \text{ ft/s})^2(300 \text{ ft}^2)} = 0.073$$

$$C_D = 0.022 + 0.119(0.073)^2 = 0.023$$

$$D = C_D qS = 0.023(1/2)(0.002048 \text{ slug/ft}^3)(987.4 \text{ ft/s})^2(300 \text{ ft}^2) = 6889 \text{ lb}$$

$$T_{\text{avail}} = T_{\text{SL}}\left(\frac{\rho}{\rho_{\text{SL}}}\right)(1 + 0.7M_\infty) = 17,500 \text{ lb}\left(\frac{0.002048}{0.002377}\right)(1 + 0.27)$$

$$= 24,577 \text{ lb}$$

$$P_s = \frac{(T - D)V}{W} = \frac{(24,577 \text{ lb} - 6889 \text{ lb})(987.4 \text{ ft/s})}{21,737 \text{ lb}} = 803.5 \text{ ft/s}$$

5.17.5 Thrust and Drag Curves

The methods used in the P_s calculation to evaluate thrust and drag at a specific Mach number and altitude can be used to generate thrust and drag curves similar to Fig. 5.13. Figure 5.49 is a curve generated using these methods for the F-16 at $h = 10,000$ ft.

5.17.6 Takeoff

An F-16 taking off in full afterburner in air-to-air combat configuration certainly meets the requirement for $T \gg D$, and so Eq. (5.56) can be used to estimate takeoff distance. The predicted $C_{L_{\max}} = 1.21$ for takeoff. Using $M = 0.15$ to calculate the average takeoff thrust,

$$T_{\text{avail}} = T_{\text{SL}}\left(\frac{\rho}{\rho_{\text{SL}}}\right)(1 + 0.7M_\infty) = 17,500 \text{ lb}(1.0)(1 + 0.105) = 19,337 \text{ lb}$$

$$s_{\text{TO}} = \frac{1.44W_{\text{TO}}^2}{\rho S C_{L_{\max}} gT} = \frac{1.44(25,223 \text{ lb})^2}{0.002377 \text{ slug/ft}^3(300 \text{ ft}^2)1.21(32.2 \text{ ft/s}^2)(19,337 \text{ lb})}$$

$$s_{\text{TO}} = 1705 \text{ ft}$$

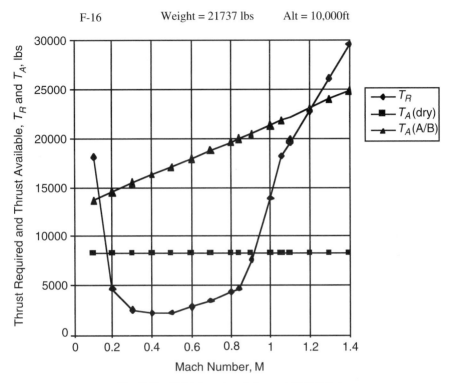

Fig. 5.49   F-16 thrust and drag at 10,000 ft.

5.17.7 Landing

When the F-16 touches down, the flight control computer automatically retracts the leading-edge flaps and flaperons. When the nosewheel is lowered to the runway, the lift generated by the aircraft is effectively reduced to zero. With landing gear extended, the F-16's $C_{D_0} = 0.05$. Using the predicted $C_{L_{max}}$ for landing of 1.37 and a landing weight of 20,000 lb,

$$0.7V_L = 0.7(1.3)V_{stall} = 0.91\sqrt{\frac{2W}{\rho S C_{L_{max}}}}$$

$$= 0.91\sqrt{\frac{2(20,000 \text{ lb})}{0.002377 \text{ slug/ft}^3(300 \text{ ft}^2)1.37}} = 202 \text{ ft/s}$$

$$D = C_D q S = C_{D_0} q S = 0.05(1/2)(0.002377 \text{ slug/ft}^3) \times (202 \text{ ft/s})^2(300 \text{ ft}^2)$$

$$= 729.9 \text{ lb}$$

$$s_L = \frac{1.69 W_L^2}{\rho S C_{L_{max}} g \, [D + \mu \, (W_L - L)]_{0.7 V_L}}$$

$$= \frac{1.69(20{,}000 \text{ lb})^2}{0.002377 \text{ slug/ft}^3 (300 \text{ ft}^2) 1.37 (32.2 \text{ ft/s}^2) \, [729.9 \text{ lb} + 0.5 \, (20{,}000 \text{ lb} - 0)]}$$

$$s_L = 2003 \text{ ft}$$

References

[1]Harned, M., "The Ramjet Power Plant," *Aero Digest*, Vol. 69, No. 1, July 1954, p. 38.

[2]Stinton, D., *The Anatomy of the Aeroplane*, American Elsevier, New York, 1966, p. 61.

[3]Goodwin, S. P., Beierle, M. T., and McLaughlin, T. E. (eds.), *Aeronautical Engineering 215 Course Booklet*, Kendall-Hunt, Dubuque, IA, 1995, pp. 27–31.

[4]Nicolai, L., *Fundamentals of Aircraft Design*, METS, Inc., Xenia, OH, 1975.

[5]McCormick, B. W., *Aerodynamics, Aeronautics, and Flight Mechanics*, Wiley, New York, 1979, p. 400.

[6]Mattingly, J. D., Heiser, W. H., and Daley, D. H., *Aircraft Engine Design*, AIAA Education Series, AIAA, New York, 1987, pp. 17–52.

Problems

Synthesis Problems

S-5.1   Brainstorm at least five ways to determine the actual drag polar of a small, hand-launched glider.

S-5.2   Brainstorm at least five different types of flight tests that could be used to determine the actual drag polar of a twin-jet subsonic military trainer aircraft.

S-5.3   Brainstorm at least five ways to increase loiter capability on a supersonic jet fighter.

S-5.4   Brainstorm at least five ways to increase supersonic cruise range of a high-speed civil transport.

S-5.5   An airplane is being designed to fly at high altitudes and low airspeeds for weeks at a time without refueling. Based on your knowledge of the relationship between induced drag and aspect ratio and the relative importance of induced and parasite drag at low speeds, would you suggest a high or low aspect ratio for the wing of this plane?

S-5.6   An airplane is being designed to fly on Mars, where the density of the atmosphere is comparable to the density at $h = 100{,}000$ ft in the standard atmosphere on Earth. What problems do you expect this aircraft to have with takeoff and landing? Brainstorm five concepts for overcoming these problems.

Analysis Problems

A-5.1 Consider an aircraft with a NACA 0009 airfoil, $e_0 = 0.95$, $C_{D_0} = 0.01$, and $R = 10$. If the aircraft is flying at 5-deg angle of attack, calculate C_L and C_D.

A-5.2 A SEPECAT Jaguar supersonic strike fighter/trainer has a mass of 10,000 kg. Its wing area is 24.2 m², span is 8.69 m, Oswald efficiency factor is 0.85, and zero-lift drag coefficient C_{D_0} is 0.024. Calculate the thrust required to fly at a velocity of 300 KTAS at (a) standard sea level and (b) an altitude of 5 km.

A-5.3 Calculate the value of $(L/D)_{max}$ for the aircraft described in problem A-5.1. What is the value of C_L for this aircraft when it achieves $(L/D)_{max}$?

A-5.4 What aerodynamic features of an airplane influence the stall speed?

A-5.5 Use the T-38 thrust and drag curves in Appendix B to answer these and all other questions about the T-38 in this chapter. Given an 8000-lb T-38 flying at 10,000 ft, determine the following: (a) thrust required T_R at Mach 0.5, (b) thrust available in military power $T_{A_{dry}}$ at Mach 0.5, (c) thrust available in maximum power $T_{A_{wet}}$ at Mach 0.5, (d) excess thrust T_X at Mach 0.5 (assume military power setting), (e) Mach number for minimum drag, (f) minimum drag, (g) minimum Mach number and what causes this limit (thrust or stall), and (h) maximum Mach number (assume maximum power setting).

A-5.6 How much faster (in feet per second) can a T-38 at maximum thrust fly than one in military thrust if both aircraft weigh 8000 lb and are at 30,000 ft in level flight?

A-5.7 (a) A T-37 has a drag polar of $C_D = 0.02 + 0.057C_L^2$, a weight of 6000 lb, and $S = 184$ ft². What is the value of L/D_{max} for this aircraft?
(b) At what equivalent airspeed would you fly for L/D_{max}?

A-5.8 (a) What is the minimum speed of a 10,000-lb T-38 at sea level in level flight? What causes this limit?
(b) What is the minimum speed of a 10,000-lb T-38 at 40,000 ft in level flight? What causes this limit?

A-5.9 Assuming thrust is proportional to density, calculate the military thrust available for a T-38 at 30,000 ft given only the sea-level value of thrust at $M = 0.7$. Compare the result to the 30,000-ft data in Appendix B using the same Mach number.

A-5.10 Given at 12,000-lb T-38 at sea level, determine the following: (a) power required at Mach 0.35, (b) power available in military power $(P_{A_{dry}})$ at Mach 0.35, (c) excess power at Mach 0.35, (d) Mach number for minimum power required, and (e) minimum power required.

A-5.11 Sketch a power-required curve and show where L/D_{max} occurs on this curve.

A-5.12   (a) An F-4 flying at L/D_{max} dumps 4000 lb of internal fuel. Should it fly faster or slower to maintain L/D_{max}?

(b) If you are flying at L/D_{max} at 10,000 ft and climb to 20,000 ft, should your true velocity be faster or slower to maintain L/D_{max}?

(c) The space shuttle extends its speed brakes fully during an approach, how does this change its true velocity for L/D_{max}?

A-5.13   Answer the following questions about a LET L23 Super Blanik sailplane, which has a mass of 500 kg. Its wing span is 16.2 m, its wing area is 19.15 m^2, its Oswald efficiency factor is 0.895, and its zero-lift drag coefficient is 0.0132.

(a) If it is flying at an altitude of 3 km over terrain with an elevation of 2 km, what is the maximum distance the glider can glide in still air, and at what speed should it initially fly to achieve this performance?

(b) What is the maximum time the Blanik can stay airborne in still air for this situation, and at what speed should it initially fly to achieve this performance?

(c) If it flies into a thermal or vertical current of air rising at 3 m/s, at what rate can the glider increase its altitude?

A-5.14   (a) What is the maximum glide range (in nautical miles) a 12,000-lb T-38 can achieve from 20,000 ft above the ground?

(b) What is the maximum glide range (in nautical miles) an 8000-lb T-38 can achieve from 20,000 ft above the ground?

A-5.15   In a steady climb, which is larger, lift or weight? Assume thrust acts along the flight path.

A-5.16   An 8000-lb T-38 is in a steady climb passing 10,000 ft at 0.5 Mach.

(a) What is its ROC using military thrust? Using maximum thrust?

(b) What is its climb angle using maximum thrust?

A-5.17   A T-38 is cruising at 20,000 ft (using "normal" power) and weighs 11,000 lb (2000 lb of this weight is usable fuel).

(a) What velocity would it fly at for maximum endurance?

(b) What is its maximum endurance (in hours)?

(c) What is its endurance (in hours) at $M = 0.5$?

(d) What is its range at $M = 0.5$?

A-5.18   A T-37 has a drag polar of $C_D = 0.02 + 0.057C_L^2$, weighs 6000 lb, (500 lb of this is usable fuel), and $c_{sl} = 0.9/h$. What is its maximum endurance at 20,000 ft standard day?

A-5.19   A T-38 is cruising at 20,000 ft standard day and weighs 11,000 lb (2000 lb of this weight is usable fuel).

(a) At what velocity should it fly for maximum range?

(b) What is its maximum range, in nautical miles?

A-5.20   Using the information in problem A-5.18 and $S = 184$ ft^2, what is the maximum range in nautical miles of the T-37 at 20,000 ft in standard day conditions.

A-5.21 A KC-135 weighing 150,000 lb has a takeoff roll of 3600 ft at sea-level density altitude.

(a) If the aircraft is loaded with 100,000 lb of fuel and all other conditions remain the same, what is its takeoff roll?

(b) If its next takeoff is made at 150,000 lb gross weight and a density altitude of 8000 ft, assuming all other factors remain the same, what is its takeoff distance?

A-5.22 A T-38 weighing 8000 lb is preparing to land using zero flaps at standard sea-level conditions. The T-38's wing area is 170 ft^2, and the coefficient of rolling friction for braking on the dry runway is 0.5. Answer the following questions:

(a) What is the T-38's stall speed? Assume the stall speed is the same as for clean configuration.

(b) What is $C_{L_{max}}$?

(c) What is the T-38's final approach airspeed?

(d) What is the T-38's landing distance? Assume that lift is reduced to zero once the T-38 lands and that $C_D = 0.05$ during the landing roll.

(e) The control tower reports that a sudden rain shower has soaked the runway, reducing the coefficient of rolling friction. How will this affect the T-38's landing distance?

A-5.23 A British Aerospace TSR.2 long-range interdiction aircraft is landing at sea level at a gross weight of 300,000 N. Its maximum lift coefficient with full-span blown flaps fully extended is 2.5. After touchdown, the pilot lowers the nose immediately to the runway and retracts the flaps to achieve zero lift for the braking roll. Using the following parameters describing the aircraft, calculate its landing distance: wing area = 65 m^2; $C_{D_0} = 0.06$; and $\mu = 0.5$.

A-5.24 A T-38 at 500 kn true airspeed with a load factor of 5 g is at 10,000 ft.

(a) What is its turn rate (degrees per second) and turn radius for a pull-up?

(b) What is its turn rate (degrees per second) and turn radius for a pull-down?

(c) What is its turn rate (degrees per second) and turn radius for a level turn?

A-5.25 An SR-71 is in a 20-deg banked level turn at Mach 3.0 and an altitude of 80,000 ft. Assume the ambient temperature is 390° R.

(a) What is the aircraft's turn radius in nautical miles?

(b) What is the aircraft's turn rate in degrees per second?

(c) If the pilot increases the bank angle to 45 deg while maintaining a level turn at Mach 3.0, what would the new turn radius be in nautical miles? What would happen to the aircraft load factor?

A-5.26 A P-51 Mustang is performing a 4-g loop at a velocity of 500 ft/s.

(a) What is the turn radius at the bottom of the loop in feet?

(b) What is the turn rate at the bottom of the loop in degrees per second?

(c) What is the turn radius at the top of the loop in feet?

(d) What is the turn rate at the top of the loop in degrees per second?

A-5.27 You are flight testing an aircraft and determine the 1-g stall speed to be 100 kn true velocity. Under the same flight conditions what would the 3-g stall speed be?

A-5.28 Find the load factor, bank angle, and turn radius for a T-41 in a level turn at a true airspeed of 120 kn and a turn rate of 5 deg/s.

A-5.29 What is the velocity and load factor a T-38 pilot should fly for the highest turn rate and lowest turn radius if $W = 12,000$ lb and the altitude is 15,000 ft?

A-5.30 (a) Sketch a typical V–n diagram. What causes each of the limits? Can any point inside the V–n diagram be sustained?
(b) Sketch how each of these limits changes with an increase in weight or an increase in altitude.

A-5.31 Refer to the T-38 V–n diagrams in Appendix B to answer the following questions:
(a) What is the maximum instantaneous load factor for a 12,000-lb T-38 at 15,000 ft and Mach 0.6?
(b) What is the maximum instantaneous load factor for a 9600-lb T-38 at sea level and Mach 0.8?
(c) What is the maximum Mach number for a 12,000-lb T-38 at sea level?
(d) What is the maximum Mach number for a 9600-lb T-38 at 15,000 ft?
(e) What is the corner velocity for a 12,000-lb T-38 at 25,000 ft?
(f) What is the corner velocity for a 9600-lb T-38 at sea level?

A-5.32 A B-52 at 20,000 ft and 200 kn true velocity has a weight of 500,000 lb and a T-38 at 10,000 ft and 500 kn true velocity has a weight of 10,000 lb. Assume standard day.
(a) Which aircraft has more energy?
(b) Which aircraft has a greater energy height?
(c) Calculate the P_s of the T-38 in maximum thrust.
(d) If the B-52 has a maximum steady climb rate of 500 ft/min, which aircraft has more specific excess power?
(e) What is the level acceleration capability of the T-38 at these conditions?

A-5.33 Given the following data for the F-4, answer the following questions: $W = 40,000$ lb; $S = 530$ ft^2; $T_{A_{\text{sea level}}} = 35,800$ lb; Mach = 1.5; $h = 20,000$ ft (standard day conditions); $C_{D_0} = 0.0224$; and $k = 0.1516$.
(a) What is the aircraft's specific excess power in straight-and-level flight?
(b) What is the highest altitude the aircraft could reach, assuming $T = D$ during the entire zoom maneuver so it can trade all of its kinetic energy for potential energy?

A-5.34 An adversary aircraft is observed with a level acceleration capability of 12 ft/s^2 using max thrust at an altitude of 10,000 ft and a velocity of 500 kn. Your aircraft has a P_s of 500 ft/s at these conditions.
(a) In level flight, can you accelerate faster than your adversary? Assume you are both at 10,000 ft and 500 kn.

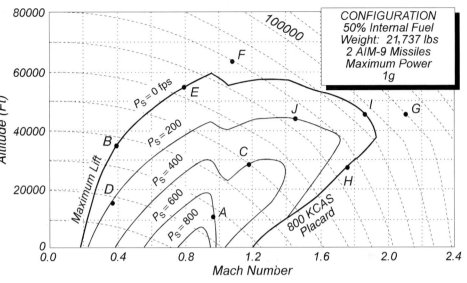

Fig. PA-5.35   F-16C specific excess power.

(b) When comparing aircraft in combat, what factors, besides P_s, should be considered?

A-5.35   Using the P_s plot in Fig. PA-5.35 answer the following questions:
(a) At what points will this aircraft stabilize in level unaccelerated flight?
(b) Sketch a possible path for this aircraft to maneuver from B to E.
(c) What is the subsonic absolute ceiling?
(d) What is the supersonic absolute ceiling?
(e) What is the maximum energy height this aircraft can obtain in sustained flight?
(f) If the aircraft performed a zoom climb from the maximum energy point, what is the maximum altitude it could reach?
(g) Can this aircraft reach point F?
(h) Can this aircraft reach point G?
(i) Sketch the minimum time to climb path from takeoff to point I.

A-5.36   Use the data provided to perform the following constraint analysis. The performance requirements are for supercruise (non-AB supersonic cruise), $M = 1.8$ at 30,000 ft; for combat turn (max AB): $5.2g$ sustained at 30,000 ft $M = 0.9$; for horizontal acceleration (max AB): $0.8M$ to $1.5M$ in 50 s at 30,000 ft; and takeoff and braking distance, $s_{TO} = S_L = 1500$ ft.
(a) Supercruise: $\alpha = 0.375$, $\beta = 1.0$, $C_{D_0} = 0.025$, $k_1 = 0.3$, determine n, V, and q.
(b) Combat turn (constant V and h): $\alpha = 0.5$, $\beta = 0.8$, $C_{D_0} = 0.013$, $k = 0.18$, $n = 5.2$, determine V and q.
(c) Horizontal acceleration: $\alpha = 0.58$, $\beta = 0.9$, $C_{D_0} = 0.022$, $k_1 = 0.23$, determine n, V, q, and dV/dt.
(d) Takeoff distance: $\alpha = 1.0$, $\beta = 1.0$, $C_{L_{maxTO}} = 1.6$, $T \gg D$, $s_{TO} = 1500$ ft.

(e) Braking distance: $\alpha = 1.0$, $\beta = 1.0$, $V_L = 1.3V_{\text{stall}}$, $C_{L_{\max}} = 2.0$, $\mu = 0.5$, $s_L = 1500$ ft.

(f) Sketch the resulting constraint diagram.

(g) What's your choice for an initial design point?

(h) Compare your choice to the historical data for similar aircraft provided in Fig. 5.35. What do you think?

(i) What constraints do you think should be relaxed?

Congratulations! You've just analyzed the F-22! See Fig. PA-36b.

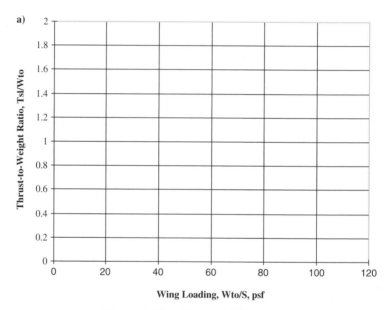

Fig. PA-5.36a   Constraint diagram.

Fig. PA-5.36b   YF-22, prototype for the F-22 advanced tactical fighter (courtesy of Lockheed Martin).

6
Stability and Control

"The balancing of a gliding or flying machine is very simple in theory. It merely consists in causing the center of pressure to coincide with the center of gravity. But in actual practice there seems to be an almost boundless incompatibility of temper which prevents their remaining peaceably together for a single instant, so that the operator, who in this case acts as peacemaker, often suffers injury to himself while attempting to bring them together."

—Wilbur Wright

6.1 Design Motivation

Simply stated, stability and control is the science behind keeping the aircraft pointed in a desired direction. Whereas performance analysis sums the forces on an aircraft, stability and control analysis requires summing the moments acting on it as a result of surface pressure and shear-stress distributions, engine thrust, etc., and ensuring those moments sum to zero when the aircraft is oriented as desired. Stability analysis also deals with the changes in moments on the aircraft when it is disturbed from equilibrium, or in other words from the condition when all forces and moments on it sum to zero. An aircraft that tends to drift away from its desired equilibrium condition, or that oscillates wildly about the equilibrium condition, is said to lack sufficient stability. The Wright brothers intentionally built their aircraft to be unstable because this made them more maneuverable. As the preceding quotation from Wilbur Wright suggests, such an aircraft can be very difficult and dangerous to fly.

Control analysis determines how the aircraft should be designed so that sufficient control authority (sufficiently large moments generated when controls are used) is available to allow the aircraft to fly all maneuvers and at all speeds required by the design specifications. Good stability and control characteristics are as essential to the success of an aircraft as are good lift, drag, and propulsion characteristics. Anyone who has flown a toy glider that is out of balance or that has lost its tail surfaces, or who has shot an arrow or thrown a dart with missing tail feathers, knows how disastrous poor stability can be to flying. Understanding stability and control and knowing how to design good stability characteristics into an aircraft are essential skills for an aircraft designer.

6.2 Language

The science of stability and control is complex, and only an orderly, step-by-step approach to the problem will yield sufficient understanding and acceptable results. This process must begin by defining quite a number of axes, angles, forces, moments, displacements, and rotations. As much as possible, these definitions will

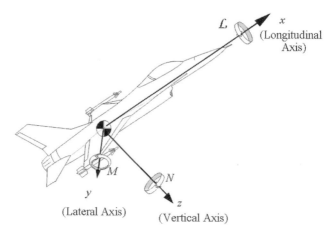

Fig. 6.1   Aircraft body axes and positive moment directions.

be consistent with those used in aerodynamic and performance analysis, but occasionally the complexity and unique requirements of stability and control problems dictate that less intuitive definitions and reference points be used.

6.2.1   Coordinate System

One of the least intuitive elements of stability and control analysis is the coordinate system as shown in Fig. 6.1. Note that the vertical z axis is defined as positive downward! The reason for this choice is a desire to have consistent and convenient definitions for positive moments. Positive moment directions are defined consistent with the right-hand rule used in vector mathematics, physics, and mechanics. This rule states that if the thumb of a person's right hand is placed parallel to an axis of a coordinate system, then the fingers of that hand will point in the positive direction of the moment about that axis. Because the moment about the aerodynamic center of an airfoil or wing was defined in Chapter 3 as being positive in a nose-up direction, the right-hand rule requires that the lateral (spanwise) axis of the aircraft coordinate system be positive in the direction from the right wing root to the right wing tip. A natural starting point for the coordinate system is the aircraft's center of gravity because it will rotate about this point as it moves through the air. The aircraft's longitudinal axis (down its centerline) is chosen parallel to and usually coincident with its aircraft reference line (defined in Chapter 4), but positive toward the aircraft's nose so that a moment tending to raise the left wing and lower the right wing is positive. This axis is chosen as the x axis to be consistent with performance analysis. Making x positive toward the front allows the aircraft's thrust and velocity to be taken as positive quantities. Because a rotation about the longitudinal axis to the right or clockwise is positive, for consistency it is desired that a moment or rotation about the aircraft's vertical axis such that the nose moves to the right be considered positive. This requires that the vertical axis be positive downward so that the right-hand rule is satisfied.

The only choice that remains is whether the lateral or vertical axis should be the y axis. The y axis is generally taken as vertical in performance analysis, but an x, y, z coordinate system must satisfy another right-hand rule in order to be consistent with conventional vector mathematics. The right-hand rule for three-dimensional orthogonal (each axis perpendicular to the others) coordinate systems requires that if the thumb of a person's right hand is placed along the coordinate system's x axis the fingers point in the shortest direction from the system's y axis to its z axis (try this on Fig. 6.1). To satisfy this right-hand rule as well as all of the previous choices for positive directions, the coordinate system's y axis must be the aircraft's lateral axis (positive out the right wing), and the z axis must be the vertical axis (positive down). A coordinate system such as this, which has its origin at the aircraft center of gravity and is aligned with the aircraft reference line and lateral axis, is referred to as a *body axis system*.

For consistency with aerodynamic analysis, the nose-up moment is labeled M. Because M is the moment about the y axis, the moment about the x axis is labeled $\mathcal{L}$ and the moment about the z axis is labeled N, to make them easier to remember. Note that the symbol $\mathcal{L}$ is used instead of L to avoid confusion with aircraft lift. Forces on the aircraft can be broken into components along the x, y, and z axes. These force components are labeled X, Y, and Z, respectively.

Note that the axis and moment symbol convention used in this book are by no means universal. The choice of letters signifying the different axes and moments is nearly universal, but the use of capitals and lower case, as well as script or italic fonts, varies significantly. One common variation is to use upper-case letters for the three moments, but then use lower-case subscripts when transforming them into moment coefficients. The student must determine the particular convention used when reading any stability and control book or paper in order to avoid confusion.

6.2.2 Degrees of Freedom

The aircraft has six *degrees of freedom*, six ways it can move. It has three degrees of freedom in *translation* (linear motion), which are orthogonal to each other. Components of its velocity along the x, y, and z axes are labeled u, v, and w. Note that lower case is used to avoid confusion with V, which typically has both u and w components. The aircraft also has three degrees of freedom in rotation, also orthogonal to each other.

6.2.3 Control Surfaces and Rotation

Figure 6.2 shows the three degrees of freedom in rotation, and the control surfaces that typically produce the moments that cause those rotations. Figure 6.2a shows rotation about the aircraft's longitudinal x axis. This motion is called *rolling*, and the maneuver is called a *roll*. Control surfaces on the aircraft's wings called *ailerons* deflect differentially (one trailing edge up and one trailing edge down) to create more lift on one wing, less on the other, and therefore a net rolling moment.

Figure 6.2b shows the aircraft in a *pitch-up* maneuver. Rotation of the aircraft about the lateral axis is called *pitching*. A control surface near the rear of the aircraft called an *elevator* or *stabilator* is deflected so that it generates a lift force,

Ailerons

a) Rolling about the *x* axis

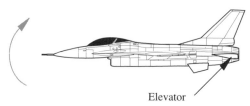

Elevator

b) Pitching about the *y* axis

Rudder

c) Yawing about the *z* axis

Fig. 6.2   Three rotations and control surfaces that produce them.

which, because of its moment arm from the aircraft center of gravity, also creates a pitching moment. An elevator is a movable surface attached to a fixed (immovable) *horizontal stabilizer*, a small horizontal surface near the tail of the aircraft, which acts like the feathers of an arrow to help keep the aircraft pointed in the right direction. A stabilator combines the functions of the horizontal stabilizer and the elevator. The stabilator does not have a fixed portion. It is said to be *all-moving*.

Figure 6.2c shows the aircraft *yawing*, rotating about the vertical axis so that the nose moves right or left. A movable surface called a *rudder*, which is attached to the aircraft's fixed *vertical stabilizer* deflects to generate a lift force in a sideways direction. Because the vertical stabilizer and rudder are toward the rear of the aircraft, some distance from its center of gravity, the lift force they generate produces a moment about the vertical axis that causes the aircraft to yaw.

6.2.4   Other Control Surfaces

A number of unusual aircraft configurations have given rise to additional types of control surfaces. These often combine the functions of two surfaces in one, and

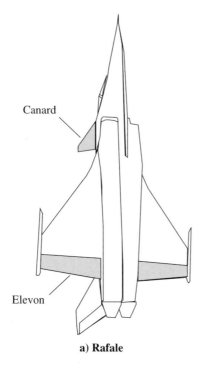

a) Rafale

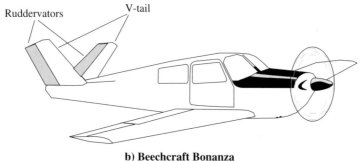

b) Beechcraft Bonanza

Fig. 6.3   Two aircraft with unusual control surfaces.

their names are created by combining the names of the two surfaces, just as the name stabilator was created by combining stabilizer and elevator. For example, the surface on the F-16 in Fig. 6.2 labeled aileron is actually a "flaperon" because it combines the functions of an aileron and a plain flap (for greater lift) in a single surface. Figure 6.3a shows the French Rafale multirole fighter aircraft. Pitch control for this aircraft is provided by *canards*, stabilators placed forward of rather than behind the wings, and *elevons*, control surfaces at the rear of the wings. Elevons move together to function as elevators and also move differentially like ailerons to provide roll control. Flying-wing aircraft, including delta-wing jet fighters such as the Mirage 2000 and Convair F-106 use elevons alone for pitch and roll control. The Vought F7U Cutlass twin-jet flying-wing fighter of the 1950s and 1960s used control surfaces exactly like elevons, but the manufacturer called them "ailevators!" The name did not find as widespread acceptance as "elevons." Figure 6.3b shows the Beechcraft Bonanza, which, unlike most aircraft with separate vertical and horizontal tail surfaces has a *V-tail*. The moveable control surfaces attached to the fixed surfaces of the V-tail are called ruddervators, because they function as elevators when moving together and rudders when moving differentially.

6.2.5 Trim

When the sum of the moments about an aircraft's center of gravity is zero, the aircraft is said to be *trimmed*. The act of adjusting the control surfaces of an aircraft so they generate just enough force to make the sum of the moments zero is called *trimming* the aircraft. The trim condition is an equilibrium condition in terms of moments. Strictly speaking, the sum of the forces acting on an aircraft does not have to be zero for it to be trimmed. For instance, an aircraft in a steady, level turn would be considered trimmed if the sum of the moments acting on it is zero, even though the sum of the forces is not.

6.2.6 Static Stability

Stability is the tendency of a system, when disturbed from an equilibrium condition, to return to that condition. There are two types of stability that must be achieved in order to consider a system stable. The first is static stability, the initial tendency or response of a system when it is disturbed from equilibrium. If the initial response of the system when disturbed is to move back toward equilibrium, then the system is said to have positive static stability. Figure 6.4a illustrates this situation for a simple system. When the ball is displaced from the bottom of the

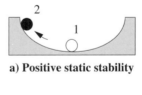

a) Positive static stability

b) Negative static stability

c) Neutral static stability

Fig. 6.4   Simple systems with positive, negative, and neutral static stability.

depression, forces resulting from the ball's weight, and the sloped sides of the depression tend to move the ball back toward its initial condition. The system is described as statically stable.

Figure 6.4b illustrates the reverse situation. When centered on the dome, the ball is in equilibrium. However, if it is disturbed from the equilibrium condition, then the slope of the dome causes the ball to continue rolling away from its initial position. This is called negative static stability because the system's initial response to a disturbance from equilibrium is away from equilibrium. The system is described as statically unstable.

Figure 6.4c shows neutral static stability. The ball on the flat surface, when displaced from equilibrium, is once again in equilibrium at its new position, so it has no tendency to move toward or away from its initial condition.

6.2.7 Dynamic Stability

The second type of stability that stable systems must have is dynamic stability. Dynamic stability refers to response of the system over time. Figure 6.5a shows the time history of a system that has positive dynamic stability. Note that the system also has positive static stability because its initial tendency when displaced from the zero displacement or equilibrium axis is to move back toward that axis. As the system reaches equilibrium, the forces and/or moments that move it there also generate momentum that causes it to overshoot or go beyond the equilibrium condition. This in turn generates forces that, because the system is stable, tend to return it to equilibrium again. These restoring forces overcome the momentum of the overshoot and generate momentum toward equilibrium, which causes another overshoot when equilibrium is reached, and so on. This process of moving toward equilibrium, overshooting, then moving toward equilibrium again is called an oscillation. If the time history of the oscillation is such that the magnitude of each successive overshoot of equilibrium is smaller, as in Fig. 6.5a, so that over time the system gets closer to equilibrium, then the system is said to have positive dynamic stability. Note that the second graph in Fig. 6.5a shows a system that has such strong dynamic stability that it does not oscillate but just moves slowly but surely to equilibrium.

The springs and shock absorbers on an automobile are familiar examples of systems with positive static and dynamic stability. When the shock absorbers are new, the system does not oscillate when the car hits a bump. The system is said to be highly damped. As the shock absorbers wear out, the car begins to oscillate when it hits a bump, and the oscillations get worse and take longer to die out as the shock absorbers get more worn out. The system is then said to be lightly damped.

A system that has positive static stability but no damping at all continues to oscillate without ever decreasing the magnitude or amplitude of the oscillation. It is said to have neutral dynamic stability because over time the system does not get any closer to or farther from equilibrium. The time history of a system with positive static stability but neutral dynamic stability is shown on the left-hand graph of Fig. 6.5b. On the right side of Fig. 6.5b is a time history of a system with neutral static and dynamic stability. When displaced from its initial condition, it

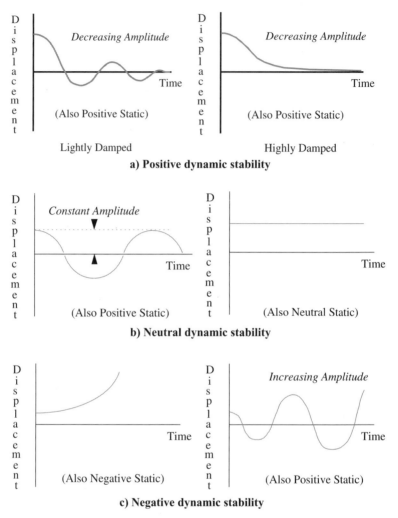

Fig. 6.5 Time histories of systems with positive, neutral, and negative dynamic stability.

is still in equilibrium, like the ball on the flat surface, and so it has no tendency to return to the zero-displacement condition.

The time histories in Fig. 6.5c are for systems with negative dynamic stability. The one on the left has negative static stability as well, so that it initially moves away from equilibrium and keeps going. The time history on the right is for a system that is statically stable, so that it initially moves toward equilibrium, but the amplitude of each overshoot is greater than the previous one. Over time, the system gets further and further from equilibrium, even though it moves through equilibrium twice during each complete oscillation.

6.3   Longitudinal Control Analysis

The analysis of the problem of adjusting pitch control to change and stabilize the aircraft's pitch attitude is called pitch control analysis or longitudinal control analysis. The term "longitudinal" is used for this analysis because the moment arms for the pitch control surfaces are primarily distances along the aircraft's longitudinal axis. Also, the conditions required for longitudinal trim (the case where moments about the lateral axis sum to zero) are affected by the airplane's velocity, which is primarily in the longitudinal direction.

The complete analysis of the static and dynamic stability and control of an aircraft in all six degrees of freedom is a broad and complex subject requiring an entire book to treat properly. A sense of how such problems are framed and analyzed can be obtained from studying the analysis of the longitudinal static stability and control problem. The longitudinal problem involves two degrees of translational freedom, the x and z directions, and one degree of freedom in rotation about the y axis.

6.3.1   Longitudinal Trim

Figure 6.6 illustrates the longitudinal trim problem for a conventional tail-aft airplane. The aircraft's center of gravity is marked by the circle with alternating black and white quarters. The lift forces of the wing and horizontal tail are shown acting at their respective aerodynamic centers. The moment about the wing's aerodynamic center due to the shape of its airfoil is also shown. The upper-case symbols L, L_t, and M_{ac} are used as in Chapter 4 for wing lift, tail lift, and wing moment respectively to indicate that they are forces and moments produced by three-dimensional surfaces, not airfoils. The horizontal tail is assumed to have a symmetrical airfoil, so that the moment about its aerodynamic center is zero. For consistency with the way two-dimensional airfoil data are presented, the locations of the wing's aerodynamic center x_{ac} and the whole aircraft's center of gravity x_{cg} are measured relative to the leading edge of the wing root. The distance of the

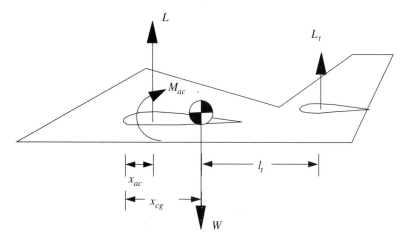

Fig. 6.6   Forces, moments, and geometry for the longitudinal trim problem.

aerodynamic center of the horizontal tail from the aircraft's center of gravity is given the symbol l_t.

Summing the moments shown in Fig. 6.6 about the aircraft's center of gravity yields

$$\sum M_{cg} = M_{ac} + L(x_{cg} - x_{ac}) - L_t l_t \tag{6.1}$$

The moments in Eq. (6.1) must sum to zero if the aircraft is trimmed. For steady flight, the forces must also sum to zero. Summing in the vertical direction,

$$\sum F_\perp = 0 = L + L_t - W \tag{6.2}$$

Together Eqs. (6.1) and (6.2) provide a system of two equations with two unknowns (because the weight is usually known and the moment about the aerodynamic center does not change with lift), which can be solved simultaneously to yield the lift required from each surface for equilibrium. In practice, the elevator attached to the horizontal tail is deflected to provide the necessary lift from the tail so that the sum of the moments is zero when the aircraft is at the angle of attack required to make the sum of the forces zero. Note that for aircraft configurations such as the one shown in Fig. 6.6, which have the horizontal tail behind the main wing, trim in level flight normally is achieved for positive values of L_t, so that the horizontal tail contributes to the total lift of the aircraft. Note also that Eqs. (6.1) and (6.2) are applicable only to the aircraft configuration for which they were derived. Similar relations can be derived for flying-wing aircraft, airplanes with canards, etc.

6.3.2 Control Authority

If an aircraft's geometry and flight conditions are known, then the lift coefficient required from the wing and pitch control surfaces can be determined using $L = C_L q S$ when Eqs. (6.1) and (6.2) are solved for L and L_t. If any of the required C_L values are greater than $C_{L_{max}}$ for their respective surfaces, then the aircraft does not have sufficient control authority to trim in that maneuver for those conditions. To remedy this situation, the aircraft designer must either increase the size of the deficient control surface or add high-lift devices to it to increase its $C_{L_{max}}$. Figure 6.7 shows a McDonnell–Douglas F-4E Phantom II multirole jet fighter. Note

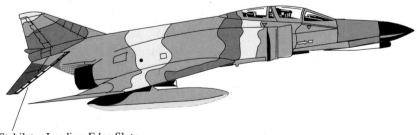

Stabilator Leading-Edge Slots

Fig. 6.7   Leading-edge slots to increase $C_{L_{max}}$ on the stabilator of the F-4E.

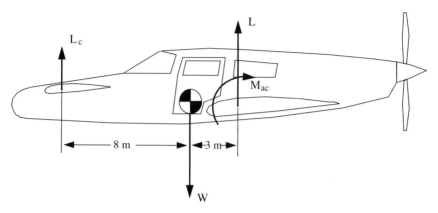

Fig. 6.8 Canard-configuration general aviation aircraft concept.

that the stabilators on this aircraft have had leading-edge slots added to them to increase their $C_{L_{max}}$ and hence their control authority.

Example 6.1

A design concept for a light general aviation aircraft uses a canard configuration as shown in Fig. 6.8. Both the wing and the canard of this aircraft have rectangular planforms. The aircraft has a mass of 1500 kg and is designed to fly as slow as 30 m/s at sea level in level flight. At this speed, its cambered main wing generates -1000 N m of pitching moment about its aerodynamic center. If the maximum lift coefficient for its canard is 1.5, how large must the canard be in order to trim the aircraft at its minimum speed?

Solution: Note that the pitching moment about the aerodynamic center is drawn nose up in Fig. 6.8 because that is the positive pitching-moment direction. The actual moment is nose down because its value is given as a negative. To trim at the specified minimum speed, the canard must generate sufficient lift so that the net moment on the aircraft measured about the center of gravity is zero. Summing the moments about the center of gravity,

$$\sum M_{cg} = 0 = -1000 \text{ N m} - L(3 \text{ m}) + L_c(8 \text{ m})$$

$$L = L_c \left(\frac{8 \text{ m}}{3 \text{ m}} \right) - \frac{1000 \text{ N m}}{3 \text{ m}} = 2.67 \, L_c - 333 \text{ N}$$

Then, summing forces in the vertical direction,

$$\sum F_\perp = 0 = L + L_c - W = L + L_c - mg$$

$$= 2.67 \, L_c - 333 \text{ N} + L_c - 1500 \text{ kg} \, (9.8 \text{ m/s}^2) = 3.67 \, L_c - 15{,}033 \text{ N}$$

$$L_c = \frac{15{,}033 \text{ N}}{3.67} = 4096 \text{ N}$$

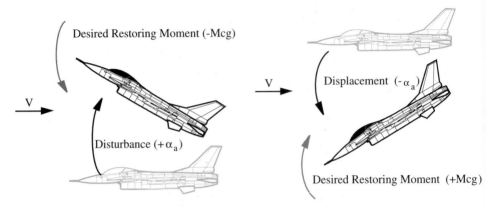

Fig. 6.9   Aircraft longitudinal static stability.

The dynamic pressure in standard sea level conditions at $V_\infty = 30$ m/s is

$$q = \tfrac{1}{2}\rho V_\infty^2 = \tfrac{1}{2}(1.225 \text{ kg/m}^3)(30 \text{ m/s})^2 = 551.3 \text{ N/m}^2$$

Then, to size the canard so that it can produce the required lift in these conditions,

$$L_c = C_{L_c} q\, S_c, \quad S_c = \frac{L_c}{C_{L_c} q} = \frac{4096 \text{ N}}{1.5\,(551.3 \text{ N/m}^2)} = 4.95 \text{ m}^2$$

6.4   Longitudinal Stability

As discussed in Sec. 6.1, adequate stability is essential to safe aircraft operations. Figure 6.9 illustrates the desired initial response of the aircraft when it is disturbed from trimmed level flight. If the disturbance causes the aircraft's angle of attack to increase, a statically stable aircraft would generate a negative pitching moment that would tend to return it to the trim condition. Likewise, if the disturbance reduced α, a statically stable aircraft would generate a positive pitching moment.

6.4.1   Static Stability Criterion

The preceding discussion of stable responses to disturbances leads to a criterion for positive longitudinal stability. This stability criterion is a condition that must be satisfied in order for an aircraft to be stable. Because, for positive static longitudinal stability, pitching moment must decrease with increasing angle of attack and increase with decreasing angle of attack, the partial derivative of the coefficient of pitching moment about the center of gravity $C_{M_{cg}} = M_{cg}/q\,Sc$, with respect to angle of attack, must satisfy

$$\frac{\partial C_{M_{cg}}}{\partial \alpha} = C_{M_\alpha} < 0 \qquad (6.3)$$

Equation (6.3) is the longitudinal static stability criterion.

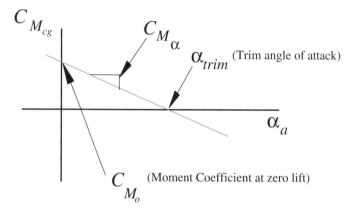

Fig. 6.10   Trim diagram.

6.4.2 Trim Diagram

A plot of pitching-moment coefficient vs angle of attack or lift coefficient reveals the relationship between static stability and trim and is usually called a *trim
diagram*. It is convenient to plot the variation of $C_{M_{cg}}$ with respect to absolute angle
of attack $\alpha_a = \alpha - \alpha_{L=0}$ because $\alpha_a = 0$ when $C_L = 0$. Figure 6.10 illustrates a
typical trim diagram for an aircraft with positive static longitudinal stability.

Note that the $C_{M_{cg}}$ vs α_a curve slope C_{M_α} is constant. This is typical at low angles
of attack. The pitching-moment coefficient at $\alpha_a = 0$ (and $C_L = 0$) is given the
symbol C_{M_0}. The angle of attack where $C_{M_{cg}} = 0$ is the *trim angle of attack* α_{trim}.
Figure 6.10 immediately makes obvious another requirement for an aircraft with
positive longitudinal stability. Because aircraft must produce lift in most cases for
equilibrium, only positive absolute angles of attack are useful as trim angles of
attack. Because $C_{M_\alpha} < 0$ for stability, it follows that C_{M_0} must be greater than 0 if
the aircraft is to trim at a useful α_{trim}. This is not strictly a stability criterion, but it
is a required characteristic of aircraft that have positive static longitudinal stability
and useful trim angles of attack.

Figure 6.11 shows $C_{M_{cg}}$ vs α_a curves for aircraft with neutral and negative static
longitudinal stability. $C_{M_\alpha} = 0$ for neutral stability, and $C_{M_\alpha} > 0$ when static stability
is negative. Clearly, for an aircraft with neutral static longitudinal stability to have
a useful α_{trim}, C_{M_0} must equal zero. Then all values of α_a are trim angles of attack.
Likewise, if an aircraft has negative static stability C_{M_0} must be less than zero for
any useful value of α_{trim}.

6.4.3 Neutral Point

The airfoils in Chapter 3 had an aerodynamic center, a point on the airfoil about
which the net aerodynamic moment did not change with angle of attack. Likewise,
a similar point can be found on an aircraft where the net total moment on the aircraft
does not change with angle of attack. This point is very useful in describing the
static stability of an aircraft.

Figure 6.12 is a drawing of an aircraft showing its c.g. and whole aircraft aerodynamic center locations. Note that if the aircraft's angle of attack increases, its

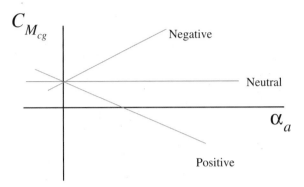

Fig. 6.11   Trim diagrams for positive, neutral, and negative static longitudinal stability.

lift will increase. This will generate a negative or nose-down moment about its c.g. because its aerodynamic center is located behind its c.g. This moment will tend to return the aircraft to its original angle of attack. Therefore, the aircraft has positive static longitudinal stability. If it is trimmed at its initial angle of attack, then its trim diagram would look like Fig. 6.10.

Now suppose the aircraft's c.g. moves aft, so that it is behind its aerodynamic center, as shown in Fig. 6.13. In this case, an increase in angle of attack and the resulting increase in lift would cause a nose-up pitching moment about the aircraft's c.g. This would tend to increase the aircraft's angle of attack further, creating more lift and more moment, and so forth, driving the aircraft farther from its original equilibrium condition. The aircraft would have negative static longitudinal stability.

If the aircraft's c.g. is located at its aerodynamic center, then by definition (of the aerodynamic center) the moment about its c.g. would not change with angle of attack. In that case, the aircraft would have neutral static longitudinal stability.

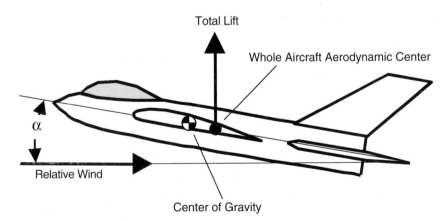

Fig. 6.12   Aircraft geometry for positive longitudinal static stability.

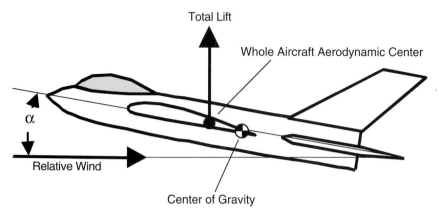

Fig. 6.13   Aircraft geometry for negative longitudinal static stability.

Figure 6.14 summarizes the effect on an aircraft's trim diagram of moving its center of gravity aft.

The whole aircraft's aerodynamic center is called its *neutral point* because placing its c.g. there gives it neutral static stability. Calling the whole aircraft's aerodynamic center its neutral point also helps avoid confusion with its wing's and horizontal tail's aerodynamic centers, which also figure prominently in stability calculations.

6.4.4   Static Margin

The relative locations of an aircraft's c.g. and its neutral point (n.p.) are used to describe its static longitudinal stability. The distance of an aircraft's c.g. ahead of its n.p., when divided by its reference chord length, is called its *static margin* (SM) If an aircraft's SM is positive, it is stable. Likewise, if its SM is negative, its c.g. is behind its n.p., and it is has negative static longitudinal stability. SM = 0 denotes an aircraft with neutral static stability.

Stated in terms of static margin, the static longitudinal stability criterion becomes $SM > 0$. Static margin is a convenient nondimensional measure of the aircraft's stability. A large static margin suggests an aircraft that is very stable and not very maneuverable. A low positive static margin is normally associated with highly

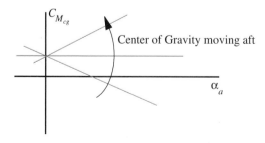

Fig. 6.14   Effect on trim diagram of moving the center of gravity aft.

Table 6.1   Static margins for several aircraft

Aircraft type	Static margin
Cessna 172	0.19
Learjet 35	0.13
Boeing 747	0.27
North American P-51 Mustang	0.05
Convair F-106	0.07
General Dynamics F-16A (early)	−0.02
General Dynamics F-16C	0.01
Grumman X-29	−0.33

maneuverable aircraft. Aircraft with zero or negative static margin normally require a computer fly-by-wire flight control system in order to be safe to fly. Table 6.1 lists static margins for typical aircraft of various types.

6.4.5 Altering Longitudinal Static Stability

The discussion of neutral point began with considering how moving the center of gravity location would change an aircraft's static longitudinal stability. To understand how other changes to an aircraft will change its stability, we must look at how each part of an aircraft contributes to its overall stability. Consider the aircraft in Fig. 6.15. The aircraft angle of attack is measured between the aircraft reference line and the freestream velocity vector V_∞. For simplicity in this analysis, the aircraft reference line is chosen to coincide with the zero-lift line of the wing and fuselage (a reference line such that when it is aligned with the freestream velocity, the wing and fuselage together produce zero lift). As a further simplification, the contribution of the horizontal tail lift to the whole aircraft lift (but not the tail's contribution to the moment) will be ignored. With these assumptions $\alpha_{L=0} = 0$ and $\alpha_a = \alpha$. At the horizontal stabilizer, the local flow velocity vector is the vector

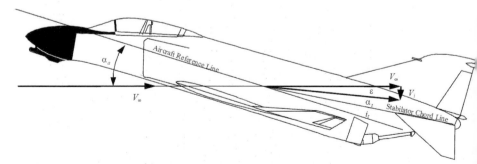

Fig. 6.15   Important velocities and angles for longitudinal stability analysis.

sum of the freestream velocity and the downwash velocity V_i. The angle between the freestream velocity and the local velocity at the tail is the downwash angle ε. The angle of attack of the horizontal tail (stabilator in this case) is labeled α_t. It is defined as the angle between the horizontal tail chord line and the local velocity vector. The angle between the horizontal tail chord line and the aircraft reference line is called the *tail incidence angle* and is given the symbol i_t.

The geometry of Fig. 6.6 was used in Sec. 6.3 in the longitudinal trim analysis. For that analysis, it was required that the moments about the aircraft's center of gravity sum to zero. The same geometry is used to determine C_{M_0}, except that the forces and moments are written in terms of nondimensional coefficients, and they do not necessarily sum to zero. The expression for C_{M_0} is obtained by dividing Eq. (6.1) by $q\,S\bar{c}$, where $\bar{c}$ is the reference chord length of the wing:

$$\frac{\sum M_{cg}}{q\,S\bar{c}} = \frac{M_{ac} + L(x_{cg} - x_{ac}) - L_t l}{q\,S\bar{c}} = C_{M_{cg}}$$

$$= C_{M_{ac}} + C_L \left(\frac{x_{cg} - x_{ac}}{\bar{c}} \right) - \frac{C_{L_t} q\, S_t l_t}{q\,S\bar{c}} \tag{6.4}$$

The following definitions are made:

$$\bar{x}_{cg} = \frac{x_{cg}}{\bar{c}}, \qquad \bar{x}_{ac} = \frac{x_{ac}}{\bar{c}}, \qquad V_H = \frac{S_t\, l_t}{S\,\bar{c}} \tag{6.5}$$

so that Eq. (6.4) becomes

$$C_{M_{cg}} = C_{M_{ac}} + C_L(\bar{x}_{cg} - \bar{x}_{ac}) - C_{L_t} V_H \tag{6.6}$$

The ratio defined in Eq. (6.5), which was given the symbol V_H, is called the *horizontal tail volume ratio* because the quantities in the numerator and the denominator of the ratio have units of volume (area multiplied by length). The lift coefficients of the wing and horizontal tail are expressed in terms of their angles of attack and lift-curve slopes:

$$C_L = C_{L\alpha}(\alpha - \alpha_{L=0}) = C_{L\alpha}\,\alpha_a, \qquad C_{L_t} = C_{L\alpha_t}(\alpha_t - \alpha_{L=0_t})$$

As with the analysis in Sec. 6.3, the horizontal tail is assumed to have a symmetrical airfoil section, so that $\alpha_{L=0_t} = 0$. Also, from Fig. 6.12, $\alpha_t = \alpha_\alpha - \varepsilon - i_t$, so that

$$C_{M_{cg}} = C_{M_{ac}} + C_{L_\alpha}\alpha_a(\bar{x}_{cg} - \bar{x}_{ac}) - C_{L\alpha_t}(\alpha_a - \varepsilon - i_t) V_H \tag{6.7}$$

Now, C_{M_0} is defined as the moment coefficient when the entire aircraft produces zero lift. For most airplanes the wing and fuselage together produce a very large proportion of the lift, and so the lift of the tail has been neglected in this analysis. With this approximation made, C_{M_0} is the moment coefficient when the wing and fuselage produce zero lift, and

$$C_{M_0} = C_{M_{ac}} - C_{L\alpha_t}(-\varepsilon - i_t) V_H = C_{M_{ac}} + C_{L\alpha_t}(\varepsilon_0 + i_t) V_H \tag{6.8}$$

where ε_0 is the downwash angle when $\alpha = 0$. This is usually a very small angle, often zero.

The moment-curve slope C_{M_α} is obtained by taking the derivative of Eq. (6.7) with respect to absolute angle of attack:

$$C_{M_\alpha} = \frac{\partial C_{M_{cg}}}{\partial \alpha_a} = \frac{\partial}{\partial \alpha_a}[C_{M_{ac}} + C_{L_\alpha}\alpha_a(\bar{x}_{cg} - \bar{x}_{ac}) - C_{L_{\alpha_t}}(\alpha_a - \varepsilon - i_t)V_H] \quad (6.9)$$

$$C_{M_\alpha} = C_{L_\alpha}(\bar{x}_{cg} - \bar{x}_{ac}) - C_{L_{\alpha_t}}\left(1 - \frac{\partial \varepsilon}{\partial \alpha}\right)V_H \quad (6.10)$$

Equations (6.8) and (6.10) give valuable insight into the influence that the wing and tail of a conventional tail-aft airplane exert on its trim diagram. Note that Eq. (6.8) reveals that, because $C_{M_{ac}}$ is normally negative or zero and ε_0 is normally very small, the incidence angle of the horizontal tail must not be zero if C_{M_0} is to be positive. Note also that i_t was defined as positive when the horizontal tail is oriented so that it is at a lower angle of attack than the main wing. This makes sense because when the main wing is producing no lift the tail, if $i_t > 0$, will be at a negative angle of attack. The lift produced by the tail in this situation would be downward, creating a nose-up pitching moment, so that $C_{M_0} > 0$.

Most conventional aircraft are designed and balanced so that their centers of gravity are aft of the aerodynamic centers of their wing/fuselage combination. For this situation, the wing term in Eq. (6.10) is positive, and because $C_{M_\alpha} < 0$ for stability, the wing tends to destabilize the aircraft. The tail term in Eq. (6.10) is negative, and so the tail must overcome the destabilizing effect of the wing in order to make the airplane stable. Expressions for C_{M_0} and C_{M_α} for other aircraft configurations can be developed using the same approach, which produced Eqs. (6.8) and (6.10).

Equation (6.10) can be used to predict how other changes in an aircraft configuration will alter its stability. For example, suppose the value of wing/fuselage

Table 6.2   Aircraft changes that affect stability

	To Increase Stability (Make C_{M_α} More Negative)	
Term	Change	How accomplished
C_{L_α}	↑ or ↓	Depends on $(\bar{x}_{cg} - \bar{x}_{ac})$ 1) Make wing more or less efficient (more or less elliptical) 2) Increase/decrease aspect ratio
$\bar{x}_{cg}$	↓	Shift weight forward
$\bar{x}_{ac}$	↑	1) Pick airfoil with more aft a.c. 2) Sweep wings
V_H	↑	1) Increase tail area or move it aft 2) Decrease wing area or chord
$C_{L_{\alpha_t}}$	↑	Make tail lift distribution more elliptical or increase apect ratio
$\dfrac{\partial \varepsilon}{\partial \alpha}$	↓	Decrease downwash by increasing aspect ratio or by placing tail above or below the plane of the wing

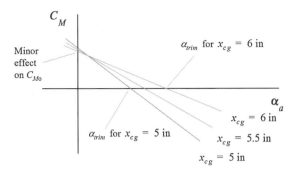

Fig. 6.16   Example trim diagram for three different c.g. positions.

lift-curve slope C_{L_α} is increased by increasing the wing aspect ratio or the wing's span efficiency factor. If, as in most conventional aircraft, the aerodynamic center of the wing/fuselage combination is forward of the aircraft center of gravity so that $x_{cg} - x_{ac} > 0$, then increasing C_{L_α} makes the wing term in Eq. (6.10) more positive. C_{M_α} therefore becomes less negative and the aircraft less stable. For an aircraft configuration where $x_{cg} - x_{ac} < 0$, on the other hand, Eq. (6.10) shows that increasing C_{L_α} increases static stability. Table 6.2 lists several other common aircraft configuration changes and the effect they have on stability.

6.4.6 Changing Trim

Changes in stability also affect trim, as indicated by Fig. 6.16. Note that most of the changes in Table 6.2 have a relatively small effect on C_{M_0}. Their primary action is to change C_{M_α}. This in turn changes the trim angle of attack where the C_M vs α lines intersect the trim diagram's horizontal axis, as shown on Fig. 6.16. Note that in the example in Fig. 6.16, moving the c.g. aft caused the reduced static stability.

Equations (6.8) and (6.10) also can be used to determine how changing horizontal tail incidence angle changes the trim diagram. This, in fact, is one of the primary uses of the trim diagram because it shows how longitudinal control of the aircraft works. Figure 6.17 is an example. Note that changing i_t changes C_{M_0} without affecting C_{M_α}. This is apparent because the C_M vs α lines remain parallel to each other. Changing i_t simply offsets the lines from each other, which in turn causes them to intersect the horizontal axis at different trim angles of attack. In other words, changing i_t changes trim angle of attack without changing stability. In general, deflecting the stabilator trailing edge up (increasing i_t) increases α_{trim}, as is the case in Fig. 6.17.

6.5 Dynamic Longitudinal Stability

Just giving an aircraft positive static longitudinal stability does not ensure that it will fly well. Analyzing its dynamic stability is a complex process that would occupy an entire textbook. However, before leaving this topic entirely, it is worth describing the primary types of dynamic longitudinal motions commonly observed in aircraft, and indicating how to change them to eliminate adverse characteristics.

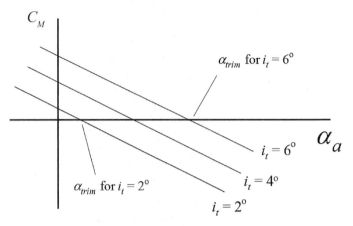

Fig. 6.17   Example trim diagram for three different tail incidence angles.

6.5.1   Longitudinal Dynamic Modes

The dynamic longitudinal motion of a statically stable aircraft follows a pattern similar to that shown in Fig. 6.5a. The graph in Fig. 6.5a could easily be a plot of an aircraft's pitch attitude as it encounters a disturbance, such as a wind gust, and then moves back toward its initial condition. However, if we look at how other parameters vary during an aircraft's response to a gust we would typically discover two separate oscillations, largely independent of each other, going on at the same time. These oscillations are called *dynamic modes*. Both longitudinal dynamic modes produce changes in pitch attitude, but in other ways they are very distinct and different from each other.

6.5.2   Short-Period Mode

The first mode that we would typically notice in an aircraft's response to a longitudinal disturbance is called the *short-period* mode. As its name implies, the short-period mode has a relatively short period and therefore a relatively high frequency. The mode involves almost exclusively an angle-of-attack oscillation, driven by the aircraft's C_{M_α}. The oscillation is usually highly damped, but only because aircraft designers take great care to make it that way. If the magnitude of its C_{M_α} is too high (SM too large, too much static stability) relative to damping effects, an aircraft's short-period mode can be lightly damped, undamped, or even dynamically unstable. Because the period of this mode is often shorter than the pilot's response time, this type of oscillation can be very difficult for a pilot to correct. In such cases, designers frequently use flight-by-wire flight control systems with artificial damping features to correct the problem.

6.5.3   Phugoid

The second longitudinal dynamic mode is the *phugoid*. This mode involves a flight-path angle/airspeed oscillation with almost no change in angle of attack. For instance, a wind gust might raise an aircraft's nose above its equilibrium condition, so that it is climbing. Unless the pilot changes the level-flight equilibrium cruise thrust setting to compensate for this climb, the aircraft's speed will decrease. As speed decreases, so does lift, and so the aircraft's nose pitches down, toward equilibrium. The aircraft is now in a descent, and so its speed increases, and so forth, oscillating above and below the level flight γ and airspeed. The pitch attitude, of course, also oscillates with the changing flight-path angle.

Although the short-period mode produces changes in pitch attitude via changes in angle of attack, without significantly affecting flight-path angle or airspeed, the phugoid reverses this, having little effect on angle of attack. The phugoid is a fairly slow, lightly damped oscillation, easily controlled by the pilot. However, in the case of free-flight gliders with no remote or onboard control the phugoid can be a problem.

The main damping of the phugoid comes from changes in drag resulting from airspeed changes. Therefore high-performance, low-drag aircraft often have very lightly damped phugoids. For free-flight gliders it is often possible to avoid a problem phugoid by simply launching the glider at its equilibrium speed and flight-path angle. This avoids exciting the oscillation as long as no wind gust or other disturbance occurs.

6.6 Lateral-Directional Stability

An aircraft that had only the three longitudinal degrees of freedom, x and z translation and rotation about the y axis, would not be very practical. It would only be able to travel in a straight line. Aircraft need to rotate about their x and z axes and translate along their y axis in order to turn and maneuver. As a result, they require stability in these additional degrees of freedom. This stability is called lateral-directional stability because it involves lateral motion (along the y axis), lateral rotation (rolling about the x axis), and directional (yawing about the z axis) rotation. These two rotations and one translation are all analyzed together because a change in any one almost always produces changes in the others.

6.6.1 Angles, Forces, and Moments

Recall that angle of attack is a primary variable in longitudinal stability analysis. The corresponding angle in lateral-directional stability is sideslip angle β. As shown in Fig. 6.18, β is the angle between the freestream velocity vector and the aircraft's x axis. A positive β occurs when the aircraft is flying as shown with "wind blowing in the pilot's right ear" because a positive yaw moment (nose right) is required to align the x axis with the velocity vector.

When an aircraft with good stability yaws, it generates yawing and rolling moments that tend to return it to equilibrium. The tendency of an aircraft, when β is disturbed from equilibrium, to produce a yawing moment that tends to reduce β back to zero is called *positive directional stability*. The corresponding tendency to return to wings level when disturbed is called *positive lateral stability*.

6.6.2 Static-Directional Stability

As shown in Fig. 6.19, positive static-directional stabilty requires that an aircraft, when β is positive, generate a right-hand or positive moment around the z axis to

Fig. 6.18   Sideslip angle.

reduce β back to zero. Aircraft designers usually achieve this by placing a vertical stabilizer on the rear fuselage, far aft of the center of gravity. When the aircraft yaws, this gives the vertical tail an angle of attack to the relative wind, and it makes lift. As seen in Fig. 6.19, this lift is in a direction that, because the vertical tail is behind the c.g., produces a moment that tends to reduce β back to zero. Recall that the moment about the z axis is identified by the symbol N, so that the yawing-moment coefficient is $C_N = N/q_\infty Sb$. Note that C_N is defined using wing span b rather than chord length as the reference length. The slope of the C_N vs β curve is C_{N_β}. Based on the preceding discussion, the criterion for positive directional stability is that $C_{N_\beta} > 0$. Figure 6.20 shows a C_N vs β plot for an aircraft with positive static-directional stability.

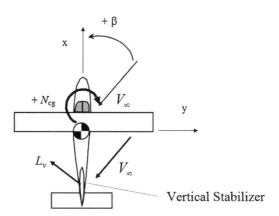

Fig. 6.19   Aircraft top view with static-directional stability geometry.

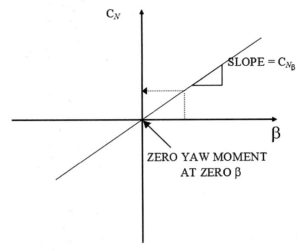

C_N

SLOPE = C_{N_β}

β

ZERO YAW MOMENT
AT ZERO β

Fig. 6.20 Yawing-moment coefficient vs sideslip angle plot for an aircraft with positive static-directional stability.

6.6.3 Static Lateral Stability

The requirement for positive static lateral stability is that an aircraft roll away from a sideslip. In other words, if an aircraft with positive static lateral stability is sideslipping to the right, so that β is positive, then it will generate a moment that will cause it to roll to the left. To understand why this is considered positive stability, see the discussion of spiral mode stability in Sec. 6.7.1. Because the positive direction for rolling about the x axis is to the right, an aircraft with positive static lateral stability will produce a negative rolling moment when its sideslip angle is positive.

Recall that the moment about the x axis is identified by the symbol $\mathcal{L}$, so that the rolling-moment coefficient is $C_{\mathcal{L}} = \mathcal{L}/(q_\infty S b)$. Note that $C_{\mathcal{L}}$, like C_N, is defined using wing span b, rather than chord length as the reference length. The slope of the $C_{\mathcal{L}}$ vs β curve is $C_{\mathcal{L}_\beta}$. Based on the preceding discussion, the criterion for positive static lateral stability is that $C_{\mathcal{L}_\beta} < 0$. Figure 6.21 shows a C_N vs β plot for an aircraft with positive directional stability.

Designers use several features of an aircraft to cause it to have $C_{\mathcal{L}_\beta} < 0$. These include wing sweep, wing position, and wing dihedral. A vertical tail that sticks high above an aircraft's c.g. also contributes to $C_{\mathcal{L}_\beta} < 0$.

To understand how wing sweep contributes to $C_{\mathcal{L}_\beta} < 0$, look at Fig. 6.18 again. The aircraft shown in Fig. 6.18 has swept wings. When the aircraft is yawed so it has a positive sideslip angle, as shown, its right wing is less swept relative to its velocity vector than its left wing. Because the right wing has less sweep relative to the left wing, it will produce more lift. Therefore, the aircraft will roll left. Another way to look at it is the right wing has a higher aspect ratio than the left wing when flying as shown and so will have a higher lift-curve slope than the left wing. This difference is shown in Fig. 6.22.

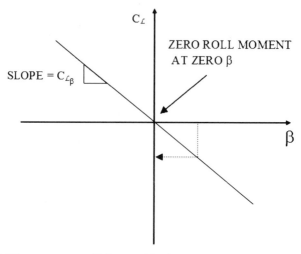

Fig. 6.21   Rolling-moment coefficient vs sideslip angle plot for an aircraft with positive static lateral stability.

Note that the effect of wing sweep on C_{L_β} increases as angle of attack and lift on the wings increases. It also goes away when the wings are making zero lift. This can be an advantage for aircraft that must make vertical climbs, where lift is zero. In a situation such as this, roll caused by sideslip is undesireable because it would cause the aircraft to make a corkscrew-type path if it yawed.

To understand how wing placement on the fuselage contributes to $C_{L_\beta} < 0$, consider Fig. 6.23. This shows *crossflow* (sideways component of airflow) around three sideslipping fuselages, each with a wing attached at a different vertical position. The high wing in Fig. 6.23a gets an upward crossflow component on its upwind wing and a downward component on its downwind wing. This causes it

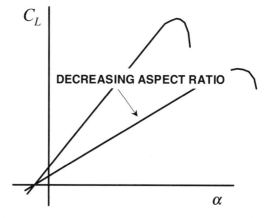

Fig. 6.22   Effect of aspect ratio on wing lift-curve slope.

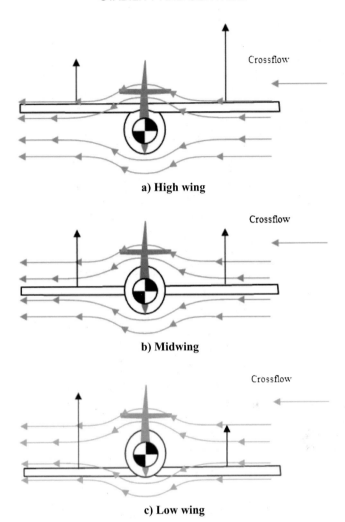

Fig. 6.23   Effect of wing position on rolling moment due to sideslip.

to roll away from the sideslip, contributing to positive static lateral stability. The midwing in Fig. 6.23b gets no effect, no net rolling moment from sideslip, neutral static lateral stability. The low wing in Fig. 6.23c gets a downward crossflow component on its upwind wing and an upward component on its downwind wing. This causes it to roll toward the sideslip, contributing to negative static lateral stability.

Wing dihedral angle is the angle in a rear view of the aircraft between its y axis and a line drawn from the middle of the wing root to the middle of the wing tip as shown in Fig. 6.24. Dihedral is commonly given the symbol Γ. Figure 6.24a shows a rear view of an aircraft with positive wing dihedral and the crossflow components due to a positive sideslip. When an aircraft wing has positive dihedral, positive sideslip angle results in a negative (left) rolling moment, a positive contribution to

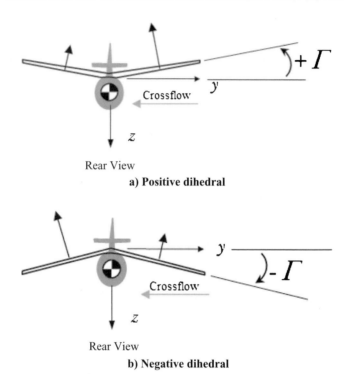

a) Positive dihedral

b) Negative dihedral

Fig. 6.24   Effect of dihedral on rolling moment due to sideslip.

static lateral stability. Figure 6.24b shows a rear view of an aircraft with negative wing dihedral and the crossflow components caused by a positive sideslip. When an aircraft wing has negative dihedral, positive sideslip angle results in a positive (right) rolling moment, a negative contribution to static lateral stability.

Figure 6.25 shows how a tall vertical tail contributes to positive static lateral stability. Because its aerodynamic center is above the aircraft's center of gravity, it generates a negative rolling moment when the aircraft has a positive sideslip angle. This is a positive contribution to static lateral stability.

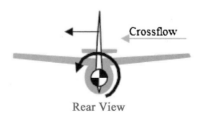

Fig. 6.25   Effect of vertical tail on static lateral stability.

6.7 Dynamic Lateral-Directional Stability

Most aircraft exhibit three distinct dynamic lateral-directional modes. These are the roll mode, the spiral mode, and the Dutch roll mode. Each mode has unique characteristics, and each is influenced in different ways by the same aircraft features that contribute to static lateral and directional stabilty. A basic understanding of these effects can be helpful in appreciating the complex interactions involved in these modes.

6.7.1 Spiral Mode

The spiral mode typically involves a gradual increase in bank angle, causing the aircraft to make a descending turn. As bank angle increases, the aircraft's nose drops further below the horizon, and its speed increases. This is sometimes called the "death spiral" because of the ultimate outcome if not corrected. The mode usually does not oscillate. In most aircraft, it happens slowly enough that a pilot can easily correct it. Many succesful modern aircraft have unstable or neutrally stable spiral modes.

The requirement for positive spiral mode stability is that an aircraft, when disturbed (rolled) from wings-level flight, should generate a moment that tends to return it to wings level. This is important for uncontrolled free-flight gliders, which rely only on their natural stability to recover from disturbances. Good lateral stability is also important for piloted aircraft, which must be flown without outside visual references (at night or in clouds), because this characteristic reduces the workload required of the pilot to keep the aircraft "shiny side up."

The actual mechanism of achieving positive spiral mode stability is slightly more complicated than static lateral stability. When most aircraft roll, their rolling moment does not usually depend much on bank angle φ. In that sense, most aircraft have neutral static stability relative to bank angle. This condition is desireable because it allows aircraft to easily maintain a required bank angle while turning. However, if an aircraft is disturbed from wings level by something other than the pilot's action, unless that disturbance puts it into a perfectly coordinated turn, the aircraft will slip toward the inside of the turn, causing it to yaw.

If this sideslip to the inside of the turn produces a rolling moment that tends to return the aircraft to the original wings-level zero-bank-angle-condition, the aircraft has positive spiral stability. If, instead, the sideslip causes the aircraft to create a rolling moment that tends to further increase its bank angle, the aircraft has negative spiral stability. The key to which of these will occur is in the relative magnitudes of the aircraft's $C_{\mathcal{L}_\beta}$ and C_{N_β}.

If $|C_{N\beta}/C_{L\beta}| > 2/3$, the aircraft's static directional stability turns it into the direction it is slipping, reducing the yaw and preventing its lateral stability from rolling it out of the turn. In this situation, the aircraft often has a very dangerous, divergent spiral mode. This is particularly true for unpiloted, free-flight gliders. To correct this situation, the designer can increase dihedral or sweep or mount the wing higher on the fuselage to increase $C_{\mathcal{L}_\beta}$. It would also be effective to reduce the size but not the height of the vertical tail or move it closer to the aircraft's center of gravity to reduce C_{N_β}.

6.7.2 Dutch Roll Mode

If $|C_{N\beta}/C_{L\beta}| < 1/3$, an aircraft's lateral stability is out of balance with and overpowers its directional stability. When a disturbance causes the aircraft to yaw, its powerful lateral stability causes it to roll away from the yaw before its directional stability can turn it into the sideslip. This triggers an oscillating rolling and yawing maneuver that, because of its side-to-side rolling motion, reminded the engineers who named it of the side-to-side rolling motion that characterized the style of famous Dutch ice skaters of the time. A Dutch roll that is lightly damped can quickly trigger airsickness in even the most stalwart airliner passengers. Its period can be shorter than the pilot's reaction time, and so it can be difficult for the pilot to correct. The designer can correct this by decreasing dihedral or sweep, or mounting the wing lower on the fuselage, to decrease $C_{\mathcal{L}_\beta}$. It would also be effective to increase the size and reduce the height of the vertical tail or move it farther from the aircraft's center of gravity to increase C_{N_β}.

6.7.3 Roll Mode

The roll mode is the aircraft's response to aileron inputs. When the pilot deflects ailerons, the resulting rolling moment causes the aircraft to begin to roll. As the roll rate increases, the motion of the wings causes an additional relative wind component that increases the angle of attack on the downgoing wing and decreases it on the upgoing wing. This, in turn, creates a moment opposite the rolling moment created by the ailerons, which gradually builds until the aircraft reaches a steady roll rate. Ideally, the roll rate should build quickly to a rate that is sufficiently rapid for the type of aircraft and its mission. The designer can correct a sluggish roll rate and slow buildup to steady-state roll rate by increasing the size and deflection of the ailerons. Likewise, the designer can correct an aircraft that is too sensitive in roll by reducing the aileron size. The use of an electronic fly-by-wire flight control system allows the designer to correct the second problem with software changes. However, software cannot create control authority that is not there, so that increasing roll rates requires more or larger surfaces.

6.7.4 Adverse Yaw and Spins

Part of the roll mode is a yaw response that is caused by differential drag on the two wings. The wing with the aileron deflected trailing-edge-down creates more lift than the one with the aileron deflected up. This in turn creates more drag on the wing with the greater lift. If the wings are long, as on a sailplane, this differential drag can create a powerful yawing moment away from the direction of the roll. At best, this can delay the turn. At worst, if done at high angle of attack near stall, it can push the aircraft into a rotating stalled condition called a spin. Spins are driven by adverse yaw and poststall lift differences. They tend to persist unless the pilot takes action to counteract the yaw and reduce the angle of attack below stall. Designers can reduce adverse yaw by using spoilers for roll control or by putting the ailerons closer to the aircraft centerline. They can also use differential ailerons that deflect trailing edge up much more than they deflect trailing edge down. This works because the trailing-edge-up deflection is so extreme that it causes flow separation and additional drag to balance the additional drag due to lift of the opposite trailing-edge-down aileron.

6.8 Chapter Summary

6.8.1 Coordinate System, Forces, and Moments

1) Axes: x positive aft, y positive out right wing, z positive down
2) Forces: X, Y, Z aligned with x, y, z axes
3) Moments: $\mathcal{L}$, M, N about x, y, z axes respectively
4) Velocity components: u, v, w along x, y, z axes, respectively

6.8.2 Rotations and Control Surfaces that Produce Them

1) Six degrees of freedom: three translation—x, y, z; three rotation—roll about x axis, pitch about y, yaw about z
2) Rolling about x axis: ailerons, elevons, flaperons, tailerons (differential tails)
3) Pitching about y axis: elevators, stabilators, elevons, canards
4) Yawing about z axis: rudders, ruddervators, V-tails, adverse yaw from ailerons

6.8.3 Trim and Stability

1) Trim: equilibrium, $\Sigma M = 0$
2) Static stability: initial tendency to return toward equilibrium when disturbed
3) Dynamic stability: tendency over time to return to equilibrium
4) Longitudinal stability: translation along x and z axes and rotation about y axis
5) Lateral-directional stability: rotation about x and z axes and translation along y axis

6.8.4 Longitudinal Trim

Free-body diagram, $\Sigma F = 0$, $\Sigma M = 0$, solve for unknowns.
Example: Douglas DC-3 in SLUF. Find lift of the tail, L_t required to trim.

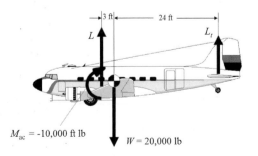

$M_{ac} = $ -10,000 ft lb

$W = 20,000$ lb

Solution:

$$\Sigma F = 0 = L + L_t - 20,000 \text{ lb}$$
$$L = 20,000 \text{ lb} - L_t$$
$$\Sigma M = 0 = L(3 \text{ ft}) - L_t(24 \text{ ft} - 10,000 \text{ ft lb})$$
$$= (20,000 \text{ lb} - L_t)(3 \text{ ft}) - L_t(24 \text{ ft} - 10,000 \text{ ft lb})$$
$$= 60,000 \text{ ft lb} - L_t(27 \text{ ft} - 10,000 \text{ ft lb})$$
$$L_t = 50,000 \text{ ft lb}/27 \text{ ft} = 1852 \text{ lb}$$

Note that this is an upward-lifting force on the tail, a typical condition for aircraft designed for efficient cruising flight.

6.8.5 Longitudinal Static Stability

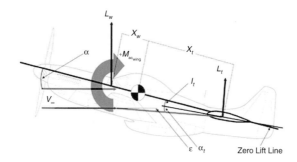

1) Tail incidence angle i_t:
   a) Angle between horizontal tail chord line and aircraft Z.L.L.
   b) Positive trailing edge up (as shown)
2) Tail angle of attack α_t:
   a) Tail sees a different angle of attack than the wing
   b) $\alpha_t = \alpha - i_t - \varepsilon$
3) Trim diagram:

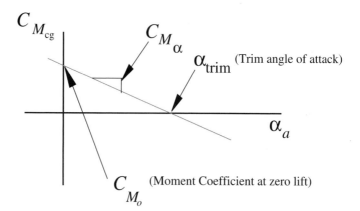

$$C_{M_0} = C_{M_{ac}} - C_{L_{\alpha_t}} (-\varepsilon - i_t) V_H = C_{M_{ac}} + C_{L_{\alpha_t}} (\varepsilon_0 + i_t) V_H$$

$$\bar{x}_{cg} = \frac{x_{cg}}{\bar{c}}, \quad \bar{x}_{ac} = \frac{x_{ac}}{\bar{c}}, \quad V_H = \frac{S_t l_t}{S \bar{c}}$$

$$C_{M_\alpha} = C_{L_\alpha}(\bar{x}_{cg} - \bar{x}_{ac}) - C_{L_{\alpha_t}}\left(1 - \frac{\partial \varepsilon}{\partial \alpha}\right) V_H$$

4) Contributions of wing and tail:

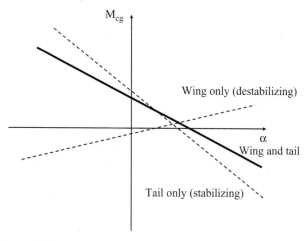

5) Changing stability:

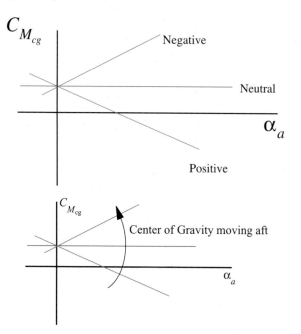

To Increase Stability (Make C_{M_a} More Negative)

Term	Change	How Accomplished
$C_{L\alpha}$	↑ or ↓	Depends on $(\bar{x}_{cg} - \bar{x}_{ac})$ 1) Make wing more or less effecient (more or less elliptical) 2) Increase/decrease aspect ratio
$\bar{x}_{cg}$	↓	Shift weight forward
$\bar{x}_{ac}$	↑	1) Pick airfoil with more aft ac 2) Sweep wings
V_H	↑	1) Increase tail area or move it aft 2) Decrease wing area or chord
$C_{L_{\alpha_t}}$	↑	Make tail lift distribution more elliptical or increase apect ratio
$\dfrac{\partial \varepsilon}{\partial \alpha}$	↓	Decrease downwash by increasing aspect ratio or by placing tail above or below the plane of the wing

6) Changing trim: By changing stability,

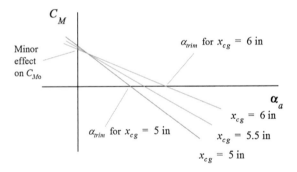

What this means physically is that first, changing stability does *not* affect C_{M_0} much, second, changing stability changes slope of the line, and third, this affects angle of attack at which $M_{cg} = 0$ (called the trim angle of attack or α_{trim}).

By changing i_t,

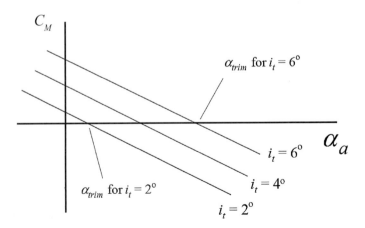

What this means physically is that 1) changing tail incidence does *not* affect stability because slope of the line does not change and 2) changing i_t changes α_{trim} by changing C_{M_o} without changing stability.

6.8.6 Lateral-Directional Static Stability

1) $C_{N_\beta} > 0$ for static directional stability:
   a) Move tail aft
   b) Make tail bigger
   c) Increase tail lift-curve slope

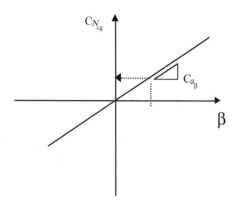

2) $C_{\mathcal{L}_\beta} < 0$ for static lateral stability:
   a) Vertical tail above c.g.
   b) Positive wing dihedral
   c) Positive wing sweep
   d) High wing placement

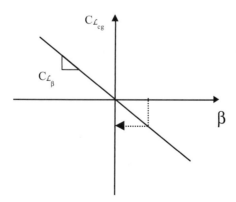

6.8.7 Aircraft Dynamic Modes of Motion

1) Longitudinal modes:
   a) *Short period*. High-frequency, heavily damped oscillation. Frequency faster than human response time, so that pilot can aggravate oscillations (PIO).

U = constant (velocity in the x body axis direction); θ and α vary (θ = pitch attitude).

b) *Phugoid.* Low-frequency, lightly damped oscillation—exchange of kinetic and potential energy. Easily controlled by pilot, though balsa gliders frequently show this mode during flight if they are not launched at their trim speed. $\alpha \approx$ constant; θ and U vary.

2) Lateral-directional modes:

a) *Dutch roll.* High-frequency, lightly damped oscillation. Frequency faster than human response time, so that pilot can aggravate oscillations (PIO). Typically objectionable when $|C_{N_\beta}/C_{L_\beta}| < 1/3$. To correct poor Dutch roll characteristics, increase directional stability, and decrease lateral stability.

b) *Spiral mode.* Nonoscillatory mode—primarily consists of a steady increase in roll angle (bank) and yaw rate if unstable. Unstable spiral will result in the aircraft or glider corkscrewing into the ground. Typically objectionable when $|C_{N_\beta}/C_{L_\beta}| > 2/3$. To correct an unstable spiral, increase roll stability, and decrease directional stability.

c) *Roll mode.* Roll-rate response to pilot's aileron command—not a factor for gliders.

6.9   More Details—Calculating Stability Parameters

In addition to giving insight into how a conventional tail-aft configuration aircraft's wing and tail contribute to its stability, and how to make changes to those components to improve stability, Eqs. (6.8) and (6.10) can be used to determine many additional stability characteristics of an aircraft. For instance, the two equations can be solved together for a given aircraft configuration's trim angle of attack. With that value determined, they can be used again to find the aircraft's trim speed.

Example 6.2

A conventional tail-aft flying model aircraft has the following characteristics: For the wing $S = 0.8\,\text{ft}^2$, $c = 0.4\,\text{ft}$, $C_{L_\alpha} = 0.078/\text{deg}$, $\varepsilon_0 = 0$, and $\partial\varepsilon/\partial\alpha = 0.2$. For the tail $S_{\text{tail}} = 0.3333\,\text{ft}^2$, $c_{\text{tail}} = 0.3333\,\text{ft}$, $C_{L_{\alpha\,\text{tail}}} = 0.068/\text{deg}$, and $i_t = 4.8\,\text{deg}$. For the airplane $x_{ac} = 0.1\,\text{ft}$, $x_{cg} = 0.2\,\text{ft}$, $l_t = 1.2\,\text{ft}$, $C_{M_{ac}wb} = -0.04$, and $W = 0.03\,\text{lb}$. What is this aircraft's trim speed (the speed at which it will fly in equilibrium) at sea level? What would happen if the aircraft were launched at 15 ft/s?

Solution: The aircraft's trim diagram will indicate its trim angle of attack and hence its trim lift coefficient. The values of C_{M_0} and C_{M_α} define the trim diagram. First, calculate V_H:

$$V_H = \frac{S_t l_t}{S\bar{c}} = \frac{0.3333\,\text{ft}^2(1.2\,\text{ft})}{0.8\,\text{ft}^2(0.4\,\text{ft})} = 1.25$$

Then

$$C_{M_\alpha} = C_{L_\alpha}(\bar{x}_{cg} - \bar{x}_{ac}) - C_{L_{\alpha_t}}\left(1 - \frac{\partial\varepsilon}{\partial\alpha}\right)V_H = C_{L_\alpha}\left(\frac{x_{cg}}{c} - \frac{x_{ac}}{c}\right) - C_{L_{\alpha_t}}\left(1 - \frac{\partial\varepsilon}{\partial\alpha}\right)V_H$$

$$= 0.078/\text{deg}\left(\frac{0.2\,\text{ft}}{0.4\,\text{ft}} - \frac{0.1\,\text{ft}}{0.4\,\text{ft}}\right) - (0.068/\text{deg})(1 - 0.2)1.25$$

$$= -0.0485/\text{deg}$$

$$C_{M_0} = C_{M_{ac}} + C_{L_{\alpha_t}}(\varepsilon_0 + i_t)V_H = -0.04 + (0.068/\text{deg})(0 + 4\,\text{deg})(1.25)$$

$$= 0.3$$

The trim diagram for this aircraft will look like Fig. 6.10, so that

$$\alpha_e = \frac{-C_{M_0}}{C_{M_\alpha}} = \frac{-0.3}{-0.0485/\text{deg}} = 6.186\,\text{deg}$$

and

$$C_L = C_{L_\alpha}\alpha_a = C_{L_\alpha}\alpha_e = (0.078/\text{deg})(6.186\,\text{deg}) = 0.482$$

The aircraft will be in equilibrium when it is at its equilibrium (trim) angle of attack and at a true airspeed such that lift equals weight, so that

$$L = W = C_L q S, \quad q = \frac{W}{C_L S} = \frac{0.03\,\text{lb}}{0.482(0.8\,\text{ft}^2)} = 0.0777\,\text{lb/ft}^2 = \tfrac{1}{2}\rho V^2$$

$$V = \sqrt{\frac{2q}{\rho}} = \sqrt{\frac{2(0.0777\,\text{lb/ft}^2)}{0.002377\,\text{slug/ft}^3}} = 8.09\,\text{ft/s}$$

If the aircraft were launched at 15 ft/s, it would still trim at α_e and the corresponding C_L so that

$$L = C_L q S = C_L \tfrac{1}{2}\rho V^2 S = 0.482\left(\frac{0.002377\,\text{slug/ft}^3}{2}\right)(15\,\text{ft/s})^2(0.8\,\text{ft}^2)$$

$$= 0.103\,\text{lb}$$

$$n = \frac{L}{W} = \frac{0.103\,\text{lb}}{0.03\,\text{lb}} = 3.44$$

Thus, if launched at $V_\infty = 15$ ft/s, the aircraft would commence a pull-up into a loop at load factor 3.44. Unless it had sufficient thrust, its speed (and the load factor) would begin to decrease as its flight-path angle increased until the aircraft either completed the loop or pitched back down at a lower speed and load factor. Eventually, after perhaps several oscillations of airspeed and flight-path angle (phugoid mode) it would stabilize at its trim airspeed, 8.09 ft/s. As long as the air density and the aircraft geometry and weight are as described, it can only be in equilibrium when flying in level flight if it is at its trim airspeed.

6.9.1 Mean Aerodynamic Chord

For untapered wings, the wing chord length is used as the reference chord length $\bar{c}$ in the expression for moment coefficient. For tapered wings, a simple average chord length is sometimes used. The most commonly used value for $\bar{c}$ is known as the mean aerodynamic chord (MAC). The MAC is a weighted average chord defined by the expression

$$\text{MAC} = \frac{1}{S} \int_{-b/2}^{b/2} c^2 \, dy \tag{6.11}$$

For untapered wings, MAC $= c$. For linearly tapered wings, Eq. (6.11) simplifies to

$$\text{MAC} = \tfrac{2}{3} c_{\text{root}} \frac{1 + \lambda + \lambda^2}{1 + \lambda} \tag{6.12}$$

6.9.2 Aerodynamic Center

The advantage of using MAC for $\bar{c}$ is that it not only is used in defining moment coefficient, but it also can be used to approximate the location of the wing's aerodynamic center. Just as the aerodynamic center of airfoils is normally located at about 0.25 c, for wings the aerodynamic center is located approximately at the quarter-chord point of the MAC for Mach numbers below M_{crit}. At supersonic speeds the aerodynamic center shifts to approximately 0.50 MAC. For swept wings the spanwise location of the MAC is important because it must be known in order to locate the wing aerodynamic center. For untapered or linearly tapered wings, the spanwise location of the MAC, y_{MAC}, is given by

$$y_{\text{MAC}} = \frac{b}{6} \frac{1 + 2\lambda}{1 + \lambda} \tag{6.13}$$

where λ is the wing taper ratio defined in Fig. 4.1. The aerodynamic center of swept wings is then, for subsonic flight, approximately located at

$$x_{\text{ac}} = y_{\text{MAC}} \tan \Lambda_{\text{LE}} + 0.25 \text{ MAC} \tag{6.14}$$

and for supersonic flight

$$x_{\text{ac}} = y_{\text{MAC}} \tan \Lambda_{\text{LE}} + 0.50 \text{ MAC} \tag{6.15}$$

where the leading edge of the wing root chord is taken as $x = 0$. Figure 6.26 illustrates this location and also demonstrates a simple graphical method for locating the MAC and aerodynamic center on linearly tapered wings.

As shown in Fig. 6.26, the graphical method for locating the MAC involves drawing the 50% chord line of the wing, then laying out lines with lengths equal to c_{root} and c_{tip} at opposite ends and alternate sides of the wing. A line is drawn connecting the endpoints of these two new lines. This third line intersects the 50% chord line of the wing at the midchord point of the MAC. The checkerboard bar, pointed at both ends, is a commonly used symbol for the MAC.

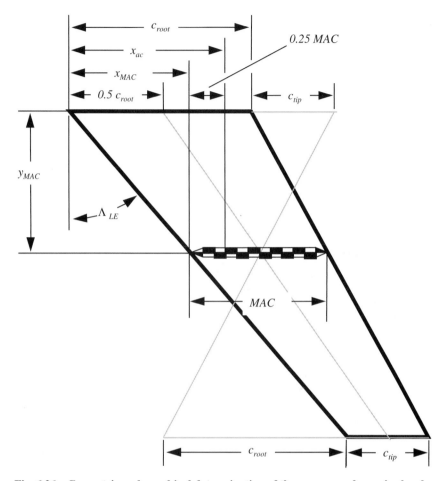

Fig. 6.26 Geometric and graphical determination of the mean aerodynamic chord.

6.9.3 Fuselage and Strake Effects

Strakes or leading-edge extensions and, to a lesser degree, fuselages tend to shift the aerodynamic center so that the location of the aerodynamic center of the wing/fuselage combination is not at the same as for the wing alone. The effect of strakes and leading-edge extensions can be estimated by treating them as additional wing panels, using Eq. (6.14) to locate the aerodynamic center of the strake by itself, then calculating a weighted average aerodynamic center location, with the areas of the strake and wing providing the weight factor:

$$x_{ac_{wing+strake}} = \frac{x_{ac_{wing}}S + (x_{ac_{strake}} - x_{ac_{wing}})S_{strake}}{S + S_{strake}} \tag{6.16}$$

The effect of the fuselage on the aerodynamic center is approximated using an expression obtained from extensive wind-tunnel testing[1]:

$$x_{ac_{wing + strake+fuselage}} = x_{ac_{wing + strake}} - \frac{l_f w_f^2 \left[0.005 + 0.111 \left(\frac{l_{ac_{wing + strake}}}{l_f} \right)^2 \right]}{S C_{L_{\alpha \, wing + strake}}} \qquad (6.17)$$

where $C_{L_{\alpha_{wing + strake}}}$ has units of 1/radians, w_f is the maximum width of the fuselage, l_f is the fuselage length, and $l_{ac_{wing + strake}}$ is the distance from the nose of the fuselage to the aerodynamic center of the wing/strake combination. From Eqs. (6.16) and (6.17) it is apparent that strakes, leading-edge extensions, and fuselages all tend to move the aerodynamic center forward. A look at Eq. (6.10) confirms that moving the aerodynamic center forward is destabilizing. As a result, increasing the size of an aircraft's fuselage and/or strakes would require a commensurate increase in the size of the horizontal tail, if the same aircraft stability is to be maintained.

6.9.4 Calculating Neutral Point Location

Recall that the neutral point is the location of the center of gravity that would cause the airplane to have neutral static longitudinal stability. From Eq. (6.10), moving the center of gravity aft (increasing $\bar{x}_{cg}$) increases the magnitude of the (destabilizing) wing term and decreases l_t and V_H, so that the aircraft becomes less stable. Neutral static stability is achieved when $C_{M_\alpha} = 0$, so that an approximate expression for the location of the neutral point can be developed by setting Eq. (6.10) equal to zero and solving for $\bar{x}_{cg}$. The expression obtained in this way is approximate if V_H is treated as constant. This is a reasonable approximation for most aircraft because l_t is usually more than 10 times greater than $x_{cg} - x_{ac}$. A change in x_{cg} has a much larger effect on the wing term of Eq. (6.10) and an almost negligible effect on the tail term. Setting Eq. (6.10) equal to zero and solving for $\bar{x}_{cg}$ yields

$$C_{M_\alpha} = 0 = C_{L_\alpha}(\bar{x}_{cg} - \bar{x}_{ac}) - C_{L_{\alpha_t}} \left(1 - \frac{\partial \varepsilon}{\partial \alpha} \right) V_H$$

$$\bar{x}_{cg(for \, C_{M_\alpha} = 0)} = \bar{x}_n = \bar{x}_{ac} + V_H \frac{C_{L_{\alpha_t}}}{C_{L_\alpha}} \left(1 - \frac{\partial \varepsilon}{\partial \alpha} \right) \qquad (6.18)$$

6.9.5 Calculating Static Margin

The definition of neutral point leads to a very convenient and commonly used alternate criteria for static longitudinal stability. It is clear from Eqs. (6.10) and (6.18) that locating the center of gravity at the neutral point gives the aircraft neutral stability, moving the center of gravity forward of the neutral point produces positive static stability, and moving the center of gravity aft of the neutral point makes the aircraft statically unstable. An alternate criterion for positive static longitudinal stability, therefore, is that the center of gravity is forward of the neutral point. This criterion is normally stated in terms of the aircraft's static margin (SM), which is defined as

$$SM = \bar{x}_n - \bar{x}_{cg} \qquad (6.19)$$

As a final comment on static margin, it is interesting to note the relationship between static margin, lift-curve slope, and moment-curve slope. An inspection

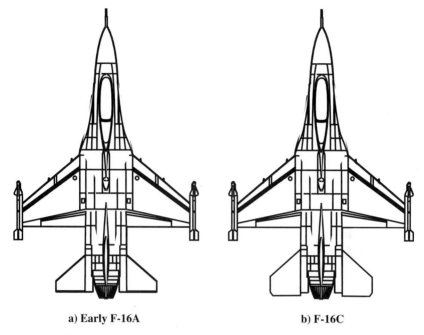

a) Early F-16A   b) F-16C

Fig. 6.27   Planform views of an early F-16A and an F-16C.

and comparison of Eqs. (6.10), (6.18), and (6.19) reveals

$$C_{M_\alpha} = -C_{L_\alpha}(\text{SM}) \tag{6.20}$$

6.10   Stability and Control Analysis Example: F-16A and F-16C

Figure 6.27 illustrates an early model F-16A and a later F-16C. The differences between the stabilators of the two aircraft are apparent. The increase in stabilator area was made to all but the earliest F-16As to increase pitch control authority. Table 6.3 lists descriptive data for each aircraft. Determine the effect of this change on the F-16's neutral point location and static margin, for both subsonic and supersonic flight.

Solution: We will run the complete analysis for the subsonic case, then repeat it for supersonic flight. The stability analysis begins by estimating the location of the aerodynamic center of the wing/strake/fuselage combination, which will be the same for both aircraft. For the F-16 wing alone,

$$\lambda = c_{\text{tip}}/c_{\text{root}} = 3.5\,\text{ft}/16.5\,\text{ft} = 0.212$$

$$\text{MAC} = \tfrac{2}{3}c_{\text{root}}\frac{1 + \lambda + \lambda^2}{1 + \lambda} = 11.4\,\text{ft}$$

$$y_{\text{MAC}} = \frac{b}{6}\frac{1 + 2\lambda}{1 + \lambda} = 5.875\,\text{ft}$$

Table 6.3   Descriptive parameters for an early F-16A and an F-16C

Item	Early F-16A	F-16C
Wing:		
S, ft^2	300	300
c_{root}, ft	16.5	16.5
c_{tip}, ft	3.5	3.5
b, ft	30 (no missiles or rails)	30 (no missiles or rails)
x of root chord leading	0	0
edge, ft	(20 ft aft of fuselage nose)	(20 ft aft of fuselage nose)
Λ_{LE}, degrees	40	40
Stabilator:		
S_t, ft^2	108	135
c_{root}, ft	10	11
c_{tip}, ft	2	3
b, ft	18	18
x of root chord leading		
edge, ft	17.5	17
Λ_{LE}, degrees	40	40
Strake (exposed):		
S_{strake}, ft^2	20	20
c_{root}, ft	9.6	9.6
c_{tip}, ft	0	0
b, ft	2	2
x of root chord leading		
edge, ft	−8	−8
Λ_{LE}, degrees (avg.)	80	80
Fuselage		
l_f	48.5	48.5
w_f	5	5
Whole airplane		
$\bar{x}$(relative to M.A.C.)	.35	.35

For subsonic flight,

$$x_{ac} = y_{MAC} \tan \Lambda_{LE} + 0.25 \text{ MAC}$$

$$= (5.875 \text{ ft}) \tan 40 \text{ deg} + 0.25(11.4 \text{ ft}) = 7.8 \text{ ft}$$

Adding the effect of the strake,

$$\lambda_{strake} = c_{tip}/c_{root} = 0 \text{ ft}/9.6 \text{ ft} = 0$$

$$\text{MAC} = \tfrac{2}{3} c_{root} \frac{1 + \lambda + \lambda^2}{1 + \lambda} = 6.4 \text{ ft}$$

$$y_{MAC} = \frac{b}{6}\frac{1+2\lambda}{1+\lambda} = 0.33 \text{ ft}$$

$$x_{ac_{strake}} = y_{MAC_{strake}} \tan \Lambda_{LE_{strake}} + 0.25 \text{ MAC}_{strake}$$

$$= (0.33 \text{ ft}) \tan 80 \text{ deg} + 0.25(6.4 \text{ ft}) = 3.5 \text{ ft}$$

but these are defined relative to the leading edge of the strake root chord, not the wing root chord. From Table 6.3, the strake root is 8 ft forward of the wing root, so that relative to the wing

$$x_{ac_{strake}} = -4.5 \text{ ft (subsonic)}$$

and

$$x_{ac_{wing+strake}} = \frac{x_{ac_{wing}} S + \left(x_{ac_{strake}} - x_{ac_{wing}}\right) S_{strake}}{S + S_{strake}} = 6.5 \text{ ft}$$

Now, adding the effect of the fuselage, using $C_{L_{\alpha wing/strake}} = 0.068/\text{deg}$ (predicted in Sec. 4.7) and the fact that the wing-root leading edge is 20 ft aft of the fuselage nose, so that $l_{ac_{wing/strake}} = 20 \text{ ft} + x_{ac_{wing/strake}}$

$$x_{ac_{wing+strake+fuselage}} = x_{ac_{wing+strake}} - \frac{l_f w_f^2 \left[0.005 + 0.111 \left(\frac{l_{ac_{wing+strake}}}{l_f}\right)^2\right]}{S C_{L_{\alpha wing+strake}}}$$

$$= 6.5 \text{ ft} - \frac{48.5 \text{ ft} (5 \text{ ft})^2 \left[0.005 + 0.111 \left(\frac{26.5 \text{ ft}}{48.5 \text{ ft}}\right)^2\right]}{300 \text{ ft}^2 (0.068/\text{deg})(57.3 \text{ deg/rad})}$$

$$= 6.4 \text{ ft}$$

Next, the aerodynamic center of the F-16A stabilator is located for subsonic flight.

$$\lambda_{stabilator} = c_{tip}/c_{root} = 2 \text{ ft}/10 \text{ ft} = 0.2$$

$$\text{MAC} = \frac{2}{3} c_{root} \frac{1+\lambda+\lambda^2}{1+\lambda} = 6.9 \text{ ft}$$

$$y_{MAC} = \frac{b}{6}\frac{1+2\lambda}{1+\lambda} = 3.5 \text{ ft}$$

$$x_{ac_{stab}} = y_{MAC_{stab}} \tan \Lambda_{LE_{stab}} + 0.25 \text{ MAC}_{stab}$$

$$= (3.5 \text{ ft}) \tan 40 \text{ deg} + 0.25(6.9 \text{ ft}) = 4.7 \text{ ft}$$

This is defined relative to the leading edge of the stabilator root chord. From Table 6.3, the stabilator root is 17.5 ft aft of the wing root, so that relative to the wing

$$x_{ac_{stab}} = 22.2 \text{ ft}$$

But the distance of interest for the stabilator is l_t, the distance from the stabilator's aerodynamic center to the aircraft center of gravity. Table 6.3 lists the center of gravity as 0.35 MAC, so that relative to the wing root

$$x_{cg} = y_{MAC} \tan \Lambda_{LE} + 0.35 \text{ MAC}$$

$$= (5.875 \text{ ft}) \tan 40 \text{ deg} + 0.35(11.4 \text{ ft}) = 8.9 \text{ ft}$$

and

$$l_t = x_{ac_{stab}} - x_{cg} = 22.2 \text{ ft} - 8.9 \text{ ft} = 13.3 \text{ ft}$$

It is now possible to calculate tail volume ratio:

$$V_H = \frac{S_t l_t}{S\bar{c}} = \frac{108 \text{ ft}^2 (13.3 \text{ ft})}{300 \text{ ft}^2 (11.4 \text{ ft})} = 0.42$$

Because the F-16's $\bar{x}_{cg}$ is specified relative to the leading edge of the MAC, it is convenient (and common) to express $\bar{x}_{ac_{wing+strake+fuselage}}$ and $\bar{x}_n$ relative to the same reference. This requires subtracting the distance between the root leading edge and the MAC leading edge from the value of x_{ac}. The expression for $\bar{x}_{ac_{wing+strake+fuselage}}$ then becomes

$$\bar{x}_{ac_{wing+strake+fuselage}} = \frac{x_{ac_{wing+strake+fuselage}}}{\bar{c}} = \frac{6.4 \text{ ft} - (5.875 \text{ ft}) \tan 40 \text{ deg}}{11.4 \text{ ft}} = 0.13$$

Recall from Sec. 4.7 that $\partial\varepsilon/\partial\alpha = 0.64$.

$$\bar{x}_n = \bar{x}_{ac_{wing+strake+fuselage}} = V_H \frac{C_{L_{\alpha t}}}{C_{L_\alpha}} \left(1 - \frac{\partial\varepsilon}{\partial\alpha}\right)$$

$$= 0.13 + 0.42 \left(\frac{0.0536}{0.0572}\right)(1 - 0.64) = 0.27$$

So that the F-16A's static margin is

$$SM = \bar{x}_n - \bar{x}$$

$$= 0.27 - 0.35 = -0.08$$

Similar calculations for the F-16C yield

$$SM = 0.36 - 0.35 = +0.01$$

Now we repeat the calculations for supersonic flight, when the aerodynamic centers of the wing and stabilator shift from the quarter-chord of their MAC to the half-chord.

$$x_{ac} = y_{MAC} \tan \Lambda_{LE} + 0.50 \text{ MAC}$$

$$= (5.875 \text{ ft}) \tan 40 \text{ deg} + 0.50 (11.4 \text{ ft}) = 10.6 \text{ ft}$$

Adding the effect of the strake,

$$x_{ac_{strake}} = y_{MAC_{strake}} \tan \Lambda_{LE_{strake}} + 0.50 \text{ MAC}_{strake}$$

$$= (0.33 \text{ ft}) \tan 80 \text{ deg} + 0.50 (6.4 \text{ ft}) = 5.1 \text{ ft}$$

But these are defined relative to the leading edge of the strake-root chord, not the wing-root chord. From Table 6.3, the strake root is 8 ft forward of the wing root, so that relative to the wing

$$x_{ac_{strake}} = -2.9\,\text{ft}$$

and

$$x_{ac_{wing+strake}} = \frac{x_{ac_{wing}} S + \left(x_{ac_{strake}} - x_{ac_{wing}}\right) S_{strake}}{S + S_{strake}} = 9.1\,\text{ft}$$

To perform the supersonic calculation for fuselage effect, supersonic lift-curve slope must be predicted. A specific Mach number must be chosen. For $M = 1.5$,

$$C_{L_\alpha} = \frac{4}{\sqrt{M_\infty^2 - 1}} = 0.051/\text{deg}$$

$$x_{ac_{wing+strake+fuselage}} = x_{ac_{wing+strake}} - \frac{l_f w_f^2 \left[0.005 + 0.111 \left(\dfrac{l_{ac_{wing+strake}}}{l_f}\right)^2\right]}{S C_{L_{\alpha\,wing+strake}}}$$

$$= 9.1\,\text{ft} - \frac{48.5\,\text{ft}\,(5\,\text{ft})^2 \left[0.005 + 0.111 \left(\frac{29.1\,\text{ft}}{48.5\,\text{ft}}\right)^2\right]}{300\,\text{ft}^2\,(0.051/\text{deg})(57.3\,\text{deg/rad})}$$

$$= 9.0\,\text{ft}$$

Next, the aerodynamic center of the F-16A stabilator is located:

$$\lambda_{stabilator} = c_{tip}/c_{root} = 2\,\text{ft}/10\,\text{ft} = 0.2$$

$$\text{MAC} = \frac{2}{3} c_{root} \frac{1 + \lambda + \lambda^2}{1 + \lambda} = 6.9\,\text{ft}$$

$$y_{MAC} = \frac{b}{6} \frac{1 + 2\lambda}{1 + \lambda} = 3.5\,\text{ft}$$

$$x_{ac_{stab}} = y_{MAC_{stab}} \tan \Lambda_{LE_{stab}} + 0.4\,\text{MAC}_{stab}$$

$$= (3.5\,\text{ft}) \tan 40\,\text{deg} + 0.50\,(6.9\,\text{ft}) = 6.4\,\text{ft}$$

This is defined relative to the leading edge of the stabilator root chord. From Table 6.3, the stabilator root is 17.5 ft aft of the wing root, so that relative to the wing

$$x_{ac_{stab}} = 23.9\,\text{ft}$$

But the distance of interest for the stabilator is l_t, the distance from the stabilator's aerodynamic center to the aircraft center of gravity. Table 6.3 lists the center of gravity as 0.35 MAC, so that relative to the wing root

$$x_{cg} = y_{MAC} \tan \Lambda_{LE} + 0.35 \, MAC$$
$$= (5.875 \, \text{ft}) \tan 40 \, \text{deg} + 0.35 \, (11.4 \, \text{ft}) = 8.9 \, \text{ft}$$

and

$$l_t = x_{ac_{stab}} - x_{cg} = 23.9 \, \text{ft} - 8.9 \, \text{ft} = 15 \, \text{ft}$$

It is now possible to calculate tail volume ratio

$$V_H = \frac{S_t l_t}{S \bar{c}} = \frac{108 \, \text{ft}^2 \, (15 \, \text{ft})}{300 \, \text{ft}^2 \, (11.4 \, \text{ft})} = 0.47$$

Recall from Sec. 4.7 that $\partial \varepsilon / \partial \alpha = 0.64$. Because the F-16's $\bar{x}_{cg}$ is specified relative to the leading edge of the MAC, it is convenient (and common) to express $\bar{x}_{ac_{wing+strake+fuselage}}$ and $\bar{x}_n$ relative to the same reference. This requires subtracting the distance between the root leading edge and the MAC leading edge from the value of x_{ac}. The expression for supersonic $\bar{x}_{ac_{wing+strake+fuselage}}$ then becomes

$$\bar{x}_{ac_{wing+strake+fuselage}} = \frac{x_{ac_{wing+strake+fuselage}}}{\bar{c}} = \frac{9 \, \text{ft} - (5.875 \, \text{ft}) \tan 40 \, \text{deg}}{11.4 \, \text{ft}} = 0.36$$

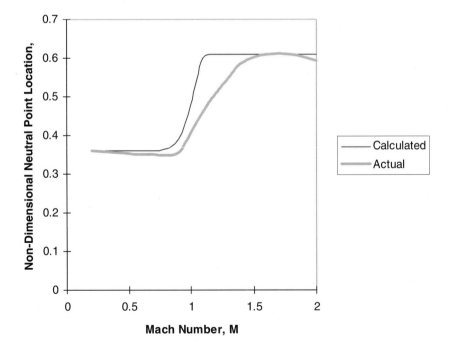

Fig. 6.28   Calculated and actual variation of F-16C neutral point with Mach number.

And for supersonic neutral point

$$\bar{x}_n = \bar{x}_{ac_{wing+strake+fuselage}} = V_H \frac{C_{L_{\alpha t}}}{C_{L_\alpha}} \left(1 - \frac{\partial \varepsilon}{\partial \alpha}\right)$$

$$= 0.36 + 0.47 \left(\frac{0.051}{0.0544}\right)(1 - 0.64) = 0.52$$

So that the F-16A's supersonic static margin is

$$\text{SM} = \bar{x}_n - \bar{x}$$

$$= 0.52 - 0.35 = +0.17$$

Similar calculations for the F-16C yield

$$\text{SM} = 0.61 - 0.35 = +0.26$$

Figure 6.28 plots neutral point locations calculated for the F-16C vs Mach number and compares them with actual values. Note that, despite the F-16's relatively complex aerodynamics, the method produced reasonably good estimates.

Reference

[1]Raymer, D. P., *Aircraft Design: A Conceptual Approach*, AIAA Education Series, AIAA, Washington, DC, 1989, Chap. 4.

Problems

Synthesis Problems

S-6.1   The YF-22 and X-31 have demonstrated the ability to maneuver at angles of attack above 60 deg. At these extreme angles, well beyond stall, conventional control surfaces sometimes lose their control authority, or even work in reverse. Brainstorm five concepts for control mechanisms that might be used to control an aircraft in pitch, roll, and yaw at very high angles of attack, up to 90 deg.

S-6.2   Flying-wing airplanes (including delta-wing jet fighters) have no canard or horizontal tail to serve as a trimming surface. They are trimmed entirely by changing the pitching-moment coefficient of the wing. This limits their ability to use highly cambered, high-lift airfoils because one of the inevitable consequences of high camber is a strong nose-down pitching moment. Brainstorm at least five ways to allow a flying wing to use a highly cambered airfoil, at least on the inner 40% of its span, but still be trimmable.

S-6.3   The area of the F-16's stabilator was increased in order to increase its pitch control authority. One of the consequences of this change was an increase in the aircraft's static margin. Brainstorm at least five ways to increase an aircraft's pitch control authority without increasing its stability.

Analysis Problems

A-6.1 Fill in the following table:

Motion	Control surface	Axis
Roll		
Pitch		
Yaw		

A-6.2 How many degrees of freedom does an aircraft have?

A-6.3 Define static and dynamic stability.

A-6.4 Explain why a weathervane is stable (points into the wind).

A-6.5 Explain the tradeoff between stability and maneuverability.

A-6.6 A conventional aircraft (tail to the rear), is in trimmed, level, unaccelerated flight. The wing is generating 40,000 lb of lift and has a moment around the aerodynamic center of −20,000 ft-lb. The aerodynamic center of the wing is located at $0.25c$, the center of gravity is located at $0.45c$, the aircraft has a chord of 5 ft, and the symmetric tail aerodynamic center is located 10 ft behind the center of gravity. What is the lift generated by the tail, and what is the weight of the aircraft? (Hint: Draw a sketch and assume thrust and all drag forces act through the center of gravity.)

A-6.7 An aircraft with a canard is in trimmed, level, unaccelerated flight. The wing is generating 40,000 lb of lift and has a moment around the aerodynamic center of −20,000 ft-lb. The aircraft has a chord of 5 ft, the aerodynamic center is located at $0.25c$, the center of gravity is located at $0.10c$, and the canard aerodynamic center is located 5 ft ahead of the center of gravity. What is the lift generated by the canard, and what is the weight of the aircraft?

A-6.8 A Su-35 has a wing, horizontal tail, and canard as shown in Fig. PA-6.8. The horizontal tail is generating 11,000 lb of lift and has an aerodynamic center located 16.2 ft behind the aircraft center of gravity. The canard is generating 4000 lb of lift and has an aerodynamic center located 14.9 ft in front of the aircraft c.g. Both the horizontal tail and canard have symmetric airfoil sections. The wing's aerodynamic center is located 3.0 ft in front of the aircraft c.g. The cambered wing is generating a moment around the aerodynamic center M_{ac} of −16,400 ft-lb.
(a) If the aircraft is in trimmed, level flight, how much lift is the wing generating?
(b) What is the weight of the aircraft?

A-6.9 (a) How would increasing the tail volume ratio change the longitudinal static stability of a conventional aircraft?
(b) How would moving the center of gravity forward change the stability of a conventional aircraft?

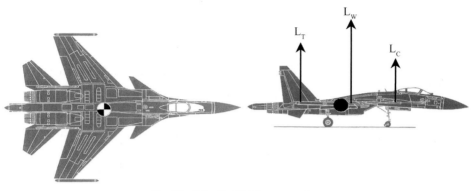

Fig. PA-6.8   Su-35 trim geometry.

(c) When an aircraft goes supersonic, its aerodynamic center shifts from $0.25c$ to $0.5c$. How would this change the stability of a conventional aircraft?

A-6.10   An aircraft has the following data: The center of gravity is located $0.45c$ behind the leading edge of the wing, the aerodynamic center of the wing body is at $0.25c$, the tail volume ratio is 0.4, the wing lift-curve slope is 0.08/deg, the tail lift-curve slope is 0.07/deg, $\partial\varepsilon/\partial\alpha = 0.3$, the tail setting angle is 3 deg, $C_{MAC} = -0.05$, and the downwash angle at zero lift is zero. The weight is 2500 lb, the wing area is 200 ft^2, and the aircraft is flying at sea-level conditions.

    (a) Calculate the neutral point.
    (b) Calculate the static margin.
    (c) Is this aircraft stable?
    (d) Calculate C_{M_α}, C_{M_0}, and α_e, and plot the aircraft's trim diagram
    (e) What is this aircraft's trimmed lift coefficient?
    (f) What is this aircraft's trim speed?

7
Structures

"Ut tensio sic vis" (As the stretch so the force).

Robert Hooke

7.1 Design Motivation

Fundamentally, an aircraft is a structure. Aircraft designers design structures. The structures are shaped to give them desired aerodynamic characteristics, and the materials and structures of their engines are chosen and shaped so they can provide needed thrust. Even seats, control sticks, and windows are structures, all of which must be designed for optimum performance.

Designing aircraft structures is particularly challenging because their weight must be kept to a minimum. There is always a tradeoff between structural strength and weight. A good aircraft structure is one that provides all of the strength and rigidity to allow the aircraft to meet all of its design requirements, but which weighs no more than necessary. Any excess structural weight often makes the aircraft cost more to build and almost always makes it cost more to operate. As with small excesses of aircraft drag, a small percentage of total aircraft weight used for structure instead of payload can make the difference between a profitable airliner or successful tactical fighter and a failure.

Designing aircraft structures involves determining the loads on the structure, planning the general shape and layout, choosing materials, and then shaping, sizing, and optimizing its many components to give every part just enough strength without excess weight. Because aircraft structures have relatively low densities, much of their interiors are typically empty space, which in the complete aircraft is filled with equipment, payload, and fuel. Careful layout of the aircraft structure ensures structural components are placed within the interior of the structure so that they carry the required loads efficiently and do not interfere with placement of other components and payload within the space. Choice of materials for the structure can profoundly influence weight, cost, and manufacturing difficulty. The extreme complexity of modern aircraft structures makes optimal sizing of individual components particularly challenging. An understanding of basic structural concepts and techniques for designing efficient structures is essential to every aircraft designer.

7.2 Solid Mechanics

The most fundamental concept that must be understood in order to design and analyze structures is the physics that governs how a solid object resists or supports a load applied to it. The study of this phenomenon is called solid mechanics or mechanics of materials. Solids are composed of molecules held together in a matrix

315

by strong intermolecular forces. When an external force is applied to a solid, the molecules in contact with the force are moved by it, causing them to move relative to other molecules in the matrix. The shift of the relative positions of the molecules in the matrix causes the magnitude of the intermolecular forces to change in a way that tends to return the molecules to their original relative positions. In this way, the applied force is propagated through the solid object as changes in intermolecular forces. If the object is free to move, the applied force will cause it to accelerate. On the molecular level, the changes in intermolecular forces cause each molecule to accelerate with the object. If the object is restrained, the restraint applies forces to the object that counters the applied force, and these are communicated to each molecule in the matrix by changes in intermolecular forces.

7.2.1 Stress and Strain

Figure 7.1 shows a restrained solid object to which an external force has been applied. The object is cut by an imaginary line so that the intermolecular forces in it can be examined. If no load is applied to the body, the intermolecular forces within it are in balance, so that there is no net force on any molecule. This is also true if the force is applied to a restrained body. However, it should be apparent that the forces between molecules on opposite sides of the imaginary cut line will not be the same when a force is applied as when there is no external force. Summing the forces on the portion of the body left of the line makes this clear:

$$\sum f = F_1 - \sum_{i=1}^{n} \Delta f_i \qquad (7.1)$$

where F_1 is the applied external force and Δf_i are the changes in the intermolecular forces between molecules on opposite sides of the line relative to the magnitudes of the forces when no external force is applied.

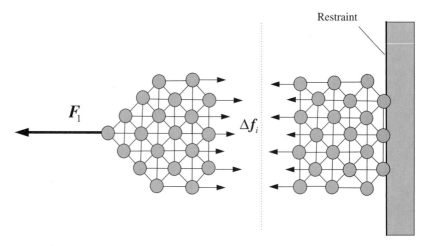

Fig. 7.1   Changes in intermolecular forces as a result of applied load.

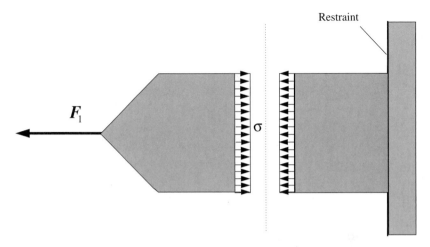

Fig. 7.2   Internal stresses in a solid object as a result of an external force.

The forces between the billions of molecules in a solid object are commonly represented as a *stress*. A stress is a measure of the total Δf_i per unit area within an object. Figure 7.2 illustrates the same situation as Fig. 7.1, but with the intermolecular forces represented as a stress. Stresses such as this are usually given the symbol σ.

Summing forces on the portion of the object left of the imaginary cut line yields

$$\sum F_x = -F_1 + \sigma\,A, \quad \text{or} \quad \sigma = F_1/A \qquad (7.1a)$$

where A is the cross-sectional area of the object where it is cut by the imaginary line.

The Δf_i are generated by shifts in the relative positions of molecules. This change in molecule positions when a force is applied causes the solid object to change shape or deform. The amount of change in an object's dimension per unit length in that direction is called *strain* and given the symbol ε. Strain is defined as

$$\varepsilon \equiv \frac{\Delta l}{l} \qquad (7.2)$$

where l is the overall length of the object and Δl is the change in its length when the force is applied.

If the force applied to the body is not too strong, then when it is removed the molecules will return to their original positions relative to each other. This process is called *elastic deformation*. Anyone who has flown on a modern jet airliner has probably noticed the elastic deformation or flexing of its wings during flight.

7.2.2 Stress–Strain Relationship: Hooke's Law

Each material has a characteristic relationship between the stress applied to it and the amount of strain it exhibits. For the situation shown in Fig. 7.2, this

Table 7.1   Values of Young's modulus
for some common aircraft materials[2,3]

Material[a]	E, psi
4340 Steel	29×10^6
Stainless steel	26×10^6
2024-T4 Aluminum	10.7×10^6
7075-T6 Aluminum	10.3×10^6
Titanium	16×10^6
Graphite/epoxy	22×10^6
Kevlar/epoxy	12×10^6
Fiberglass/epoxy	5×10^6
Aircraft spruce	1.3×10^6
Balsa wood	1×10^6

[a]The numbers associated with some materials
designate particular alloys and heat treatment.
Values for composite materials are based on
unidirectional layup with 60% of fiber contents.

relationship is given as

$$\varepsilon = \sigma/E \qquad (7.3)$$

where E is the *modulus of elasticity* characteristic of each material. The modulus of elasticity is also referred to as *Young's modulus* after the English engineer Thomas Young who suggested the concept in 1807. Equation (7.3) is often referred to as *Hooke's law* after another Englishman, Robert Hooke, who observed in 1678, "Ut tensio sic vis" (as the stretch so the force).[1] E is a measure of the stiffness of a material. Materials with very high values of E change their dimensions relatively little when a force is applied. Table 7.1 lists values for E for some common aircraft materials.

Some of the materials in Table 7.1 (graphite/epoxy, Kevlar®/epoxy, and fiberglass/epoxy) are known as composite materials or just composites. Composites are composed of very strong fibers embedded in a softer material. The fibers give the composite very high stiffness, whereas the softer material gives it toughness and a rigid shape that the fibers by themselves would not have. Actually, wood is a naturally occurring composite material, with strong cellulose fibers held together by softer material. Modern man-made composites imitate many of wood's good characteristics, but have greater strength and stiffness.

Composite fibers can be woven into cloth or mat to give the material good strength in all directions and good resistance to shear, or they can be all placed parallel to each other (unidirectional, like wood) to produce the greatest strength in one direction. The fibers can also be placed in the composite in layers, with the fibers in each layer oriented differently. In this way, the strength and bending characteristics of the composite can be tailored to the needs of a particular design application. Composite materials have great potential for significantly reducing the weight and cost of aircraft structures, if new design and fabrication methods can be developed that allow composites to perform to their full potential.

7.2.3 Plastic Deformation

If the force applied to the solid is strong enough, it can cause the molecules to move so far from their original relative positions that some intermolecular forces with other molecules become weaker and others that originally were weak become stronger. As a result, when the external force is removed, the molecules might not return to their original relative positions, but might remain in some new configuration in which they are in equilibrium. This process, called *plastic deformation*, causes the shape of the solid to change permanently. Plastic deformation of an aircraft structure can seriously affect its ability to function properly. The maximum structural limits on aircraft are always set to avoid plastic deformation of the structure. The stress beyond which a material will undergo plastic deformation is called its *yield strength* σ_y, and the load limit for a structure beyond which it will be permanently deformed is called its *yield limit*.

7.2.4 Failure

Very strong forces applied to a solid might cause some molecules to move so far from their neighbors that the intermolecular forces between them disappear and the object develops cracks or even breaks apart. This situation is called *structural failure*. Failure of aircraft structures frequently results in complete destruction of the aircraft. Maximum structural limits on aircraft are set with a factor of safety relative to the loads that would cause structural failure. This factor of safety is usually 1.5 for aircraft, so that if a load factor (see Sec. 5.12 and 5.13) of 12 would produce structural failure the maximum allowable load factor for the aircraft would be 8. Loads beyond which structural failure will occur are called *ultimate loads*, and the maximum stress that a material can endure without failure is its *ultimate strength* σ_u. Table 7.2 lists yield strengths and ultimate strengths for some common

Table 7.2   Values of yield strength and ultimate strength for some common aircraft materials[2–4]

Material[a]	Yield strength, psi	Ultimate strength, psi
4340 Steel	163,000	180,000
Stainless steel	165,000	190,000
2024-T4 Aluminum	42,000	57,000
7075-T6 Aluminum	70,000	78,000
Titanium	143,000	157,000
Graphite/epoxy	170,000	170,000
Kevlar®/epoxy	160,000	160,000
Fiberglass/epoxy	60,000	60,000
Aircraft spruce	9,400	9,400
Balsa wood	3,500	3,500

[a]The numbers associated with some materials designate particular alloys and heat treatment. Values for composite materials based on unidirectional layup with 60% of fiber contents.

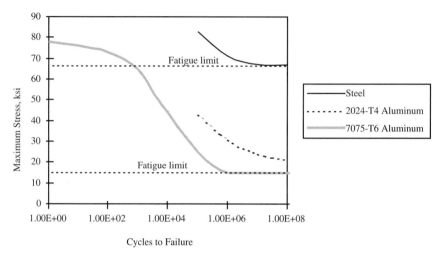

Fig. 7.3   Fatigue life as function of maximum cyclic load for typical aircraft metals.

aircraft materials. Note that the very strong fibers in composite materials prevent them from yielding significantly before the fibers break and the materials fail.

7.2.5 Fatigue

Many materials, especially metals, will develop cracks and eventually fail after many cycles of having loads applied and removed without ever being stressed beyond their ultimate strength. This process of developing cracks due to cyclic loading is called *fatigue*. Figure 7.3 shows a typical relationship between maximum loads and the number of cycles a material can endure before developing fatigue cracks.

Two of the metals in Fig. 7.3 have *fatigue limits*, stress levels below which the metals will not develop cracks no matter how many cycles of loading they undergo. For some metals such as steel, this fatigue limit is quite high, and useful aircraft structures can be designed so that their fatigue limit is never exceeded. For other materials such as many aluminum alloys, however, the fatigue limit is so low that structures designed to never exceed the fatigue limit are not practical. As a result, these structures must be designed to a specific service life, usually designated in terms of a maximum number of flight hours. For example, the original design service life of the Cessna T-37 jet trainer was 5000 h, which many aircraft reached after only 10 years. Structural strengthening has allowed many of these aircraft to serve three times as long. Aircraft that have exceeded their design service life are sometimes still flown without modification, but care must be taken to periodically inspect their structure for developing cracks. Some aircraft are equipped with devices that record the loads applied to the aircraft in order to more accurately predict and monitor structural fatigue.

Composite materials are not free from fatigue problems. One of the most serious fatigue failures in composites is called *delamination*. Most composites are made up of many layers. The strong fibers frequently run only within layers, so that the layers are held to each other only by softer material. Delamination occurs when minor damage or a manufacturing defect causes a crack to develop between layers and then grow during many cycles of loading. Developing methods of reliably detecting and repairing delamination and other damage in composite structures is one of the keys to unlocking the full potential composites have for saving weight and cost.

7.3   Types of Stress

The force applied in Fig. 7.2 tended to stretch or elongate the object. This type of load on a structure is called a *tensile load*. Structures can be stressed in other ways, however. Figure 7.4a shows a *compressive load* (one that tends to compress the structure), and Fig. 7.4b shows a load that produces *shear stress* in a structure. A shear stress tends to move different parts of a structure in opposite directions. Consistent with the symbol used for aerodynamic shear stresses caused by friction, structural shear stresses are given the symbol τ.

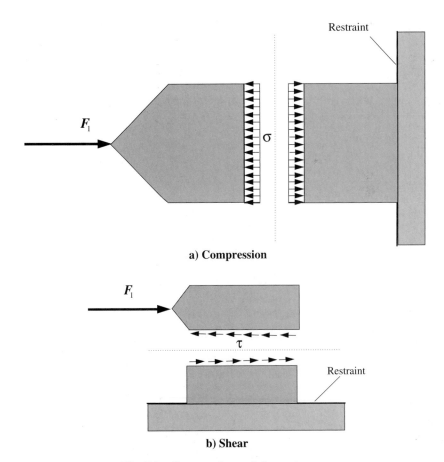

a) Compression

b) Shear

Fig. 7.4   Compression and shear stresses.

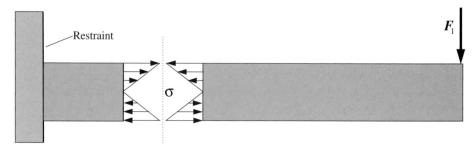

Fig. 7.5   Beam with a bending load.

Hooke's law for a structure loaded in compression is the same as for tensile loads. For shear, however, a different form of Eq. (7.3) is used:

$$\gamma = \tau/G \qquad (7.4)$$

where γ is the shear strain and G is Young's modulus for shear, also called the *modulus of rigidity*. The magnitude of the modulus of rigidity for most materials is less than half the magnitude of their modulus of elasticity.

Figure 7.5 shows an object under a bending load. This is really the same situation as in Fig. 7.4b, rotated 90 deg to the right, but the object is a long, slender beam, such as might be found in an aircraft wing. Because of the beam's shape, the force applied at its end creates a very strong moment. The stresses that result from this bending moment are much greater than the shear stresses caused by the force in the object. Figure 7.5 shows that the stresses in the beam caused by the moment are tensile at the top of the beam decreasing to zero midway in the beam cross section and reaching a maximum compressive stress at the bottom of the beam. The midway point in the beam cross section where compressive and tensile stresses are zero is called the beam's *neutral axis*. If the beam's cross-sectional shape is symmetrical about a plane running down the length of the beam, then its neutral axis lies on this plane of symmetry.

The magnitude of the compressive or tensile stress for any vertical y position in the beam cross section is given by

$$\sigma = \frac{M\,y}{I} \qquad (7.5)$$

where M is the moment in the beam as a result of to the load and I is the *area moment of inertia* of the beam's cross-sectional shape defined by

$$I \equiv \int^{A} y^2 \, \mathrm{d}A \qquad (7.6)$$

Note that Eq. (7.6) applies only to beam cross sections that are symmetrical about the y axis. For the more general case of asymmetrical cross sections, see Ref. 2. Clearly, from Eq. (7.5) and Fig. 7.5, the greatest tensile and compressive stresses in the beam are at the top and bottom of the beam cross section, farthest from the neutral axis. Shear stresses as a result of the load are also present and should not be ignored.

7.4 Loads

An aircraft structure must be designed to withstand a large number of different types of loads as shown in Fig. 7.5. Some of these loads, such as catapult, towing, arresting, external stores, and landing gear loads, are applied to the structure at a few discreet locations. These are referred to as point loads or concentrated loads. Others, primarily the aerodynamic loads, are the result of pressures and shear stresses distributed over the aircraft surface, and hence called distributed loads. These loads are not distributed uniformly, and the locations on aircraft surfaces where maximum pressure loads occur change as flight conditions change. Many of these loads are unsteady, conducive to structural fatigue. Before detail design of an aircraft structure can occur, the maximum magnitudes and frequencies of application of these many loads that the aircraft must sustain in order to meet the design requirements must be determined.

7.4.1 Point Loads

Maximum magnitudes of most concentrated loads can be determined fairly easily. Maximum catapult and arresting gear loads are determined using Eqs. (5.56) and (5.57). The takeoff distance equation is modified to include the catapult force F_{cat} with the takeoff thrust:

$$s_{TO} = \frac{1.44\, W_{TO}^2}{\rho S C_{L_{max}}\, g\, (T_{SL} + F_{cat})} \qquad (7.7)$$

$$F_{cat} = \frac{1.44\, W_{TO}^2}{\rho S C_{L_{max}}\, g\, s_{TO}} - T_{SL} \qquad (7.8)$$

Similarly, the landing distance equation is modified by replacing the relatively small aerodynamic and rolling-friction forces with the arresting gear force F_{arr}:

$$s_L = \frac{1.69\, W_L^2}{\rho S C_{L_{max}}\, g\, F_{arr}} \qquad (7.9)$$

$$F_{arr} = \frac{1.69\, W_L^2}{\rho S C_{L_{max}}\, g\, s_L} \qquad (7.10)$$

Towing loads are naturally significantly less than catapult loads. Assuming a maximum towing load equal to 50% of the aircraft's maximum takeoff weight would allow for the possibility of towing on steep inclines and uneven surfaces. Rolling friction and braking loads are given by

$$F_{brake} = \mu\, N \qquad (7.11)$$

where $\mu < 0.1$ for a free-rolling wheel and $\mu < 0.6$ for a braking wheel and N is the portion of the aircraft weight carried by each landing gear. Aircraft are normally designed to place 10–12% of the aircraft weight on the nose gear and 88–90% distributed evenly between or among the main gear. However, the nose-down pitching moment created by braking will increase the nose-gear load, depending on the aircraft geometry. Likewise, any small turns made during braking can place

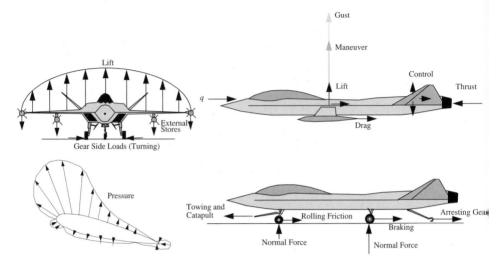

Fig. 7.6   Some of the many loads on an aircraft structure.

more than 50% of the aircraft weight on one main gear. As a general rule for maximum rolling friction and braking load calculations, allow for the possibility of 50% of the total weight being on the nose gear and 100% on each main gear. Maximum *side loads* (depicted in Fig. 7.6) during turns are of approximately the same magnitude as braking loads. Landing loads will be significantly higher than this.

Worst-case landing loads require that the landing gear be able to sustain the forces necessary to absorb all of a maximum specified sink rate in the length of the landing-gear *stroke* as depicted in Fig. 7.7. This can produce forces on each landing gear that are four or more times the total aircraft landing weight, especially for carrier-based aircraft. Landing gear for land-based aircraft are generally designed for a maximum sink rate on landing of 12 ft/s, whereas carrier-based aircraft are designed for 24 ft/s sink rates. During a bad landing with significant bank and/or sideslip, a large portion of this load can be side load. Designers must make a

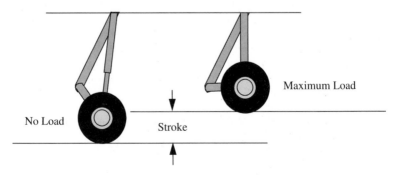

Fig. 7.7   Landing-gear stroke.

tradeoff between the extra weight associated with additional structural strength and heavier landing gear with longer stroke.

Point loads on the structure from externally or internally mounted stores, engines, equipment, passengers, and payload are simply the weight of the item and any pylons, seats, mounting brackets, etc. multiplied by the maximum load factor that the aircraft will sustain when these items are carried. The load as a result of drag of any external stores and the thrust of engines must also be considered, especially if they are hung on long pylons that cause the forces to produce strong twisting moments on the structure.

Many aerodynamic loads also apply point loads to portions of the structure. This occurs typically because lifting and control surfaces, though they sustain distributed aerodynamic loads, are designed to attach to the rest of the aircraft structure at only a few points. These surfaces essentially collect the distributed load and concentrate it into a few points. The maximum aerodynamic load on wing-to-fuselage attachment points results from the maximum lift the aircraft is designed to generate. For an aircraft designed to sustain a load factor of nine, the design maximum lift can be more than nine times the weight (depending on the magnitude and direction of trim lift generated by control surfaces). The drag and pitching moment generated by the wing in this condition also add to the load on the fuselage.

Control surfaces place similar lift, drag, and pitching-moment loads on an aircraft. In addition, control sticks, linkages, cables, and actuators place loads on the aircraft structure caused by the aerodynamic resistance to control surface movements that they must overcome. Attachment points for control systems must be very rigid to avoid sloppiness in the controls, which reduces control effectiveness.

7.4.2 Distributed Loads

Not all distributed loads on an aircraft structure are aerodynamic. Fuel is often carried within the wings in rubber bladders, or in integral fuel tanks portions of the structure that have been sealed against leaks. Other types of liquids (water, fire-retardant chemicals, insect spray, etc.) can also be carried. Some of these liquids might be under pressure, adding to the magnitude of the distributed load. Typically, though, pressurized liquids and gases are carried in separate pressure vessels, which load the aircraft structure at discrete points. In addition to liquids, some other types of cargo (grain, gravel, etc.) also place distributed loads on the structure.

7.4.3 Aerodynamic Loads

Pressure loads are generally of much greater magnitude than aerodynamic loads caused by shear. The highest air pressures are generally at stagnation points, where at high speeds and low altitudes total pressures can be over 3000 psf. Low static pressures in regions of high flow velocities also place distributed loads on the structure.

Even when the pressure and shear loads on an airfoil are represented as lift-and-drag point loads at the airfoil's center of pressure, they still must be considered as a distributed load across the span of the wing. Figure 7.8a shows a typical spanwise lift distribution. Drag and pitching moment also have spanwise distributions. These distributions typically have their maximum magnitudes when the aircraft is

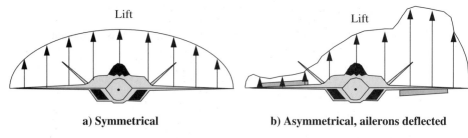

a) Symmetrical   b) Asymmetrical, ailerons deflected

Fig. 7.8   Symmetrical and asymmetrical spanwise lift distributions.

maneuvering at its maximum design load factor at low altitude and high speed. If the aircraft is banking or rolling, the lift distribution is no longer symmetrical, and the wing generating the most lift often has a peak in the lift distribution near the deflected aileron. Figure 7.8b shows this situation. For asymmetrical maneuvers such as this, the maximum load factor limit is set by the maximum structural load that can be sustained by the most heavily loaded wing.

7.4.4   Gust Loads

For some aircraft, particularly transports, design maximum aerodynamic loads are not caused by maneuvering but result from encounters with *gusts* or air turbulence. Gusts result from uneven heating of the Earth's surface, which produces strong vertical air currents and winds in the atmosphere. Gusts also result from the strong trailing vortex systems shed in the wakes of large aircraft. The strongest gust load that aircraft are designed to sustain is one caused by an aircraft flying into a strong vertical air current, which abruptly changes its angle of attack and the lift it is producing. Airline passengers are frequently reminded of the effects of vertical gusts when the pilot turns on the "Fasten Seat Belts" sign.

Figure 7.9 illustrates an aircraft that has abruptly encountered a vertical air current. The current is called a sharp-edged gust because it does not have reduced velocities around its edges. This would never happen in nature, but most standards for gust loading are specified in terms of an equivalent sharp-edged gust. An equivalent sharp-edged gust actually has higher velocities in the center and lower ones at the edge, but it is modeled for analysis purposes as having uniform velocities that abruptly stop at its edge and that produce approximately the same effect on the aircraft as a real gust. Figure 7.9 shows the worst situation in terms of structural loads.

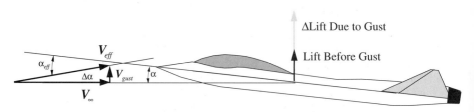

Fig. 7.9   Aircraft encountering an updraft.

The aircraft is in level flight generating lift equal to its weight when it encounters a gust, which is pure updraft. (The air currents in it are oriented perpendicular to the horizon and moving upward.) The aircraft's velocity vector is horizontal and much greater than the gust velocity, so that the primary effect of the vertical gust as it adds vectorially to the horizontal velocity vector is to change the direction of the effective freestream velocity vector V_{eff} (just as downwash does). The change in the effective freestream direction changes the aircraft's effective angle of attack α_{eff}. The effect of an updraft is to increase α_{eff} and therefore increase lift. This sudden increase in lift can cause very heavy loads on the aircraft's wing structure.

The gust velocity is small relative to V_∞; otherwise, it would cause such a large change in angle of attack that the wing would stall. For a relatively small V_{gust}, the magnitude of the change in angle of attack $\Delta\alpha$ is given by

$$\Delta\alpha = \tan^{-1}\left(V_{\text{gust}}/V_\infty\right) \approx V_{\text{gust}}/V_\infty \qquad (7.12)$$

and the change in lift coefficient is

$$\Delta C_L = C_{L_\alpha}\Delta\alpha = C_{L_\alpha}\left(V_{\text{gust}}/V_\infty\right) \qquad (7.13)$$

The change in lift is

$$\Delta L = \Delta C_L qS = C_{L_\alpha}\left(V_{\text{gust}}/V_\infty\right)1/2\,\rho\,V_\infty^2 S = \frac{C_{L_\alpha}\rho\,V_{\text{gust}}V_\infty S}{2} \qquad (7.14)$$

and the change in load factor is

$$\Delta n = \frac{\Delta L}{W} = \frac{C_{L_\alpha}\rho\,V_{\text{gust}}V_\infty S}{2W} = \frac{C_{L_\alpha}\rho\,V_{\text{gust}}V_\infty}{2\,(W/S)} \qquad (7.15)$$

Assuming the load factor prior to encountering the gust is 1, the maximum load factor during the encounter is

$$n_{\text{gust}} = 1 + \Delta n = 1 + \frac{C_{L_\alpha}\,\rho\,V_{\text{gust}}V_\infty}{2\,(W/S)} \qquad (7.16)$$

In some design specifications and regulations, Eq. (7.16) is modified with a gust alleviation factor K_g, which accounts for the fact that true sharp-edged gusts do not exist and the actual response of the aircraft is less than predicted by the preceding analysis:

$$n_{\text{gust}} = 1 + \Delta n = 1 + \frac{K_g\,C_{L_\alpha}\,\rho\,V_{\text{gust}}V_\infty}{2\,(W/S)} \qquad (7.17)$$

An aircraft is just as likely to encounter a downdraft as an updraft when flying through turbulent air. A downdraft is a vertical air current like an updraft, except that the direction of the airflow is downward. The reaction of an aircraft to a downdraft is similar to an updraft, but because V_{gust} is negative the second term in Eq. (7.17) is negative as well. The first term in Eq. (7.17) remains positive, so that the magnitude of n_{gust} is less. However, because most aircraft have lower negative structural limits encountering a downdraft could still be a problem.

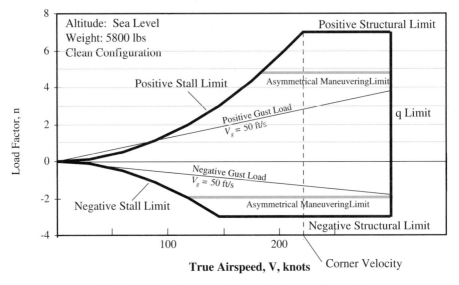

Fig. 7.10   V–n diagram with gust loads superimposed.

7.4.5 V–n Diagram

The V–n diagram discussed in Sec. 5.13 is often used to summarize the loads that the aircraft is designed to withstand and to verify that gust encounters within the aircraft operating envelope will not cause gust factors that exceed structural limits. Figure 7.10 is an example of this. Note that asymmetrical maneuvering load limits and gust loads have been added to the diagram.

Figure 7.10 is for a military jet trainer, and so maximum structural limits are quite high. However, if n_{max} were 3 instead of 7, the aircraft would be gust limited at speeds above 270 kn. This means that if the aircraft flew faster than 270 kn and encountered a 50-ft/s equivalent sharp-edged gust it would exceed its structural limits. This is a common situation for many light aircraft and airliners.

7.5 Structural Layout

Conceptual design of aircraft structures requires deciding where major structural members will be placed within the aircraft. These are critical decisions because a misplaced member can cause many headaches later in the design process. Careful placement of structures can save significant structural weight and can greatly simplify manufacturing, operation, and maintenance of the aircraft.

Most aircraft structural components have names that were borrowed from ship structures. Figure 7.11 illustrates the major types of aircraft structural components. The main load-bearing members in the wing are called *spars*. Spars are strong beams that run spanwise in the wing and carry the force and moments as a result of the spanwise lift distribution. The chordwise pressure and shear distributions on each airfoil are carried to the spars by the wing skin and airfoil-shaped structural frames called *ribs*. The ribs help the wing keep its airfoil shape, and together with

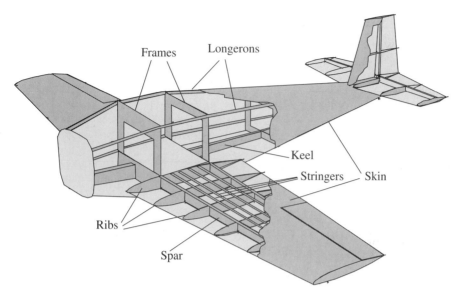

Fig. 7.11   Types of structural components.

the skin and spars form tubes and boxes that resist wing twisting or torsion. The pressure and shear distributions on the wing skin are collected by the ribs and transmitted to the spars. The loads on most ribs are relatively small, though some might carry concentrated loads from landing gear, engines, or external stores. Wing skins are usually quite thin, and so they frequently have additional stiffeners or *stringers* attached to them. Stringers help transmit the skin surface loads to the ribs and spars, and they help keep the skin from bending too much under load. Structural components of stabilizers and control surfaces are given the same names as similar components in wings.

Fuselages also have structural beams, frames, skins, and stiffeners. Fuselage frames are sometimes called *bulkheads*, and they typically run perpendicular to the longitudinal axis. Fuselage beams are called *longerons*, except for the center beam, which is called a *keel*. Keels are used primarily on carrier-based navy aircraft because they need a strong structure to which they can attach catapult bars and arresting gear tail hooks. Other types of aircraft have fewer beams in their fuselages. A *monocoque* fuselage has no keel, longerons, or stringers at all, but gets all of its bending and torsional stiffness from the tubes and boxes formed by its skin and frames. A *semimonocoque* fuselage is more common. It has some stringers and longerons to stiffen the skin, but is otherwise similar to the monocoque design.

Every structure design problem is different, but the following general guidelines suggest pitfalls to avoid and goals to strive for when laying out an aircraft structure:

1) Never attach anything to skin alone. Even thick aluminum skin has relatively little strength against point loads perpendicular to its surface. Pylons, landing gear, control surfaces, etc. must be attached through the skin to major structural components (spars, ribs, frames, keels, etc.) within the structure.

2) Do not pass structural members through air inlets, passenger cabins, cargo bays, etc.

3) Major load-bearing members such as spars should carry completely through a structure. Putting unnecessary joints at the boundaries of fuselages, nacelles, etc. weakens the structure and adds weight.

4) Whenever possible, attach engines, equipment, landing gear, systems, seats, pylons, etc. to existing structural members. Adding structures to beef up attachment points adds weight. Plan the positions of major structural members so that as many systems as possible can be attached to them, and so the structures can carry as many different loads as is practical.

5) Design redundancy into your structures so that there are multiple paths for loads to be transmitted. In this way, damage or failure of a structural member will not cause loss of the aircraft.

6) Mount control surfaces and high-lift devices to a spar, not just the rear ends of ribs.

7) Structural layout is a very creative process. Innovation can often save complexity, weight, and cost. Follow the suggestions on creative problem solving in Chapter 1.

Figure 7.12 illustrates the early stages of a conceptual structural layout for a jet fighter. Positions of the multiple wing spars have been designated, including a short stub spar at the rear near the root to be used as a mounting point for the

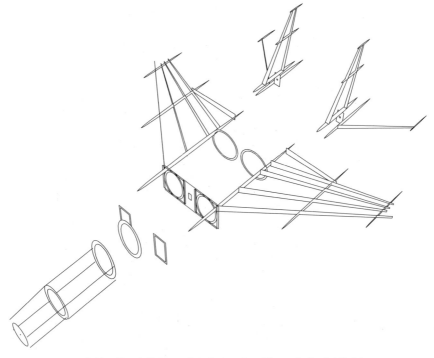

Fig. 7.12   Partially completed structural layout of a jet fighter.

wing flaps. Fuselage frames have been placed so that they line up with some of the spars to serve as attachment points. More will be added until all the spars have structure, not just skin, to attach to. The frames have cutouts in them to designate the paths of the air intake ducts, which must be kept clear of all structures. In the forward fuselage area, some frames and longerons have been placed to define the clear space reserved for the cockpit. The farthest forward frame is positioned to serve as a mounting point for the radar antenna. As the layout progresses, more structural members will be added, along with representations of all of the nonstructural components of the aircraft, until a complete model of the aircraft interior is built.

7.6 Materials

Aircraft materials have progressed tremendously from the early days of "bamboo, burlap, and bailing wire." The modern aircraft designer has a variety of high-performance materials to choose from. The goal is to produce a structure that has sufficient strength and stiffness for a minimum weight, cost, and manufacturing effort. Two of the parameters to be considered when selecting materials, therefore, are strength-to-weight ratio $\sigma_u/(\rho g)$ and stiffness-to-weight ratio $E/(\rho g)$. These two parameters are often referred to as structural efficiency. Values for typical aircraft materials are summarized in Table 7.3.

Note that, with the exception of the highest-performing composites, the structural efficiencies of all of the materials in Table 7.3 are reasonably close to each other. This correctly suggests that all of the materials are suitable for use in aircraft. Other considerations therefore become the deciding factors in selecting which materials are used.

Aluminum alloys are by far the most popular materials in most modern aircraft. Although they have lower structural efficiency values than steel, they are also

Table 7.3   Strength-to-weight and stiffness-to-weight ratios for typical aircraft materials[2-4]

Material[a]	ρ, slug/in.^{3}	$\sigma_u/(\rho g)$, 10^3 in.	$E/(\rho g)$, 10^6 in.
4340 Steel	0.00879	636	163
Stainless steel	0.00888	190	165
2024-T4 Aluminum	0.00311	570	107
7075-T6 Aluminum	0.00314	772	103
Titanium	0.00497	981	100
Graphite/epoxy	0.00174	3040	393
Kevlar®/epoxy	0.00155	3200	240
Fiberglass/epoxy	0.00201	1230	77
Spruce wood	0.00048	584	81
Balsa wood	0.00016	679	194

[a]Numbers associated with some materials designate particular alloys and heat treatment. Values for composite materials are based on a unidirectional layup with 60% fiber content.

less dense. This is an advantage when they are used for wing spars, etc., which must sustain bending loads. Given two beams of the same length, cross-sectional shape, and weight, one of steel and one of aluminum, the aluminum beam will be able to sustain greater bending loads. This is true because its cross-sectional dimensions and area moment of inertia I in Eq. (7.6) will be greater. Aluminum alloys are also preferred for their superior resistance to corrosion and the ease with which they can be shaped. Their main disadvantages are their relatively low melting temperatures and fatigue limits. Aluminum–lithium alloys offer equal strength but 10% lower weight than the more traditional aluminum–copper alloys. Some aluminum–magnesium alloys have also been used in the past, but their susceptibility to burning makes them unpopular.

Steel is commonly used in aircraft structures such as engine mounts and firewalls, which must sustain moderate temperatures for long periods and withstand the heat of a fuel fire for a short period without failing. Steel is also commonly used for landing gear and structural joints, which must sustain intense, cyclic loads. In general, steel is an appropriate choice of material when temperature, loading conditions, or volume limits make aluminum impractical.

Titanium has very good heat resistance, but it is expensive and requires special manufacturing methods and equipment. Its use in aircraft is normally reserved for high-heat areas around engine exhausts and the leading edges of supersonic aircraft wings. A large portion of the skins and internal structure of the Lockheed YF-12 (Fig. 1.1) and SR-71 high-altitude Mach 3+ aircraft are made of titanium.

Wood is still a popular material for some light aircraft, and balsa wood is used widely in flying model aircraft and in some very lightweight aircraft such as those powered by human muscle or solar energy. Wood's very low heat resistance and susceptibility to dry rot has limited its use in large aircraft. The largest aircraft ever built with an all-wood structure was the famous Hughes H.2 Hercules flying boat, better known as the *Spruce Goose*.

Composites have great potential, and they are quite popular for difficult-to-form nonstructural shapes such as wheel fairings and engine cowlings. Their use in primary structures is rapidly increasing, as industry learns to design more effectively to exploit their strengths. Composites will comprise 35% of the structure of the F-22 and 40% of the Euro-Fighter Typhoon.

Composites require completely different manufacturing, maintenance, and repair tools and methods, a major expense for an aircraft manufacturer that has previously worked with metals. Their susceptibility to delamination has caused designers to use much higher factors of safety than for metals. As better design, manufacturing, and inspection methods are developed, the use and performance of composites will continue to increase. Some composites pose environmental hazards on disposal or when they burn. This issue must also be resolved before composite use can become widespread.

7.7   Component Sizing

Once the aircraft structure has been laid out and materials chosen, the detail design of the structure can begin. This process includes selecting the shapes of each structural member and sizing it to ensure adequate strength. Because aircraft structures are so complex, and every load is borne by more than one member,

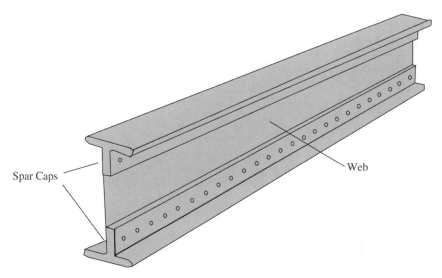

Fig. 7.13   Parts of a built-up spar.

detailed analysis of the entire structure is very difficult and time consuming. The following discussion of a simple beam-bending problem will explain in general how the sizing process works and some of the factors that must be considered.

7.7.1   Choosing Structural Shapes

Before a structural member can be sized, its cross-sectional shape must be chosen. For beams such as wing spars, a simple rectangular cross section is sometimes used. For the same cross-sectional area and weight per unit span, however, C- or I-shaped cross sections will have higher values of I because they have more of their area farther from their neutral axes where the stresses are higher. I-shaped cross sections are very common choices for aircraft spars. They can be extruded whole or built up from pieces. As shown in Fig. 7.13, the top and bottom portions of the spar are called *spar caps*, and the relatively thin sheet of material connecting them is called the *web*. Spar caps are primarily loaded in tension and compression whereas the web is designed primarily to resist shear.

7.7.2   Sizing to Stress Limits

Once the cross-sectional shape of a spar is chosen, the shape's area moment of inertia can be determined. Then the spar can be sized to withstand the expected design loads. As discussed in Sec. 7.4, and shown in Fig. 7.6, spars will typically have both point loads and distributed loads. The spar cross section must be sized so that the bending moment from 1.5 (factor of safety) times the maximum design loads will not cause the tension and compression stresses in the spar caps to exceed the ultimate stress of the material from which they are made. If the aircraft's design maximum load factor is 8, then the point load from an external store hanging from a pylon attached to the spar that must be considered is the weight of store and pylon multiplied by 12. The moment at a given point on the spar due to that point load is the load multiplied by its moment arm to the point.

For distributed loads such as the spanwise lift distribution, the moment is determined by integration:

$$M = \int_0^b l\,(x - x_o)\,\mathrm{d}x \tag{7.18}$$

where l is 1.5 times the design maximum lift per unit span (airfoil lift) at the spanwise location x, and x_o is the spanwise location about which the moments are being summed. Note that for these spar calculations, a coordinate system is chosen with x running spanwise and y vertical to be consistent with common practice. Because an easily integrated algebraic expression for the spanwise lift distribution is not normally available, trapezoid rule or Simpson's rule numerical integration can be used to approximate the moment.

Once the total moment at a given spanwise location on the spar is known, Eq. (7.5) is solved for the required area moment of inertia for the spar cross section at that point:

$$I = \frac{M\,y}{\sigma_u} \tag{7.19}$$

Consider first a spar with a rectangular cross section, as shown in Fig. 7.14. This is a common section shape for wooden spars (in the Piper Cub, for example). Note that the grain (fibers) in the wood are oriented spanwise, for maximum strength in tension and compression. For this shape, Eq. (7.6) can be integrated algebraically as

$$I \equiv \int^A y^2\,\mathrm{d}A = \frac{wh^3}{12} \tag{7.20}$$

where w is the width of the base of the rectangular cross section and h is its height, as depicted in Fig. 7.14.

Because the spar must fit within the wing, the shape and size chosen for the wing's airfoil determine the maximum possible height of the spar. As shown in Fig. 7.14, the maximum y distance from the neutral axis in the section is just 50% of h, so that Eq. (7.20) can be combined with Eq. (7.19) and solved for the required

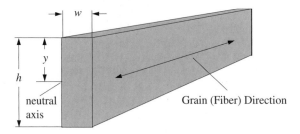

Fig. 7.14   Rectangular cross-section wooden spar.

cross-section width:

$$w = \frac{M(h/2)}{\sigma_u \left(h^3/12\right)} = \frac{6M}{\sigma_u h^2} \tag{7.21}$$

Ideally, Eq. (7.21) should be evaluated at each point along the span, and the dimensions of the spar changed accordingly. In practice, especially for wooden spars that are milled from larger stock, manufacturing is much simpler if a single size of cross section is used across the entire span. For this reason, such spars are primarily used in wings with a taper ratio of unity. This section is sized for the point on the span where 150% of design maximum loads produces the greatest moment.

Next consider the built-up spar in Fig. 7.13. Its height is also determined by the thickness of the wing it must fit inside. As a simplifying approximation, assume the web contributes very little to the magnitude of I, and that each spar caps' contribution to I can be modeled as its area A_c, multiplied by the square of a characteristic distance y_c from the neutral axis,

$$I \equiv \int^A y^2 \, dA \approx 2A_c \, y_c^2 \tag{7.22}$$

The skin attached to the top and bottom of the spar may be thick enough to also contribute significantly to I, so that A_c becomes the area of the spar cap and skin, and y_c may be approximately equal to 50% of h. For this case, combining Eq. (7.22) with Eq. (7.19) and solving for the required area of spar caps and skin yields

$$A_c = \frac{M \, h/2}{2\sigma_u \, (h/2)^2} = \frac{M}{\sigma_u h} \tag{7.23}$$

A built-up spar such as this is much easier to design to fit inside a tapered wing, because to do so only requires cutting the web to the appropriate shape before the spar is assembled. Ideally, Eq. (7.23) would be evaluated everywhere along the span to obtain a continuous function describing the variation of the required spar cap and skin area. In practice, it is normally sufficient to evaluate Eq. (7.23) at enough discrete points along the span to adequately describe the variation. The spar caps may be extrusions that do not taper unless additional machining is done to them. Doublers, additional strips of material, can be added to either side of the spar cap when the spar is assembled to increase area where the moment is greatest. For a typical wing loaded as shown in Fig. 7.6, this maximum moment will likely occur at the wing root.

Sizing of a built-up spar is not complete until the required web thickness is determined. Webs must primarily resist shear, both vertical shear resulting from the load and horizontal shear caused by compression at the top of the spar and tension at the bottom. These stresses are relatively small compared to the stresses in the spar caps, however, and the webs can be quite thin. Because they must primarily carry shear stresses, webs made of composite materials should have their fibers in a mesh or with multiple layers in which each layer has fibers oriented 90 deg or 45 deg relative to fibers in adjacent layers. Wooden webs are normally made of plywood with the grain in each ply 90 deg from the grain in adjacent layers for the same reason. If a web must be made of wood with grain in a single direction (as with balsa wood sheets for built-up model airplane spars), the grain should be oriented vertically, perpendicular to the spanwise direction and the grain in the spar caps. This allows the spar caps to carry the vertical shear perpendicular to their grain while the web carries the horizontal shear perpendicular to its grain, so that shear does not tend to separate the relatively weak lateral bonds between fibers.

7.7.3 Sizing to Fatigue Limits

Sizing of structural members made of metals must be modified slightly to check fatigue limits. This normally involves simply reevaluating Eq. (7.23) using a stress below the ultimate stress and only using 100% of the design loads. The stress to be used would be chosen from a chart like Fig. 7.3 so as to give the desired number of cycles to failure. The required area calculated in this manner would be compared with that from the ultimate stress calculation and the larger of the two used for sizing. A structure sized by either failure or fatigue stress limits should also be checked in a similar manner to be sure it does not exceed its plastic deformation limits, using yield stresses and 100% of the design load.

7.7.4 Sizing to Deflection Limits

In some cases, structural members are sized not by failure limits but by elastic deformation limits. In the case of spar bending, this limit would normally be specified by a maximum deflection limit under the design load. The general expression for deflection of a beam requires an integration of strain as a result of shear and moment over the entire span and can be quite complex. However, for untapered spars with constant cross-sectional shape closed-form expressions can be integrated for simple loading cases. Figure 7.15 illustrates three of these that are most useful in approximating spar loading and deflections. In addition, more complex loadings can be approximated as summations of several different simpler loadings. The resulting deflections are approximated as the sum of the deflections due to each of the simpler loadings.

7.7.5 Buckling

Long, slender structural members and thin skins will fail under compressive stresses well below ultimate. This failure results from a structural instability that causes the structure to bend in the middle when loaded axially in compression. Figure 7.16 illustrates buckling of a pin-jointed slender column and a thin skin. The expression for the critical load F_{cr}, which will cause a pin-jointed slender column to buckle, is

$$F_{cr} = \frac{\pi^2 E I}{b^2} \tag{7.24}$$

where b is the length of the column. Equation (7.24) was derived by the 18th-century Swiss mathematician Leonhard Euler and, like Eq. (3.3), is named for him. There is seldom any confusion between the two equations, however, because they are generally used in very different contexts. Similar expressions exist for buckling of structures with different shapes and end conditions.

Buckling of skins can seriously degrade the strength and the aerodynamics of an aircraft's wings. To minimize this effect, stiffeners or stringers are often attached

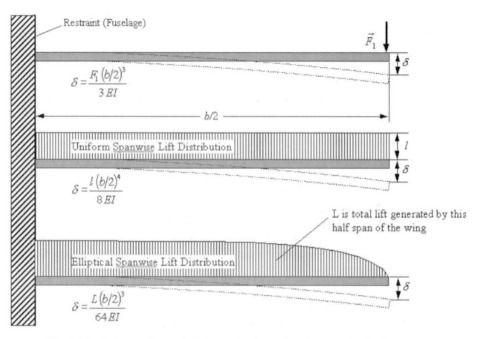

Fig. 7.15 Untapered spar deflections for three simple spanwise loading cases.

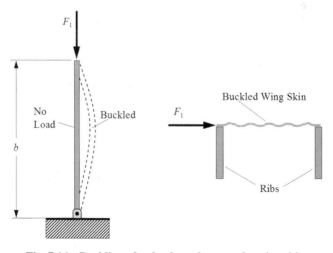

Fig. 7.16 Buckling of a slender column and a wing skin.

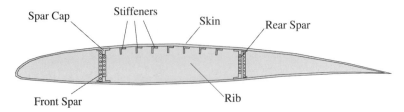

Fig. 7.17   Cross-section view of a wing panel.

to the skin to increase I and raise the critical load. Many modern aircraft skins are milled from much thicker slabs of material, with the stiffeners milled in place, rather than fastened on later. Figure 7.17 illustrates wing skin stiffeners.

7.7.6 Design Considerations

The preceding sections on structure fundamentals should give the aircraft designer a better understanding of some of the tradeoffs that are frequently made between aerodynamics, structural weight, and manufacturability. Decisions about the shape of a wing are good examples. First, consider the thick, untapered wings of the Piper Cherokee 140 and 180 light airplanes. These were undoubtedly designed this way to save on manufacturing costs, because all of the ribs in an untapered wing are identical and the spars can be made from aluminum extrusions with no additional machining. Older Cherokees never quite performed as well as their chief rival, the Cessna 172, however, and so in the 1970s the Cherokee Archer and Warrior appeared with longer-span, tapered outer wing panels. These were more expensive to manufacture, but the improved performance as a result of higher e values and lower induced drag of the tapered wings made the change worth the effort.

It was mentioned in Chapter 4 that the highest values of span efficiency factor e for a linearly tapered wing are achieved for taper ratios near 0.3. However, the F-16 was designed with a taper ratio of 0.2, and many Boeing commercial transports have taper ratios even lower than that. Why? Note the difference between the bending displacement formula in Fig. 7.15 for the rectangular and elliptical spanwise load distributions. The deflection for the elliptical load is much less because the load is carried further inboard, where it has a smaller moment arm. Tapering a wing has the same effect on spanwise lift distributions. Lower taper ratios move the lift further inboard and save on structural weight. This structural weight savings makes up for the increased induced drag and allows each aircraft to fly its mission for less total fuel used. Jet fighter aircraft wings with the lowest weight per unit area are delta wings, which have taper ratios near zero.

Now consider again the case of the Hawker Typhoon and Tempest fighters of World War II, which were discussed in Sec. 4.6. The thicker wing of the Typhoon, in addition to generating more lift, gave its spar a much higher value of I, so that it could be lighter while achieving the same strength as the Tempest's wing. Indeed, the Typhoon's empty weight was some 450 lb less than the nearly identical Tempest (the Tempest also had a slightly longer fuselage), whose thinner wing let it fly faster before encountering shock-induced separation. Because both aircraft had the same maximum takeoff weight, the Typhoon's structural weight savings

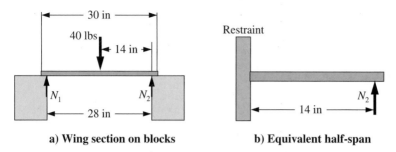

a) Wing section on blocks   b) Equivalent half-span

Fig. 7.18   Wing-section geometry for structural test and analysis.

let it carry more payload, a good feature for an aircraft whose low maximum speed relegated it to ground-attack duties. As maximum speeds of military aircraft have continued to increase, wings have become progressively thinner, and heavier.

7.8   Structural Sizing Example

The Society of Automotive Engineers (SAE) sponsors an annual contest for university students. The object of the SAE contest is to design and build a radio-controlled aircraft that will lift the maximum possible payload weight using a specified piston engine. (Similarly the AIAA Foundation sponsors a Design/Build/ Fly competition as well as other student design competitions.) In the past, aircraft with empty weights less than 6 lb have lifted payloads weighing nearly 30 lb! Naturally, these aircraft have very good structural efficiencies.

Suppose that, as a first step in designing an aircraft for this contest, it is desired to build a sample wing section that will be subjected to a load-bearing test. The goal of the test is for the wing section to support a 40-lb load at its center span when placed on blocks 28 in. apart, as depicted in Fig. 7.18. The challenge is to size the wing structure to carry the load, but at minimum weight. A 12% thick airfoil with a 10-in. chord is specified for this test wing panel.

A simple construction method involves cutting the wing shape from Styrofoam, inserting a rectangular-section balsa wood spar into the shape at its maximum thickness point, and covering the section with a thin plastic skin. The weights of the foam and skin are negligible, so that the sizing task comes down to designing the lightest-weight balsa spar that can carry the load.

Figure 7.18b shows half of the span of the wing with the forces drawn on it. The symmetry of the problem makes the midspan point on the wing act as if it were fastened rigidly to a restraint. Summing forces and moments for the whole wing in Fig. 7.18a shows that the point loads on the spar as a result of the support blocks are each 20 lb located 14 in. from the midspan. This produces a maximum bending moment at the midspan point of 280 in pounds. The maximum height of the spar is 1.2 in. (12% of the 10-in. chord). Using Eq. (7.21) and the value of σ_u for balsa wood from Table 7.2, the width of the spar must be

$$w = \frac{6M}{\sigma_u h^2} = \frac{6(280 \text{ in.} \cdot \text{lb})}{(3500 \text{ lb/in.}^2)(1.2 \text{ in.})^2} = 0.33 \text{ in.}$$

If a maximum deflection limit of 0.5 in. at the midspan were imposed, the first equation in Fig. 7.15 can be solved for the required width of the spar:

$$I = \frac{wh^3}{12} = \frac{F\,(b/2)^3}{3E\delta}$$

$$w = \frac{4F(b/2)^3}{h^3E\delta} = \frac{4(20\,\text{lb})(14\,\text{in.})^3}{(1.2\,\text{in.})^3\,\left(1{,}000{,}000\,\text{lb/in.}^2\right)(0.5\,\text{in.})}$$

$$w = 0.37\,\text{in.}$$

7.9 Weight Estimates

Once the structure is designed and sized, its weight and center of gravity must be determined. The weight of each member is its volume multiplied by its material density and the acceleration of gravity. The center of gravity of a member composed of a single material is located at the centroid of its volume, which can be determined by integration, by published closed-form solutions (for standard shapes), or by various graphical methods. For members of uniform cross section down their length, their center of gravity is at their midspan. The weights of all of the members and the moments of the weights about some arbitrary reference point are summed, then the total moment is divided by the total weight to determine the center of gravity of the whole structure. For the balsa wood spar sized in Sec. 7.8, using the density of balsa wood from Table 7.3,

$$W = \rho g(whb) = (0.00016\,\text{slug/ft}^3)(32.2\,\text{ft/s}^2)(0.37\,\text{in.})(1.2\,\text{in.})(30\,\text{in.})$$

$$= 0.0686\,\text{lb.}$$

Its center of gravity is at its midspan.

7.10 Finite Element Analysis

An introduction to aircraft structural design methods would not be complete without mentioning finite element analysis. This form of analysis uses the power of modern computers to predict stresses and deflections in very complex structures. The basic method involves dividing the structure into thousands, even millions of tiny structural elements that are linked to each other at nodes or junctions at their corners. Hooke's law is written in matrix form for each element, and the condition is enforced that the displacement of a node shared by two elements must be the same in the statement of Hooke's law for both elements. In this way, a huge matrix of equations describing the stress–strain relationships and enforced equalities of displacements for shared nodes is constructed. For most complex structures, this matrix does not have a single solution. The methods of calculus of variations (optimization theory) are used to determine a solution to the matrix that minimizes the total strain energy of the structure.

Finite element analysis is the method of choice for structural design. It has given engineering vast new capabilities for optimizing structures, saving weight, and saving money. It can truly be said that without this powerful tool current and future generations of aircraft would be less capable, more expensive, and more susceptible to unexpected structural failures.

7.11 Chapter Summary

1) Axial stress:

$$\sigma = F_1/A$$

2) Bending stress:

$$\sigma = \frac{M\,y}{I}$$

3) Area moment of inertia:

$$I \equiv \int^A y^2 \, dA = \frac{wh^3}{12}$$

for rectangular cross section.

4) Hooke's law:

$$\varepsilon = \sigma/E$$

5) Buckling:

$$F_{cr} = \frac{\pi^2 E I}{b^2}$$

6) Fatigue and fatigue limit:

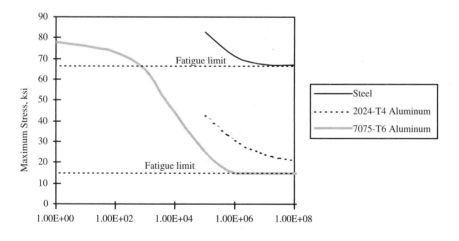

7) Gust response:

$$n_{gust} = 1 + \Delta n = 1 + \frac{K_g C_{L_\alpha} \rho V_{gust} V_\infty}{2(W/S)}$$

8) Gust response on V–n diagram:

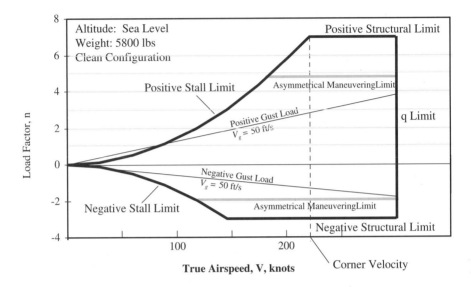

References

[1]Raymer, D. P., *Aircraft Design: A Conceptual Approach*, AIAA Education Series, AIAA, Washington, DC, 1989, Chap. 14.

[2]Peery, D. J., *Aircraft Structures*, McGraw–Hill, NewYork, 1950, p. 279.

[3]Higdon, A., Ohlsen, E. H., Stiles, W. B., and Weese, J. A., *Mechanics of Materials*, Wiley, New York, 1967, Chaps. 1 and 2.

[4]Niu, M. C. Y., *Airframe Structural Design*, Conmilit Press, Ltd., Hong Kong, 1993, Chap. 4.

Problems

Synthesis Problems

S-7.1   During the design of a new business jet, the aerodynamics group proposes placing streamlined fuel tanks on the aircraft's wing tips to reduce the plane's induced drag. Would this change increase or decrease the bending loads on the aircraft's wing root structure?

S-7.2   During the design of a new jet fighter, the weapons group proposes moving the main landing gear from the fuselage to the wings to make room for a larger internal weapons bay. Would this change increase or decrease the bending loads on the aircraft's wing-root structure?

S-7.3   During the design of a replacement for the F-111 deep interdiction strike fighter, the aerodynamics group proposes changing to a variable-sweep wing like the F-111's. They argue that this change would allow the aircraft to meet its takeoff and landing distance requirements (with the wings unswept) with a much smaller wing, which would give the aircraft less cruise drag, and that with the wings swept back the aircraft would have a higher maximum speed and a smoother ride in turbulence. What effect do you think this change would have on the aircraft's structural weight?

S-7.4 Brainstorm five lightweight materials and structures concepts for a flying radio-controlled aircraft with a high-aspect-ratio wing to compete in the SAE Aero Design contest as described in Sec. 7.8. Especially consider lightweight alternatives for landing gear, rear fuselage, wing, and engine mounts.

Analysis Problems

A-7.1 A jet airliner is being designed to have a nearly elliptical spanwise lift distribution on its wings. The design maximum lift distribution has a total lift of 250,000 lb per wing, which creates the same moment at the wing root as if it were a point load located 25 ft from the root. A tapered built-up wing spar is to be used which has a maximum height of 2 ft at the wing root. If the wing spars are to be made of 7075-T6 aluminum and a factor of safety of 1.5 is to be used, and if the maximum allowable deflection at the tip of the 65-ft-long wing under the design load is 4 ft, what is the minimum acceptable area of the spar caps and skins at the wing root.

A-7.2 For the jet airliner analyzed in problem A-7.1, a decision is made to move the aircraft's twin engines from being mounted on either side of the rear fuselage to being mounted on pylons under the wings. If each engine, nacelle, and pylon weighs 7000 lb, and they are to be mounted 20 ft out from the wing root, how will this affect the sizing of the wing-root spar caps? Assume the design load factor is 3.0.

A-7.3 Straighten out the wire in a paper clip, then cyclically bend a portion of the wire through 90 deg and straighten it again repeatedly until it breaks. How many bending cycles did the wire sustain before it failed? Now try the same experiment, using a vice and a pair of pliers to put a sharper bend in the wire than could be made with your fingers. How many cycles to failure this time? Explain the difference.

A-7.4 The B-52 was designed as a high-altitude bomber, but the advent of radar and surface-to-air missiles forced the U.S. Air Force to change B-52 tactics to include flying long portions of a typical bombing mission at low altitudes, below radar coverage. As a result, B-52 wings began developing cracks. Why do you think this happened?

8
Sizing

"The outside has to be bigger than the inside"
Howard W. Smith and Robert Burnham

8.1 Design Motivation

The last major task in defining a conceptual aircraft configuration is *sizing*, determining how large the aircraft must be to carry enough fuel and payload to perform the design mission(s). This task includes, but is more than, making sure everything the aircraft must carry inside it will fit. Because the fuel required to fly the mission depends on the size of the aircraft and the size of the aircraft depends in part on how much fuel it must carry, the sizing problem must be carefully formulated and solved in order to obtain a useful result. Aircraft size has a very profound effect on cost, and cost is often one of the most important constraints on a new aircraft design. In the 1950s and 1960s, aircraft performance requirements were often the primary design drivers, especially for military aircraft, and higher-than-planned costs were often accepted by customers in order to get the performance they desired. In the present day, cost is far more important to customers, and performance requirements are frequently revised in order to allow a new aircraft design to meet its cost goal.

Consider, for instance, the U.S. Air Force F-15 and F-16 tactical fighter aircraft shown in Fig. 8.1. Both aircraft have approximately the same P_s and performance characteristics at all but high Mach numbers. (The F-15's variable engine inlets give it more thrust and higher P_s and V_{max} above $M = 1.8$). The larger F-15 has a higher payload and a higher maximum range, but it is also more expensive. Because of the cost advantage, the U.S. Air Force bought many more F-16s than F-15s (2358 F-16s vs 807 F-15s as of 1996). During the 1991 Persian Gulf War, the F-16's short range made it useful for missions to Kuwait and (with extensive tanker aircraft support) southern Iraq, but relatively few longer missions. Most of the long-range missions into central Iraq were flown primarily by F-15s, F-117s, and F-111s, all of them larger and more expensive aircraft but with the range and payload to perform the required missions with lower requirements for tanker support. U.S. Air Force requirements for a replacement for the F-16 include significantly greater range but at an affordable cost. Careful sizing analysis and prudent use of advanced technology are required when developing an aircraft to meet these strongly contradictory requirements.

Cost has always been a primary design driver for commercial aircraft. However, a large part of an airliner's total cost is not in its initial purchase price, but in its operating costs. Because larger aircraft carry more passengers and fuel, they can

Fig. 8.1   F-15 (left) and F-16 (right) tactical fighters (U.S. Air Force photograph).

fly farther and generate more revenue before they have to make expensive stops for fuel. Of course, not all airline routes are long enough and have sufficient passenger traffic to justify a large airliner. Most airlines therefore purchase a mix of sizes of airliners, each one suited to a particular type of route and passenger load. In these times of intense airline competition, airline purchasing agents must choose carefully the performance requirements they set for new airliners, and aircraft designers must make very accurate sizing analyses so that the aircraft they design meet but do not exceed the required range, payload, and cost.

8.2   Internal Layout

The first step in sizing analysis is to determine the weight and volume of the payload, systems, and crew that must be carried to perform the design mission(s). These normally are set by the customer and do not change with changing aircraft size. Many of these components must be carried in particular portions of the aircraft. For example, weather radar must be carried in a forward-facing portion of the aircraft, the forward fuselage or a pod on the wing. The pilot(s) normally require(s) good visibility forward and downward, and so the cockpit or flight deck (with some notable exceptions) normally is placed high and near the aircraft nose. Passengers need easy access to normal and emergency exit doors. Heavy payloads, especially those that will be dropped, offloaded, or jettisoned during flight, should be placed close to the aircraft's center of gravity. Engines must have unobstructed pathways for air into their inlets. Components that require maintenance should be easily accessible (by someone standing on the ground, if possible). The locations of these

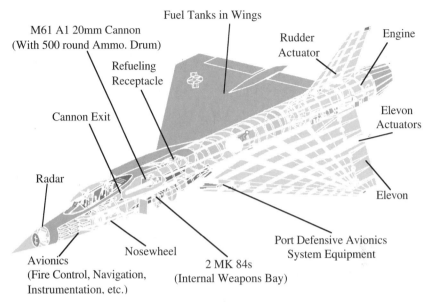

M61 A1 20mm Cannon
(With 500 round Ammo. Drum)

Fuel Tanks in Wings

Refueling
Receptacle

Rudder
Actuator

Engine

Cannon Exit

Elevon
Actuators

Radar

Elevon

Avionics
(Fire Control, Navigation,
Instrumentation, etc.)

Nosewheel

2 MK 84s
(Internal Weapons Bay)

Port Defensive Avionics
System Equipment

Fig. 8.2   Internal arrangement drawing (drawn by Robert Bodwell).

fixed components and required payloads within the aircraft should be determined in general early in the conceptual design process. An internal arrangement or internal layout drawing is used to illustrate and deconflict the locations of these various components. Figure 8.2 is an example of this type of drawing.

8.3   Structures and Weight

Design, analysis, and sizing of an aircraft's structure is normally left for the detail design phase. For conceptual design, it is sufficient to estimate the weight, center of gravity, and volume of the structure, and to ensure the structure is planned carefully so that it will be strong, light, inexpensive, buildable, and maintainable. As discussed in Chapter 7, planning of the structure at the conceptual design level involves deciding what materials will be used, where the major structural components will be placed within the aircraft volume and where major systems, engines, payload, etc. will be attached to the structure. The major structural members are often shown on an internal arrangement drawing along with the systems, cockpit, engines, etc., which must attach to them. A drawing of this sort allows the designer to ensure that nonstructural components are placed so that they do not interfere with the efficient design and placement of the structure.

Initial estimates of structural weight, center of gravity, and volume are based on historical data for similar aircraft. A crude but useful first estimate of the structural weights of major aircraft components is based on a historical average value of weight per unit reference area of each type of component. For wings and control surfaces, the reference area is the surface planform area. For fuselages, external fuel tanks, nacelles, and pods, the reference area is the component's wetted area. Table 8.1 lists the average weight per area of components for several types of

Table 8.1   Average weights per unit area and center of gravity locations for several aircraft components

Component	Jet fighter, lb/ft^2	Jet transport, lb/ft^2	Light airplane, lb/ft^2	Reference area, ft^2	Center of gravity location
Wing	7.0[a]	10.0	2.5	S	0.4 MAC
Fuselage	4.8[a]	5.0	1.4	$S_{\text{wet fus}}$[b]	0.4–0.5 length
Vertical tail	6.0	6.0	2.0	S_v[c]	0.4 MAC
Horizontal tail	4.0	6.0	2.0	S_t	0.4 MAC

[a] Add 25% to the weight of fuselages and wings for airplanes designed to operate from aircraft carriers
[b] Wetted area of the fuselage.
[c] Planform area of the vertical tail.

aircraft. The weights of other components are more easily and accurately predicted as weight fractions (for landing gear and miscellaneous systems) and weight-to-thrust ratios (for engines). Table 8.2 lists typical values.

These weight prediction methods provide quite accurate results for many cases. However, predicting wing weight for jet-powered aircraft using this method often results in significant errors. This is true because a jet fighter or transport airplane wing's weight is profoundly affected by its thickness-to-chord ratio, taper ratio, and aspect ratio, and those characteristics vary significantly from aircraft to aircraft. Because wing weight is a significant part of the weight of an airplane's structure, a large error in its prediction can yield unreasonable sizing results. For a more accurate wing weight estimate, the following formula based on a more detailed analysis of historical data[2] for land-based jet fighter and transport aircraft is used

$$\frac{W_{\text{wing}}}{S} = 0.04 \frac{n_{\text{max}}^{0.2}\ \mathcal{R}^{1.8}\ (1+\lambda)^{0.5}}{(t/c)^{0.7}\cos\Lambda_{\text{LE}}} \quad (8.1)$$

Table 8.2   Average weight fractions and center of gravity locations for several aircraft components

Component	Jet fighter, lb/lb	Jet transport, lb/lb	Light airplane, lb/lb (ex. eng.)	Multiplier, lb (hp for SHP)	Center of gravity location
Landing gear	0.033	0.043	0.057	W_{TO}	15% nose gear,
(carrier capable)[a]	(0.047)				85% main gear
Uninstalled engine	0.1	0.2	1.8 lb/hp[b]	T_{SL}[b]	0.4–0.5 length
Installed engine	0.13	0.26	2.0 lb/hp[b]	T_{SL}[b]	0.4–0.5 length
All else. misc.	0.17	0.17	0.1	W_{TO}	0.4–0.5 length

[a] Landing gear on aircraft capable of operating from an aircraft carrier are heavier, because they must stand up to much higher landing forces (harder landings) plus the stresses of catapult launches, which are transmitted through a linkage on the nose gear.
[b] Light aircraft engine weights are based on rated sea level SHP, in units of horsepower instead of thrust.

where n_{max} is the design load factor limit of the aircraft and t/c is the ratio of maximum thickness to chord of the wing's airfoil.

The wing weights for jet aircraft designed to operate from aircraft carriers average approximately 25% more than what is predicted by Eq. (8.1). This additional structural weight is partially caused by requirements for wing folding mechanisms and partially caused by the extra structural strength needed to resist the acceleration and deceleration forces from catapult launches and arrested landings.

Modern developments in materials and fabrication technology have made it possible to build structures that are significantly lighter. The new materials are mostly composites, made up of two or more materials such as fiberglass cloth and epoxy resin. New fabrication methods are being developed that produce components from fewer parts. This saves on the costs of the components. The full benefit of this technology has yet to be realized in aircraft design and construction. At present, because industry is still learning how to use advanced materials the presence of such materials in a structure can save from 0 to 10% of the structural weight and not save at all on cost. There is good reason to expect that when this design and fabrication technology is mature it will save 20% or more on structural weight and as much as 50% on the cost of some components.

Example 8.1

A new fighter aircraft is being designed to have a maximum load factor limit of 9, a wing area of 300 ft^2, a span of 30 ft, and a taper ratio of 0.21. If the airfoil selected for the wing is 4% thick, estimate the weight of the wing structure. How would your estimate for the wing structure weight change if a 6% thick airfoil were used for the wing?

Solution: The wing structure weight is predicted using Eq. (8.1). The aspect ratio is

$$(30 \text{ ft})^2/300 \text{ ft}^2 = 3$$

$$\frac{W_{wing}}{S} = 0.04\frac{n_{max}^{0.2} \, \mathcal{R}^{1.8}(1+\lambda)^{0.5}}{(t/c)^{0.7} \cos \Lambda_{LE}} \quad 0.04 \, \frac{9^{0.2} \, 3^{1.8}(1+0.21)^{0.5}}{(0.04)^{0.7} \cos 40 \text{ deg}} = 6.13 \text{ lb/ft}^2$$

$$W_{wing} = \frac{W_{wing}}{S}S = 6.13 \text{ lb/ft}^2 \, (300 \text{ ft}^2) = 1839 \text{ lb}$$

If the wing thickness is increased to 6%,

$$\frac{W_{wing}}{S} = 0.04\frac{n_{max}^{0.2} \, \mathcal{R}^{1.8}(1+\lambda)^{0.5}}{(t/c)^{0.7} \cos \Lambda_{LE}} \quad 0.04 \, \frac{9^{0.2} \, 3^{1.8}(1+0.21)^{0.5}}{(0.06)^{0.7} \cos 40 \text{ deg}} = 4.62 \text{ lb/ft}^2$$

$$W_{wing} = \frac{W_{wing}}{S}S = 4.62 \text{ lb/ft}^2 \, (300 \text{ ft}^2) = 1385 \text{ lb}$$

The structural sizing and weight estimate examples in Secs. 7.8 and 7.9 make it clear why making a wing thicker allows it to support the same load (primarily bending moment) with less structural weight. Because a thicker wing allows a taller spar with a larger value of I, less material can be used in the spar without exceeding the design stress of the material.

8.4   Geometry Constraints

Aircraft size can be constrained for a variety of reasons. Airliners must comply with maximum length and wing-span limits in order to use the passenger loading jetways at most airport terminals. Fighter and attack aircraft might need to fit inside existing hangers and/or bomb-resistant shelters. Carrier-based aircraft must meet strict dimensional constraints, which often require them to have wing panels and stabilizers that can be folded during transit on elevators and while parked on the flight deck and the hangar deck. Extremely large aircraft might be limited by runway and taxiway widths. Smaller aircraft might be expected to be partially disassembled and transported in railroad cars or transport aircraft, or even towed as a trailer to the owner's home and kept in a garage.

8.4.1   Volume

As the title to the AIAA paper[3] quoted at the beginning of this chapter suggests, the aircraft must be large enough to contain all of the structures, passengers, crew, engines, systems, payload, fuel, etc. required to fly the design mission(s). The volumes of payloads are normally known or specified by the customer. The passenger load and seating density are also determined by the customer. Crew station design allows some flexibility, but crew space much smaller than 60 ft^3/person is generally unacceptable. The most common types of aircraft structures occupy on the average approximately 15% of the internal volume of lifting and control surfaces and 8% of the internal volume of fuselages, external fuel tanks, and nacelles (engine pods). Table 8.3 lists typical volume allowances for some major classes of items inside an aircraft.

The volumes of all items should be determined by the most accurate means available. The data listed in Table 8.3 should be used only in the absence of more accurate estimates. For instance, when a concept for the cockpit of a jet fighter is drawn as part of its internal arrangement drawing, the proposed cockpit volume can be calculated. This calculated volume should thereafter be used in preference to the rule of thumb listed on Table 8.3.

Subtracting the volumes of all of the required components from the total internal volume of the aircraft yields the volume remaining for fuel tanks. Approximately 10% of a fuel tank's volume must be allowed for the thickness of its skin, plus fuel quantity measuring probes, etc. An additional 5% is subtracted from the volume if the tanks are filled with reticulated foam. The foam helps eliminate the danger of fire as a result of battle damage in the fuel tanks of combat aircraft. The internal arrangement drawing may also reveal that some of the aircraft's internal volume is unusable for fuel tanks, either because it is too far from the aircraft's center of gravity or too close to the engine or other incompatible systems.

8.4.2   Landing Gear

The requirement to have good ground handling characteristics imposes strict constraints on the location and geometry of landing gear. Although designers can

Table 8.3 Volume allowances for some aircraft components and payloads

Component	Factor	Multiplier
Engines:		
Nonafterburning turbojets	0.03 ft^3/lb	T_{SL}
Afterburning turbojets and turbofans	0.0325 ft^3/lb	T_{SL}
Nonafterburning low-bypass-ratio turbofans	0.035 ft^3/lb	T_{SL}
High-bypass-ratio turbofans	0.05 ft^3/lb	T_{SL}
Turboshaft	0.2 ft^3/SHP	SHP_{SL}
Reciprocating	0.3 ft^3/SHP	SHP_{SL}
Transport airplane passengers (includes overhead bins):		
Economy class, 32-in. seat pitch	38 ft^3/person	number of passengers
Tourist class, 34-in. seat pitch	45 ft^3/person	number of passengers
Domestic first class, 38-in. seat pitch, 150% wide	70 ft^3/person	number of passengers
International first class, 41-in. seat pitch, 150% wide	75 ft^3/person	number of passengers
Transport airplane exits, lavatories, galleys, closets	8.5 ft^3/person	number of passengers
Transport airplane baggage/cargo compartment(s)	9.5 ft^3/person	number of passengers
Crew stations:		
fighter	80 ft^3/person	number of crew members
transport	120 ft^3/person	number of crew members
general aviation [pilot(s) and passengers]	40 ft^3/person	number of occupants

choose from a great variety of landing-gear configurations, the most commonly used is the tricycle style. Tricycle landing gear comprise one steerable nose landing gear and two nonsteering main-gear, like the wheels of a child's tricycle. Unlike a child's tricycle, most of the weight of an aircraft's landing gear must be placed on the main-gear. This is because the aircraft must be able to rotate for takeoff. The aircraft's c.g. must not be too far ahead of the main gear, or the moment required to raise the aircraft's weight as it rotates using the main-gear pivot point will exceed the capabilities of the aircraft's pitch control surfaces. In addition, aircraft with too much of their weight carried by their nose gear tend to tip sideways or forward easily while taxiing. On the other hand, aircraft with too little weight on their nose gear are difficult to steer because their nose wheel does not have enough weight on it to get good traction. As a rule of thumb, tricycle landing gears should be designed so that 80–90% of the aircraft's weight is carried by their main gear and 10–20% is carried by their nose gear.

Figure 8.3 shows another consideration and constraint on placement of tricycle landing gear. If an aircraft's c.g. is too close to its main landing gear, it can actually

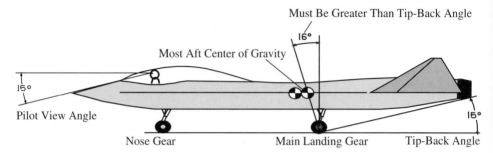

Fig. 8.3   Aircraft tip-back geometry.

move behind the wheels as they serve as the pivot point when the aircraft rotates. If this happens, the aircraft will become a "tail-sitter." Avoid this problem by making sure the tip-back angle shown in Fig. 8.3 is less than the angle ahead of vertical measured from the main-gear pivot point to the most aft position of the aircraft's c.g.

A further complication to the tip-back problem occurs occasionally when aircraft must be parked outdoors. The Martin B-57 Canberra, a subsonic twin-jet medium bomber of the 1950's, experienced this difficulty. This aircraft had fairly large tail surfaces, and its two-man crew was carried far forward in its nose. When left outside during heavy snowfall with no crew onboard, this aircraft frequently experienced so much buildup of snow on its tail surfaces that, without the ballast of crew members in its nose, its center of gravity shifted to very near the location of its main landing gear. In this condition, a slight gust of wind, such as is often readily available in a snow storm, would cause it to tip back and sit on its tail!

Example 8.2

Achieving good ground-handling qualities for an aircraft with tricycle landing gear requires placing the landing gear so that approximately 15% of the aircraft's weight is carried by the nose gear and 85% is carried by the main gear. For the aircraft in Fig. 8.4, determine the required values for x_1 and x_2 in order to achieve the required weight distribution. Assume the distance between the nose gear and the main gear is 10 m.

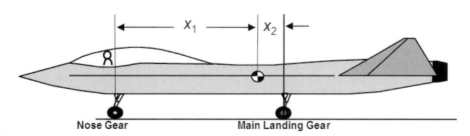

Fig. 8.4   Tricycle landing-gear weight distribution geometry.

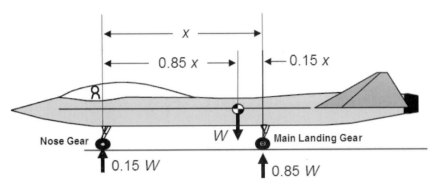

Fig. 8.5 Required geometry for proper tricycle landing-gear weight distribution.

Solution: This is a simple static equilibrium problem, similar to the trim problems of Chapter 6. We simply use the desired weight distribution percentages and sum the moments about the center of gravity. We then set the sum of the moments to zero because the aircraft must be in static equilibrium in this configuration. We use the fact that $x_1 + x_2 = 10$ m as the second equation needed to solve for the unknowns.

$$x_1 + x_2 = 10 \text{ m}$$
$$x_1 = 10 \text{ m} - x_2$$
$$\sum M = 0 = x_1(15\%) - x_2(85\%) = (10 \text{ m} - x_2)(15\%) - x_2(85\%)$$
$$0 = (150 \text{ m}) - 100x_2$$
$$x_2 = 1.5 \text{ m}$$
$$x_1 = 10 \text{ m} - x_2 = 10 \text{ m} - 1.5 \text{ m} = 8.5 \text{ m}$$

So to get 15% of the weight on the nose gear, the distance from the c.g. to the nose gear must be 85% of the total distance between the nose gear and the main gear. And to get 85% of the weight on the main gear, the distance from the main gear to the c.g. must be 15% of the total distance between the nose gear and the main gear. Figure 8.5 illustrates this geometry.

8.4.3 Rollover Criteria

Landing gear must comply with yet another geometry constraint, this one influencing the way the aircraft handles when making turns while taxiing. Figure 8.6 shows the geometry of interest. The engineer must construct an auxiliary view of the aircraft such that the viewing direction aligns one main gear with the nose gear. This view essentially looks down the pivot line about which the aircraft would tip if it were going to roll over, as it might if it turns a corner too fast. For good ground handling in turns, the rollover criteria states that the angle in this auxiliary view between the ground and a line drawn from the nose-gear–main-gear pivot line to the aircraft's center of gravity must not exceed 63 deg.

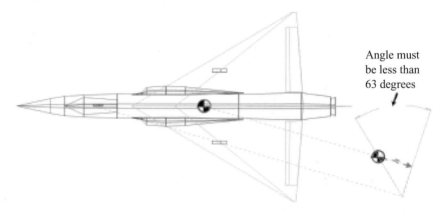

Angle must
be less than
63 degrees

Fig. 8.6   Rollover criteria geometry.

As shown in Fig. 8.6, one way to construct this auxiliary view is to start with a planform view and draw a line through the nose-gear and main-gear contact points with the ground. Then, draw a parallel line through the aircraft's c.g. Now, move a comfortable distance along the two lines away from the original drawing, and draw a line perpendicular to the first two. This new line is the ground plane in the auxiliary view. Now starting at the intersection of this new ground plane line with the line through the c.g., measure along the c.g. line toward the original drawing the distance of the c.g. above the ground. You have now constructed an auxiliary view of the aircraft's c.g. Starting at the intersection of the landing-gear line with the new ground plane, draw a line to the c.g. in this auxiliary view. Now measure the angle between this line and the ground plane in this view. If the angle is less than 63 deg, the aircraft meets the rollover criteria. Note that this angle is reduced to 54 deg for carrier-based aircraft because they need added rollover resistance on pitching, rolling carrier decks.

8.4.4 Landing-Gear Internal Volume Considerations

A further constraint on the placement of landing gear occurs if they are re-tractable. Because the main landing gear must be quite close to the aircraft's c.g. and the wing, fuel tanks, and expendable payload must be near the c.g. as well, internal volume near the c.g. is at a premium. Finding a space into which the main landing gear can be retracted is typically not a trivial task. Because the vicinity of the wing is also where most aircraft have their maximum cross-sectional area, enlarging cross-sectional area to give more internal volume in this vicinity can be very detrimental, especially if the aircraft must be capable of supersonic flight. Clever aircraft designers over the years have worked out some truly ingenious solutions to this perennial problem.

8.5 Mission Analysis

To determine how much fuel an aircraft must carry to fly a mission, we modify the constraint analysis master equation presented in Chapter 5. We break the design mission down into segments or legs and specialize the master equation for each type of leg. We need to model two basic types of mission segments. The first type assumes engines running at a set throttle level (full military power, full afterburner, 50% afterburner, 90% of full military power, etc.) and the aircraft's energy changing depending on the difference between thrust and drag. This mission leg type includes steady climbs, level accelerations, and climbing accelerations. The second mission leg type is characterized by the aircraft's throttle being adjusted until thrust just equals drag. This mission leg type includes cruise, loiter, combat air patrol, and sustained turns. Takeoff is modeled as a special type of acceleration. Descents and landings are not modeled because fuel consumption is low and usually no credit is given for the distance traveled.

There are practical reasons for ignoring the distance covered in a descent. Commercial and private aircraft frequently make descents when approaching an airport. Other traffic around the airport might force the airplane to fly in directions perpendicular to or even away from the desired landing field. Maneuvering the aircraft to land heading into the wind can also negate any distance gained in the descent. Combat aircraft often make descents very rapidly with very little forward distance traveled to deceive enemy radars. In general, because operational considerations make it impossible to guarantee any net range gained as a result of the descent, this maneuver is usually ignored. Landings are not modeled because the fuel actually used for landing is fuel set aside as reserve fuel and modeled as a loiter leg at the end of the mission.

8.5.1 Climbs and Accelerations

Most aircraft spend relatively little time and fuel accelerating. More time and fuel are typically spent climbing. Because we specify the engine throttle setting for these maneuvers, we need only use the constraint analysis master equation to solve for the time required to climb or accelerate. We then multiply thrust-specific fuel consumption (TSFC) for the maneuver by the thrust used and time required. First, we rearrange Eq. (5.75) to calculate the rate of climb and acceleration:

$$\frac{T_{\text{SL}}}{W_{\text{TO}}} = \frac{\beta}{\alpha} \left\{ \frac{q}{\beta} \left[\frac{C_{D_0}}{(W_{\text{TO}}/S)} + k_1 \left(\frac{n\beta}{q} \right)^2 \left(\frac{W_{\text{TO}}}{S} \right) \right] + \frac{1}{V} \frac{dh}{dt} + \frac{1}{g} \frac{dV}{dt} \right\} \quad (5.75)$$

$$\frac{1}{V} \frac{dh}{dt} + \frac{1}{g} \frac{dV}{dt} = \frac{\alpha}{\beta} \frac{T_{\text{SL}}}{W_{\text{TO}}} - \frac{q}{\beta} \left[\frac{C_{D_0}}{(W_{\text{TO}}/S)} + k_1 \left(\frac{n\beta}{q} \right)^2 \left(\frac{W_{\text{TO}}}{S} \right) \right] \quad (8.2)$$

Note that a simplified drag polar, with $k_2 = 0$, was used, to be consistent with the approximations made in constraint analysis in Chapter 5. Constraint analysis is used to select the design wing loading and thrust-to-weight ratio prior to performing mission analysis. Therefore, all quantities in Eq. (8.2) are known for specified flight conditions, provided the value of β from the previous mission leg is known. Both types of maneuvers are characterized by significant changes in many of the parameters in Eq. (8.2). However, if we break up long climbs and accelerations into short segments we can achieve acceptable accuracy by calculating an average rate of climb or acceleration. For the case of a steady climb, Eq. (8.2) simplifies to

$$\frac{dh}{dt} = V \left[\left(\frac{\alpha}{\beta} \frac{T_{\text{SL}}}{W_{\text{TO}}} \right) - \frac{q}{\beta} \left(\frac{C_{D_0}}{W_{\text{TO}}/S} \right) - \frac{n^2 \beta}{q} k_1 \left(\frac{W_{\text{TO}}}{S} \right) \right] \quad (8.3)$$

To estimate the time required to climb from one altitude to another, the rate of climb and flight conditions are used for the altitude midway between the initial altitude and the final altitude of the climb. With the average rate of climb known, the time to climb is given by

$$t_{\text{climb}} = \frac{(h_{\text{final}} - h_{\text{initial}})}{dh/dt} \tag{8.4}$$

The accuracy of this time-to-climb estimate is improved if the climb is broken up into smaller segments and a calculation made for the conditions and the altitude change in each segment. More complex and potentially more accurate estimates can also be used, but this average value method is acceptable for an initial estimate, and it becomes quite exact when the climb is broken into very many small segments, as might be easily done in a computer-aided analysis.

The fuel used to climb is given by multiplying the time to climb by the TSFC (calculated using the flight conditions at the average altitude and the TSFC models of Chapter 5) multiplied by the thrust. However, thrust is not known explicitly, only the thrust-to-weight ratio, and so only the fuel weight fraction for the climb can be calculated:

$$\frac{W_{f_{\text{climb}}}}{W_{\text{TO}}} = \alpha \frac{T_{\text{SL}}}{W_{\text{TO}}} c_t \, t_{\text{climb}} \tag{8.5}$$

For the case of acceleration in level flight, Eq. (8.2) simplifies to

$$\frac{dV}{dt} = g \left[\left(\frac{\alpha}{\beta} \frac{T_{\text{SL}}}{W_{\text{TO}}} \right) - \frac{q}{\beta} \left(\frac{C_{D_0}}{W_{\text{TO}}/S} \right) - \frac{n^2 \beta}{q} k_1 \left(\frac{W_{\text{TO}}}{S} \right) \right] \tag{8.6}$$

The approximate time to accelerate, using the average value method, is

$$t_{\text{accel}} = \frac{(V_{\text{final}} - V_{\text{initial}})}{dV/dt} \tag{8.7}$$

and the fuel fraction used to accelerate is

$$\frac{W_{f_{\text{accel}}}}{W_{\text{TO}}} = \alpha \frac{T_{\text{SL}}}{W_{\text{TO}}} c_t \, t_{\text{accel}} \tag{8.8}$$

For a climbing acceleration, the total leg time used to calculate fuel burn is taken as the sum of the times calculated by Eqs. (8.4) and (8.8).

8.5.2 Takeoff

The takeoff problem is a special case of the acceleration problem, with slightly different drag terms. The initial velocity for this problem is zero, and the final velocity is V_{TO}. Recall that in Chapter 5 the takeoff acceleration was given as

$$\frac{dV}{dt} = \frac{g \, [T - D - \mu(W - L)]}{W} \tag{5.48}$$

which, for aircraft that make little or no lift prior to rotation, simplifies to

$$\frac{dV}{dt} = g\left[\frac{(T-D)}{W} - \mu\right]$$

$$\frac{dV}{dt} = g\left[\left(\frac{\alpha}{\beta}\frac{T_{SL}}{W_{TO}}\right) - \frac{q}{\beta}\left(\frac{C_{D_0}}{W_{TO}/S}\right) - \mu\right] \tag{8.9}$$

For nearly constant acceleration (usually a good approximation for jet aircraft), the time to accelerate is

$$t_{TO} = V_{TO}\bigg/\frac{dV}{dt} \tag{8.10}$$

where

$$V_{TO} = 1.2\sqrt{\frac{2W_{TO}}{\rho S C_{L_{max}}}} \tag{5.51}$$

The fuel fraction used to take off is then

$$\frac{W_{f_{TO}}}{W_{TO}} = \alpha\frac{T_{SL}}{W_{TO}}c_t\,t_{TO} \tag{8.11}$$

Note that the discussion of takeoff performance in Sec. 5.11 recommended calculating drag at $0.7\,V_{TO}$. Therefore, the value of q in Eq. (8.9) should be evaluated for this speed. Also note that the assumption of zero lift prior to rotation is a conservative one, especially on soft or grassy fields. The actual takeoff time and distance can be slightly reduced to less than what is assumed in Eqs. (8.9–8.11) by causing the aircraft to produce a small fraction of its maximum lift coefficient during the takeoff roll.

8.5.3 Level Cruise and Turns

The maneuvers or mission segments that are sustained with thrust less than the maximum available require slightly different expressions for their fuel used. For these situations, we solve the Breguet endurance equation, Eq. (5.36), for fuel used.

$$E = \Delta t = \frac{1}{c_t}\frac{C_L}{C_D}\,\ln\left(\frac{W_1}{W_2}\right)$$

$$c_t\Delta t\bigg/\left(\frac{C_L}{C_D}\right) = \ln\left(\frac{W_1}{W_2}\right)$$

$$\exp\left[c_t\Delta t\bigg/\left(\frac{C_L}{C_D}\right)\right] = \left(\frac{W_1}{W_2}\right)$$

$$W_2 = W_1\exp\left\{-c_t\Delta t\bigg/\left(\frac{C_L}{C_D}\right)\right\} \tag{8.12}$$

so

$$\frac{W_f}{W_{TO}} = \frac{W_1}{W_{TO}} \left(1 - \exp\left\{-c_t \Delta t \Big/ \left(\frac{C_L}{C_D}\right)\right\}\right)$$

$$\frac{W_f}{W_{TO}} = \beta \left(1 - \exp\left\{-c_t \Delta t \Big/ \left(\frac{C_L}{C_D}\right)\right\}\right) \tag{8.13}$$

For loiter, combat air patrol, or endurance problems, loiter time is specified, so that Eqs. (8.12) and (8.13) can be used to solve for the fuel used. Note that $n = 1$ for steady, level flight in loiter and cruise problems. When loitering, the aircraft is normally flown at the speed for maximum endurance at the specified altitude. This is not necessarily true for a combat air patrol, where higher speeds can be used to reduce vulnerability to attack from unexpected directions. For a cruise problem, the time to travel a given range R is given by

$$t_{cruise} = R/V_{cruise}$$

where V_{cruise} is the velocity chosen for the cruise segment. Velocity for a cruise leg is often chosen as the velocity for maximum range at that altitude, but some other speed and/or altitude might need to be used to meet the customer's requirements. Once the cruise altitude and velocity are chosen, the fuel used to cruise is

$$\frac{W_f}{W_{TO}} = \beta \left(1 - \exp\left\{-c_t R \Big/ \left[V_{cruise}\left(\frac{C_L}{C_D}\right)\right]\right\}\right) \tag{8.14}$$

8.5.4 Best Cruise Mach/Best Cruise Altitude

It was shown in Sec. 5.10 that if the altitude and airspeed for a cruise leg are not constrained the absolute best range can be obtained (provided sufficient thrust is available at a reasonable TSFC) by flying the leg at $M = M_{crit}$ at an altitude where the velocity for M_{crit} is also the velocity for $(L/D)_{max}$. This condition was labeled best cruise Mach/best cruise altitude (BCM/BCA) in Chapter 5. For this condition $(L/D)_{max}$ is given by Eq. (5.23), and Eq. (8.14) becomes

$$\frac{W_f}{W_{TO}} = \beta \left(1 - \exp\left\{-c_t R \Big/ \left[V_{cruise}\left(\frac{1}{2\sqrt{kC_{D_0}}}\right)\right]\right\}\right) \tag{8.15}$$

8.5.5 Turns

The time required to perform a turn depends on the rate of turn ω and the number of degrees to be turned, which is given the symbol $\Delta\Psi$. The rate of turn is given by Eq. (5.61), so that the time to turn is

$$t_{turn} = \frac{\Delta\Psi}{\omega} = \frac{\Delta\Psi V}{g\sqrt{n^2 - 1}} \tag{8.16}$$

and the turn fuel weight fraction is

$$\frac{W_f}{W_{\text{TO}}} = \beta \left(1 - \exp \left\{ -c_t \frac{\Delta \Psi V}{g \sqrt{n^2 - 1}} \middle/ \left(\frac{C_L}{C_D} \right) \right\} \right) \qquad (8.17)$$

8.5.6 Total Mission Fuel Weight Fraction

Now that mathematical models have been developed for each of the types of mission segments it is possible to perform a complete mission analysis to estimate the total fuel weight fraction for the mission. The fuel fraction for each mission leg is calculated in the order the mission is flown. The fuel weight fraction used on each leg is subtracted from β for that leg to yield the β for the next leg. If the mission consisted of a takeoff, acceleration, climb, cruise, loiter, and descent/landing, this sequence of calculations would be

$$\beta_{\text{takeoff}} = 1 \quad (\text{because } W = W_{\text{TO}})$$

$$\beta_{\text{accel}} = \beta_{\text{takeoff}} - \frac{W_{f_{\text{takeoff}}}}{W_{\text{TO}}}$$

$$\beta_{\text{climb}} = \beta_{\text{accel}} - \frac{W_{f_{\text{accel}}}}{W_{\text{TO}}}$$

$$\beta_{\text{cruise}} = \beta_{\text{climb}} - \frac{W_{f_{\text{climb}}}}{W_{\text{TO}}}$$

$$\beta_{\text{loiter}} = \beta_{\text{cruise}} - \frac{W_{f_{\text{cruise}}}}{W_{\text{TO}}}$$

$$\frac{W_{f_{\text{mission}}}}{W_{\text{TO}}} = \frac{W_{f_{\text{takeoff}}}}{W_{\text{TO}}} + \frac{W_{f_{\text{accel}}}}{W_{\text{TO}}} + \frac{W_{f_{\text{climb}}}}{W_{\text{TO}}} + \frac{W_{f_{\text{cruise}}}}{W_{\text{TO}}} + \frac{W_{f_{\text{loiter}}}}{W_{\text{TO}}}$$

Note that no fuel burn or distance traveled is calculated for the descent and landing. In practice, the fuel planned for loitering is actually used for the descent and landing. This fuel is called a fuel reserve, and it is a mandatory part of the planning for nearly every flight. The fuel quantity is normally specified either as the fuel required to loiter for a certain time or as a percentage of the total mission fuel.

Example 8.3

A new business jet design concept has a clean (gear and flaps retracted) subsonic (below $M_{\text{crit}} = 0.78$) drag polar of $C_D = 0.02 + 0.06 C_L^2$, a $C_{L_{\text{max}}}$ for takeoff and landing of 1.6, a C_{D_0} for takeoff and landing of 0.04, and a design point of $T_{\text{SL}}/W_{\text{TO}} = 0.35$ and $W_{\text{TO}}/S = 65$ lb/ft^2. Its design mission consists of takeoff at sea level, acceleration to $M = 0.5$, climb to $h = 40{,}000$ ft, cruise at $M = M_{\text{crit}}$ for 2000 n miles, then descent and landing with enough fuel to loiter at $h = 10{,}000$ for 20 min. What is the fuel fraction for this design concept and this mission? Assume $c_{t_{\text{sea level}}} = 0.8/\text{h}$.

Solution: For the takeoff $\rho = 0.002377$ slug/ft^3, $a = 1116.2$ ft/s, $\alpha = 1, \beta = 1$, and $\mu = 0.03$,

$$V_{TO} = 1.2\sqrt{\frac{2W_{TO}}{\rho\,SC_{L_{max}}}} = 1.2\sqrt{\frac{2(65\text{ lb/ft}^2)}{0.002377\text{ slug/ft}^3(1.6)}} = 221.8\text{ ft/s}$$

$$0.7\,V_{TO} = 0.7(221.8) = 155.3\text{ ft/s}$$

$$q\quad\text{at}\quad 0.7\,V_{TO} = \frac{1}{2}\rho V^2 = \frac{1}{2}(0.002377\text{ slug/ft}^2)(155.3\text{ ft/s})^2 = 28.67\text{ lb/ft}^2$$

$$t_{TO} = V_{TO} \bigg/ \frac{dV}{dt} = 221.8\text{ ft/s}^2 \bigg/ \left\{g\left[\left(\frac{\alpha\,T_{SL}}{\beta\,W_{TO}}\right) - \frac{q}{\beta}\left(\frac{C_{D_0}}{W_{TO}/S}\right) - \mu\right]\right\}$$

$$t_{TO} = 221.8\text{ ft/s}^2 \bigg/ \left\{32.2\text{ ft/s}^2\left[(0.35) - (28.67\text{ lb/ft}^2)\left(\frac{0.04}{65\text{ lb/ft}^2}\right) - 0.03\right]\right\}$$

$$t_{TO} = 22.78\text{ s}$$

$$\frac{W_{f_{TO}}}{W_{TO}} = \alpha\frac{T_{SL}}{W_{TO}}c_t\,t_{TO} = (0.35)\left(\frac{0.8}{h}\right)(22.78\text{ s})\left(\frac{1\text{ h}}{3600\text{ s}}\right) = 0.00177$$

For the acceleration,

$$\rho = 0.002377\text{ slug/ft}^3 \quad a = 1116.2\text{ ft/s} \quad \alpha = 1$$

$$\beta = 1 - 0.00177 = 0.9982$$

and the acceleration is basically from V_{TO} to $M_\infty = 0.5$ or

$$V_\infty = M_\infty a = 0.5(1116.2\text{ ft/s}) = 558.1\text{ ft/s}$$

The average velocity during the acceleration is

$$\frac{558.1\text{ ft/s} + 221.8\text{ ft/s}}{2} = 390\text{ ft/s}$$

and so the average q is

$$q = \frac{1}{2}\rho V_\infty^2 = \frac{1}{2}(0.002377\text{ slug/ft}^2)(390\text{ ft/s})^2 = 180.7\text{ lb/ft}^2$$

$$\frac{dV}{dt} = g\left[\left(\frac{\alpha\,T_{SL}}{\beta\,W_{TO}}\right) - \frac{q}{\beta}\left(\frac{C_{D_0}}{W_{TO}/S}\right) - \frac{n^2\beta}{q}k_1\left(\frac{W_{TO}}{S}\right)\right]$$

$$= 32.2\text{ ft/s}^2\left[\left(\frac{0.35}{.9985}\right) - \frac{180.7\text{ lb/ft}^2}{0.9985}\left(\frac{0.02}{65\text{ lb/ft}^2}\right)\right.$$

$$\left.- \frac{0.9985}{180.7\text{ lb/ft}^2}(0.06)(65\text{ lb/ft}^2)\right] = 8.8\text{ ft/s}^2$$

The approximate time to accelerate, using the average value method, is

$$t_{accel} = \frac{(558.1 \text{ ft/s} - 221.8 \text{ ft/s})}{8.8 \text{ ft/s}^2} = 38.2 \text{ s}$$

and the fuel fraction used to accelerate is

$$\frac{W_{f_{accel}}}{W_{TO}} = 0.35(0.8/\text{h})\,(38.2 \text{ s})\left(\frac{1 \text{ h}}{3600 \text{ s}}\right) = 0.0029$$

For the climb, the average altitude is 20,000 ft, where $\rho = 0.001267$ slug/ft^3 and $a = 1036.9$ ft/s, so that

$$V_\infty = M_\infty a = 0.5(1036.9 \text{ ft/s}) = 518.5 \text{ ft/s}$$

$$\beta = 0.9982 - 0.0029 = 0.9953$$

$$n = 1$$

$$\alpha = (\rho/\rho_{SL}) = (0.001267 \text{ slug/ft}^3/0.002377 \text{ slug/ft}^3) = 0.533$$

$$c_T = c_{T\text{sea level}}(a/a_{SL}) = 0.8/\text{h}\,(1036.9 \text{ ft/s})/(1116.2 \text{ ft/s}) = 0.743/\text{h}$$

$$q = \frac{1}{2}\rho V_\infty^2 = \frac{1}{2}(0.001267 \text{ slug/ft}^2)(518.5 \text{ ft/s})^2 = 170.3 \text{ lb/ft}^2$$

$$\frac{dh}{dt} = V\left[\left(\frac{\alpha}{\beta}\frac{T_{SL}}{W_{TO}}\right) - \frac{q}{\beta}\left(\frac{C_{D_0}}{W_{TO}/S}\right) - \frac{n^2\beta}{q}k_1\left(\frac{W_{TO}}{S}\right)\right]$$

$$= 518.5 \text{ ft/s}\left\{\left[\frac{0.533}{0.9956}(0.35)\right] - \frac{170.3 \text{ lb/ft}^2}{0.9956}\left(\frac{0.02}{65 \text{ lb/ft}^2}\right)\right.$$

$$\left. - \frac{0.9953}{170.3 \text{ lb/ft}^2}0.06(65 \text{ lb/ft}^2)\right\} = 69.1 \text{ ft/s}$$

$$t_{climb} = \frac{(h_{final} - h_{initial})}{dh/dt} = \frac{(40{,}000 \text{ ft} - 0)}{69.1 \text{ ft/s}} = 579 \text{ s}$$

$$\frac{W_{f_{climb}}}{W_{TO}} = \alpha\frac{T_{SL}}{W_{TO}}c_T t_{climb} = (0.533)(0.35)(0.743/\text{h})(579 \text{ s})\left(\frac{1 \text{ h}}{3600 \text{ s}}\right) = 0.022$$

For the cruise, the altitude is 40,000 ft, where $\rho = 0.000587$ slug/ft^3 and $a = 968.1$ ft/s, so that

$$V_\infty = M_\infty a = 0.78(968.1 \text{ ft/s}) = 755.1 \text{ ft/s} = 446.8 \text{ kn}$$

$$\beta = 0.9953 - 0.022 = 0.9733 \quad n = 1$$

$$c_T = c_{T\text{sea level}}(a/a_{SL}) = 0.8/\text{h}\,(968.1 \text{ ft/s})/(1116.2 \text{ ft/s}) = 0.694/\text{h}$$

$$q = \frac{1}{2}\rho V_\infty^2 = \frac{1}{2}(0.000587 \text{ slug/ft}^2)(755.1 \text{ ft/s})^2 = 167.4 \text{ lb/ft}^2$$

$$C_L = \frac{nW}{qS} = \frac{n\beta}{q}\frac{W_{TO}}{S} = \frac{0.9733}{167.4 \text{ lb/ft}^2}(65 \text{ lb/ft}^2) = 0.378$$

$$C_D = C_{D_0} + kC_L^2 = 0.02 + 0.06(0.378)^2 = 0.0286$$

The time to travel 2000 n miles is $R/V_\infty = 2000$ n miles/446.8 n miles/h = 4.48 h so that

$$\frac{W_f}{W_{TO}} = \beta\left(1 - \exp\left\{-\left[c_t \Delta t \Big/ \left(\frac{C_L}{C_D}\right)\right]\right\}\right)$$

$$= 0.9953\left(1 - \exp\left\{-\left[(0.694)(4.48)\Big/\left(\frac{0.378}{0.0286}\right)\right]\right\}\right) = 0.209$$

Finally, for the loiter (because descents and landings are not modeled) the altitude is 10,000 ft, where

$$\rho = 0.001756 \text{ slug/ft}^3 \qquad a = 1077.4 \text{ ft/s} \qquad n = 1 \qquad \beta = 0.9733 - 0.209$$

$$= 0.764$$

$$c_T = c_{T\text{ sea level}}(a/a_{SL}) = 0.8/\text{h}(1077.4 \text{ ft/s})/(1116.2 \text{ ft/s}) = 0.772/\text{h}$$

The loiter would be flown at maximum endurance airspeed, where

$$\left(\frac{C_L}{C_D}\right)_{max} = \frac{1}{2\sqrt{kC_{D_0}}} = \frac{1}{2\sqrt{0.06(0.02)}} = 14.4$$

The loiter time is 20 min = 0.33 h, and so

$$\frac{W_f}{W_{TO}} = \beta\left(1 - \exp\left\{-\left[c_t \Delta t \Big/ \left(\frac{C_L}{C_D}\right)\right]\right\}\right)$$

$$= 0.764(1 - \exp\{-[(0.772)(0.33)/14, 4]\}) = 0.013$$

and

$$\frac{W_{f\text{mission}}}{W_{TO}} = \frac{W_{f\text{takeoff}}}{W_{TO}} + \frac{W_{f\text{accel}}}{W_{TO}} + \frac{W_{f\text{climb}}}{W_{TO}} + \frac{W_{f\text{cruise}}}{W_{TO}} + \frac{W_{f\text{loiter}}}{W_{TO}}$$

$$= 0.0018 + 0.0029 + 0.022 + 0.209 + 0.013 = 0.248$$

8.6 Sizing Equation

Once the mission fuel fraction is known, the aircraft can be sized. The sizing problem is solved by using the fact that the total weight of the aircraft is the sum of all its parts and contents, and that sum must fit the design point specified by constraint analysis:

$$W_{TO} = W_{\text{airframe}} + W_{\text{payload}} + W_{\text{engine}} + W_{f\text{mission}} \qquad (8.18)$$

$$\frac{W_{TO}}{S} = \frac{W_{\text{airframe}}}{S} + \frac{W_{\text{payload}}}{S} + \frac{W_{\text{engine}}}{S} + \frac{W_{f\text{mission}}}{S} \qquad (8.19)$$

where the payload weight includes pilot(s), passengers, cargo, weapons, avionics, sensors, etc. required for the design mission, and the systems weight includes the landing gear and "all else" weights from Table 8.2. Each term in Eq. (8.19) can be thought of as a *wing loading portion*, that portion of the total aircraft design wing loading which is contributed by a particular item. The airframe weight includes the weight of all structures plus landing gear and systems. Structural weights of most of the aircraft's components (which include estimates of electrical and hydraulic systems in those components) are estimated as functions of surface areas using the parameters in Table 8.1, so that the total wing loading portion for the structure can be represented as

$$\frac{W_{\text{structure}}}{S} = \frac{W_{\text{wing}}}{S} + \frac{W_{\text{horiz.tail}}}{S_t}\frac{S_t}{S} + \frac{W_{\text{vert.tail}}}{S_v}\frac{S_v}{S} + \frac{W_{\text{fuselage}}}{S_{\text{wet}_{\text{fuselage}}}}\frac{S_{\text{wet}_{\text{fuselage}}}}{S} \qquad (8.20)$$

For a given aircraft configuration, the ratios of surface areas do not change as the size of the aircraft increases or decreases, and so Eq. (8.20) will hold for the aircraft regardless of its size as long as its configuration does not change. Unfortunately, its configuration will change with its overall size because the volume reserved for cockpits, passenger cabins, avionics, and payloads will not change as the aircraft size increases or decreases. A volume check after the aircraft is sized may force changes in the configuration. The method for handling this situation will be described after the explanation of the sizing equation is complete.

The weight of the airframe includes, in addition to component structures and systems weight, the weight of landing gear and all else, which includes fixtures and attachment points for cargo and passenger seats, connections for lavatories and galleys (the lavatories and galleys themselves are counted as payload), systems such as air conditioning/pressurization, breathing oxygen, emergency hydraulic and electrical systems, brake controls, etc. not included in the component weights. Required mission systems like reconnaissance packages, sensors, onboard entertainment, weapons, and targeting systems are all counted as payload. According to Table 8.2, both landing gear and all-else weights are given as fractions of takeoff gross weight, so that multiplying their weight fractions by the aircraft total wing loading gives their wing loading portions.

$$\frac{W_{\text{gear}}}{S} = 0.033\frac{W_{\text{TO}}}{S}$$

or, for carrier-based aircraft,

$$\frac{W_{\text{gear}}}{S} = 0.047\frac{W_{\text{TO}}}{S}$$

and for all aircraft,

$$\frac{W_{\text{all else}}}{S} = 0.17\frac{W_{\text{TO}}}{S}$$

$$\frac{W_{\text{airframe}}}{S} = \frac{W_{\text{structure}}}{S} + \frac{W_{\text{gear}}}{S} + \frac{W_{\text{all else}}}{S} \qquad (8.21)$$

The wing loading portion for the engine is determined by multiplying the engine weight-to-thrust-ratio from Table 8.2 by the aircraft's design point thrust-to-weight ratio and wing loading from constraint analysis:

$$\frac{W_{engine}}{S} = \frac{W_{engine}}{T_{SL}} \frac{T_{SL}}{W_{TO}} \frac{W_{TO}}{S} \tag{8.22}$$

Likewise, the wing loading portion for the mission fuel is calculated by multiplying the mission fuel weight fraction by the design wing loading:

$$\frac{W_{f_{mission}}}{S} = \frac{W_{f_{mission}}}{W_{TO}} \frac{W_{TO}}{S} \tag{8.23}$$

At this point, the only unknown in Eq. (8.19) is the sized wing area. Solving for this yields:

$$S = W_{payload} \Big/ \left(\frac{W_{TO}}{S} - \frac{W_{airframe}}{S} - \frac{W_{engine}}{S} - \frac{W_{f_{mission}}}{S} \right) \tag{8.24}$$

The result of the calculation in Eq. (8.24) is final provided the aircraft configuration does not need to change after it is sized. As already mentioned, the fact that the volume occupied by the payload, passengers, crew, etc. does not change when the airplane is resized might require that the relative sizes of the fuselage and wings, for instance, change when the airplane size changes. At best, this effect will be small, and the results of an initial sizing calculation will be sufficiently accurate. At worst, the newly sized aircraft will have to be redrawn and new structural estimates made. A new drag polar will also need to be calculated for the new configuration, and a new mission fuel fraction calculated. These new results are used in another sizing calculation, and the process is repeated until it converges to an aircraft size, which satisfies internal volume and weight sizing requirements.

8.7   Weight Fraction Method

An alternative formulation of the sizing equation is also popular. It is obtained by dividing Eq. (8.19) by the design point wing loading to obtain

$$1 = \frac{W_{airframe}}{W_{TO}} + \frac{W_{payload}}{W_{TO}} + \frac{W_{engine}}{W_{TO}} + \frac{W_{f_{mission}}}{W_{TO}} \tag{8.25}$$

which expresses the weights of the components in terms of weight fractions. For this formulation, mission fuel and miscellaneous systems weight fractions are used without modification, and the engine weight fraction is obtained by multiplying the engine weight-to-thrust ratio from Table 8.2 by the design point thrust-to-weight ratio. The airframe weight fraction is obtained by dividing (8.21) by the design point wing loading to obtain

$$\frac{W_{airframe}}{W_{TO}} = \frac{W_{structure}/S}{W_{TO}/S} + 0.033 + 0.17 \tag{8.26}$$

Equation (8.25) is then solved for W_{TO}. The sized aircraft reference wing area is then obtained by dividing W_{TO} by the design point wing loading. The methods embodied in Eqs. (8.19) and (8.25) are equivalent. The choice of formulation is left to the user.

Example 8.4

The business jet design concept analyzed in Example 8.3 has a structural wing loading portion of 20 lb/ft^2 and a payload weight of 3000 lb. How large must this aircraft be to fly the design mission?

Solution: The other wing loading portions are first calculated:

$$\frac{W_{\text{airframe}}}{S} = \frac{W_{\text{structures}}}{S} + \left(\frac{W_{\text{lndg gear}}}{W_{\text{TO}}} + \frac{W_{\text{misc}}}{W_{\text{TO}}}\right)\frac{W_{\text{TO}}}{S}$$

$$= 20 \text{ lb/ft}^2 + (0.043 + 0.17)(65 \text{ lb/ft}^2) = 33.8 \text{ lb/ft}^2$$

$$\frac{W_{\text{engine}}}{S} = \frac{W_{\text{engine}}}{T_{\text{SL}}}\frac{T_{\text{SL}}}{W_{\text{TO}}}\frac{W_{\text{TO}}}{S} = (0.26)(0.35)(65 \text{ lb/ft}^2) = 5.915 \text{ lb/ft}^2$$

$$\frac{W_{f_{\text{mission}}}}{S} = \frac{W_{f_{\text{mission}}}}{W_{\text{TO}}}\frac{W_{\text{TO}}}{S} = (0.268)(65 \text{ lb/ft}^2) = 17.4 \text{ lb/ft}^2$$

Then

$$S = W_{\text{payload}} \Big/ \frac{W_{\text{TO}}}{S} - \frac{W_{\text{airframe}}}{S} - \frac{W_{\text{engine}}}{S} - \frac{W_{f_{\text{mission}}}}{S}$$

$$= \frac{3000 \text{ lb}}{(65 \text{ lb/ft}^2) - (33.8 \text{ lb/ft}^2) - (5.915 \text{ lb/ft}^2) - (17.4 \text{ lb/ft}^2)} = 380 \text{ ft}^2$$

and the sized aircraft gross weight is

$$W_{\text{TO}} = S\frac{W_{\text{TO}}}{S} = (380 \text{ ft}^2)(65 \text{ lb/ft}^2) = 24{,}730 \text{ lb}$$

8.8 Weight and Balance

Once the aircraft is sized and the required weight of fuel is known, a more detailed arrangement of internal components can be worked out. The final step in sizing ensured that sufficient internal volume existed so that all items required to be carried internally would fit, but the specific locations of each component also are important. This is because the location of the aircraft's center of gravity must be controlled so that at all times throughout the mission, as fuel is burned and payload offloaded or expended, the aircraft's static margin will remain at acceptable values. As discussed in Sec. 8.4, c.g. location is also very important to landing-gear weight distribution, tipback, and rollover criteria, and ground handling in general.

The aircraft center of gravity is determined by summing the component weights and moments about some reference location, then dividing the total moment by the total weight. This calculation must be completed for each possible aircraft weight and loading configuration. If required loading conditions are found that place the aircraft center of gravity such that its static margin is outside acceptable limits, then the fuel and payload locations within the aircraft probably should

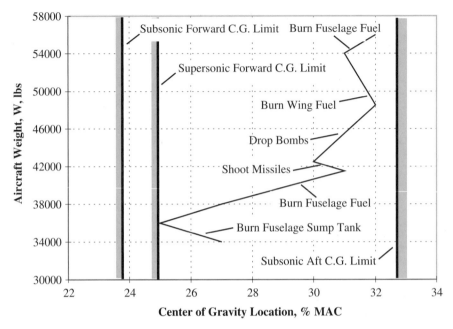

Fig. 8.7   Center-of-gravity excursion diagram for a multirole fighter.

be rearranged to alter the center of gravity shifts that result from their offloading or expenditure.

8.8.1   Center-of-Gravity Excursion Diagram

A convenient tool for verifying that the aircraft center of gravity remains within acceptable limits throughout the mission is called the c.g. excursion diagram. Figure 8.7 is an example of such a diagram. Aircraft static margin limits are translated into center of gravity limits and plotted on the diagram. These limits can be different for supersonic flight, and for certain circumstances they can be set by control authority limits rather than stability limits. All appropriate c.g. limits are shown as boundaries on the diagram. The actual variation of the c.g. throughout the mission is then plotted on the diagram to verify compliance with the limits and give the operators of the aircraft an indication of how much latitude they have in varying the loading condition to meet alternate mission requirements.

8.8.2   Center-of-Gravity Excursion Control

Many aircraft, especially aerial tankers and supersonic delta-winged aircraft such as the Convair F-106 and the Anglo-French Concorde supersonic transport, have systems for redistributing their fuel load during flight. Tankers must carry a great deal of fuel, and this frequently requires that some fuel tanks be located far from the plane's center of gravity, where offloading of the fuel will produce

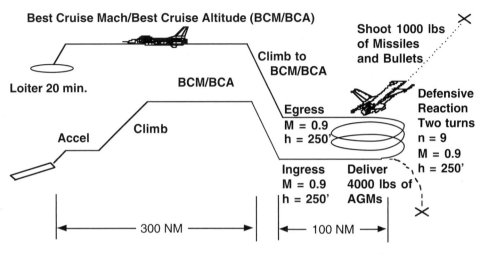

Fig. 8.8   Typical design mission for a multirole fighter.

large changes in the c.g. location. Supersonic aircraft experience large changes in their static margins between subsonic and supersonic flight as their wing aerodynamic centers shift from 0.25 mean aerodynamic chord (MAC) to approximately 0.5 (MAC). To combat these problems, these aircraft are equipped with fuel transfer systems that can pump fuel to different parts of the aircraft as needed to maintain acceptable c.g. locations.

8.9   Mission Analysis and Sizing Example

For an example of how the sizing equation is used in conceptual aircraft design, consider a conceptual design for a multirole fighter aircraft with aerodynamic characteristics similar to the F-16. Figure 1.6 illustrated a typical design mission for such an aircraft. Figure 8.8 is identical to Fig. 1.6, but with specific values for mission leg lengths, altitudes, and speeds specified. These parameters for the mission are listed in Table 8.4.

The lift and drag data for the F-16 predicted in Sec. 4.9 and the engine data specified in Sec. 5.16 will be used as models of the aircraft and engine in this mission analysis, except that during the first half of the mission the plane will be carrying two AIM-9 and two AIM-120 air-to-air missiles and two MK-84 2000-lb bombs on external rails and pylons. Modeling these external stores and suspension equipment as simple shapes to determine their wetted area, then adding this to the aircraft total wetted area and multiplying by C_{fe} yields a new value for $C_{D_0} = 0.028$, which must be used until those items are expended halfway through the mission. Assume that a design point of $W_{TO}/S = 100$ lb/ft^2 and $T_{SL}/W_{TO} = 0.9$ has been selected and that the required mission payload weighs 5000 lb. Then, for the takeoff we observe that for the business jet in Example 8.4 with $T_{SL}/W_{TO} = 0.35$ the drag and rolling friction were only 15% of the value of the thrust. We can therefore say confidently that for the F-16 making a maximum afterburner takeoff so that

Table 8.4   Multirole fighter strike mission parameters

Leg no.	Name	Altitude, ft	Mach no.	n	ΔV, ft/s	Δh, ft	Distance, n mile	Time, h	ΔW, lb
1	Take off		0.2	1	?	0	0	?	?
2	Acceleration		0.4	1	446.6	0	?	?	?
3	Climb	20,000	0.6	1	0	BCA-0	?	?	?
4	BCM/BCA	BCA	M_{crit}	1	0	0	300[a]	?	?
5	Ingress	250	0.9	1	0	0	100	?	?
6	Bomb	—	—	—	—	—	—	—	− 4,000
7	Turn	250	0.9	9	0	0	0	720 deg/ω	?
8	Shoot	—	—	—	—	—	—	—	− 1,000
9	Egress	250	0.9	1	0	0	100	?	?
10	Climb	BCA	0.6	1	0	BCA-0	?	?	?
11	BCM/BCA	BCA	M_{crit}	1	0	0	300[b]	?	?
12	Loiter	20,000	M for $(L/D)_{max}$	1	0	0	0	0.333	?

[a]Includes distance to climb on leg 3.   [b]Includes distance to climb on leg 10.

$T_{SL}/W_{TO} = 0.9$ the drag and rolling friction can be ignored. For this case, Eq. (8.9) simplifies to

$$\frac{dV}{dt} = g\frac{\alpha T_{SL}}{W_{TO}} = 32.2 \text{ ft/s}^2\,(0.9) = 28.98 \text{ ft/s}^2$$

$$V_{TO} = 1.2\sqrt{\frac{2W_{TO}}{\rho\,SC_{L_{max}}}} = 1.2\sqrt{\frac{2(100 \text{ lb/ft}^2)}{0.002377 \text{ slug/ft}^3(1.2)}} = 317.7 \text{ ft/s}$$

$$t_{TO} = V_{TO}\Big/\frac{dV}{dt} = 317.7 \text{ ft/s}/28.98 \text{ ft/s}^2 = 11 \text{ s}$$

$$\frac{W_{f_{TO}}}{W_{TO}} = \alpha\frac{T_{SL}}{W_{TO}}c_T\,t_{TO} = 0.9(2.46/h)(1.0)(11 \text{ s})(1 \text{ h}/3600 \text{ s}) = 0.0068$$

$$\beta_2 = \beta_1 - 0.0068 = 0.9932$$

Now for the acceleration, $q = 237$ lb/ft^2 at the average Mach number of 0.4.

$$\frac{dV}{dt} = g\left[\left(\frac{\alpha}{\beta}\frac{T_{SL}}{W_{TO}}\right) - \frac{q}{\beta}\left(\frac{C_{D_0}}{W_{TO}/S}\right) - \frac{n^2\beta}{q}k_1\left(\frac{W_{TO}}{S}\right)\right]$$

$$= 32.2 \text{ ft/s}^2\left[\left(\frac{1.0}{0.9932}0.9\right) - \frac{237 \text{ lb/ft}^2}{0.9932}\left(\frac{0.028}{100 \text{ lb/ft}^2}\right)\right.$$

$$\left. - \frac{0.9932}{237 \text{ lb/ft}^2}0.117\left(100 \text{ lb/ft}^2\right)\right]$$

$$= 25.4 \text{ ft/s}^2$$

$$t_{\text{accel}} = \frac{(V_{\text{final}} - V_{\text{initial}})}{dV/dt} = (446.6 \text{ ft/s})/(25.4 \text{ ft/s}^2) = 17.5 \text{ s} = 0.0049 \text{ h}$$

$$\frac{W_{f_{\text{accel}}}}{W_{\text{TO}}} = \alpha \frac{T_{\text{SL}}}{W_{\text{TO}}} c_T \frac{a}{a_{\text{SL}}} t_{\text{accel}} = 1.0(0.9)2.46/\text{h}(1.0)(0.0049) \text{ h} = 0.0108$$

$$\beta_3 = \beta_2 - 0.0108 = 0.9932 - 0.0108 = 0.9824$$

For the climb at $M = 0.6$, average altitude $= 20,000$ ft, $q = 178$ lb/ft^2, and

$$\alpha = T_A/T_{\text{SL}} = (\rho/\rho_{\text{SL}})^{0.7} = 0.64$$

$$\frac{dh}{dt} = V\left[\left(\frac{\alpha}{\beta}\frac{T_{\text{SL}}}{W_{\text{TO}}}\right) - \frac{q}{\beta}\left(\frac{C_{D_0}}{W_{\text{TO}}/S}\right) - \frac{n^2\beta}{q}k_1\left(\frac{W_{\text{TO}}}{S}\right)\right]$$

$$= 0.6(1036.9 \text{ ft/s})\left[\left(\frac{0.64}{0.9824}0.9\right) - \frac{245 \text{ lb/ft}^2}{0.9824}\left(\frac{0.028}{100 \text{ lb/ft}^2}\right)\right.$$

$$\left. - \frac{1.0(0.9824)}{245 \text{ lb/ft}^2}0.117\left(100 \text{ lb/ft}^2\right)\right]$$

$$= 294 \text{ ft/s}$$

The climb is to BCA, but that altitude is not actually known until the wing loading after the climb is known. As an approximation to get things started, assume the climb is to 40,000 ft. Then, if BCA ends up being significantly different from that the climb can be recalculated:

$$t_{\text{climb}} = \frac{(h_{\text{final}} - h_{\text{initial}})}{dh/dt} = \frac{40,000 \text{ ft}}{294 \text{ ft/s}} = 136 \text{ s} = 0.038 \text{ h}$$

$$\frac{W_{f_{\text{climb}}}}{W_{\text{TO}}} = \alpha \frac{T_{\text{SL}}}{W_{\text{TO}}} c_T \frac{a}{a_{\text{SL}}} t_{\text{climb}} = 0.64(0.9)(2.46/\text{h})\frac{1036.9}{1116.4}0.038/\text{h} = 0.05$$

$$\beta_4 = \beta_3 - 0.05 = 0.9824 - 0.05 = 0.9324$$

The distance covered during the climb is the average velocity multiplied by the time to climb:

$$s_{\text{climb}} = (0.6)(1036.9 \text{ ft/s})(136 \text{ s})(1 \text{ n mile}/6080 \text{ ft}) = 14 \text{ n mile}$$

So, at the start of BCA $\beta = 0.9324$, and $W/S = 93.24$ lb/ft^2. $M_{\text{crit}} = 0.86$. For $(L/D)_{\text{max}}$,

$$C_L = \sqrt{\frac{C_{D_0}}{k_1}} = \sqrt{\frac{0.028}{0.117}} = 0.49$$

$$L = W = C_L q S,$$

and so

$$q = (W/S)/C_L = 93.24 \text{ lb/ft}^2/0.49 = 190.6 \text{ lb/ft}^2$$

Because BCA will undoubtedly occur in the constant-temperature portion of the stratosphere, $a = 968.1$ ft/s and $V = Ma = 0.86 (968.1 \text{ ft/s}) = 832.6$ ft/s $= 492.7$ kn. Then,

$$q = \tfrac{1}{2}\rho V^2 = 190.6 \text{ lb/ft}^2$$

and so

$$\rho = 2q/V^2 = 2(190.6 \text{ lb/ft}^2)/(832.6 \text{ ft/s})^2 = 0.00055 \text{ slug/ft}^3$$

which occurs at $h = 41{,}300$ ft in the standard atmosphere. Of course, the altitude for BCA will increase as fuel burns off. The total range for this cruise segment plus the initial climb is 300 n mile, and so the cruise range is $300 - 14$ n mile $= 286$ n mile,

$$n = 1$$

$$c_T = c_{T\text{sea level}}(a/a_{SL}) = 0.8/\text{h}(968.1 \text{ ft/s})/(1116.2 \text{ ft/s}) = 0.694/\text{h}$$

$$C_D = C_{D_0} + kC_L^2 = 0.028 + 0.117(0.49)^2 = 0.056$$

The time to travel 286 nm is $R/V_\infty = 286$ n mile/492.7 n mile/h $= 0.58$ h, and so

$$\frac{W_f}{W_{TO}} = \beta \left(1 - \exp\left\{-\left[c_t \Delta t \Big/ \left(\frac{C_L}{C_D}\right)\right]\right\}\right)$$

$$= 0.9324 \left(1 - \exp\left\{-\left[(0.694)(0.58)\Big/ \left(\frac{0.49}{0.056}\right)\right]\right\}\right) = 0.0419$$

$$\beta_5 = \beta_4 - 0.0419 = 0.9324 - 0.0419 = 0.8905$$

No benefit or penalty is given for the descent to the ingress altitude. For the ingress at $h = 250$ ft, $a = 1115$ ft/s, $V = Ma = 0.9 (1115 \text{ ft/s}) = 1003.5$ ft/s $= 593.8$ kn and $\rho = 0.002360$ slug/ft^3, so that

$$q = 1/2 (0.00236 \text{ slug/ft}^3)(1003.5 \text{ ft/s})^2 = 1188 \text{ lb/ft}^2$$

$$n = 1$$

$$c_T = c_{T\text{sea level}}(a/a_{SL}) = 0.8/\text{h}(1115 \text{ ft/s})/(1116.2 \text{ ft/s}) = 0.799/\text{h}$$

$$C_L = \frac{nW}{qS} = \frac{n\beta}{q}\frac{W_{TO}}{S} = \frac{0.8905}{1188 \text{ lb/ft}^2}(100 \text{ lb/ft}^2) = 0.0749$$

$$C_D = C_{D_0} + kC_L^2 = 0.028 + 0.117(0.0749)^2 = 0.0287$$

The time to travel 100 n mile is $R/V_\infty = 100$ n mile / 593.8 n mile/h $= 0.168$ h,

so that

$$\frac{W_f}{W_{TO}} = \beta\left(1 - \exp\left\{-\left[c_t \Delta t \Big/ \left(\frac{C_L}{C_D}\right)\right]\right\}\right)$$

$$= 0.8905\left(1 - \exp\left\{-\left[(0.799)(0.168)\Big/\left(\frac{0.749}{0.0287}\right)\right]\right\}\right) = 0.045$$

$$\beta_6 = \beta_5 - 0.045 = 0.8905 - 0.045 = 0.8455$$

At this point in the mission, 4000 lb of weapons are expended, but the weight *fraction* of the weapons is not known. An assumption must be made, so that the analysis can proceed, with the intent that once the aircraft is sized the analysis must be repeated with the appropriate weapons weight fraction. Weapons weighing 4000 lb represent approximately 14% of the weight of an F-16 configured this way, and so assume

$$\beta_7 = \beta_6 - 0.14 = 0.8455 - 0.14 = 0.7055$$

The design mission now requires two complete defensive turns at ingress flight conditions and $n = 9$. Undoubtedly the afterburner will have to be used to sustain these turns, so that $c_t = 2.45/\text{h}$. The bombs have been dropped, but the missiles are still carried, so that $C_{D_0} = 0.0198$.

$$t_{\text{turn}} = \frac{\Delta\Psi V}{g\sqrt{n^2 - 1}} = \frac{4\pi(1003.5 \text{ ft/s})}{32.2 \text{ ft/s}^2\sqrt{80}} = 43.8 \text{ s} = 0.0122 \text{ h}$$

$$q = \tfrac{1}{2}(0.00236 \text{ slug/ft}^3)(1003.5 \text{ ft/s})^2 = 1188 \text{ lb/ft}^2$$

$$n = 9$$

$$c_T = c_{T\,\text{sea level}}(a/a_{SL}) = 0.8/\text{h}\,(1115 \text{ ft/s})/(1116.2 \text{ ft/s}) = 0.799/\text{h}$$

$$C_L = \frac{nW}{qS} = \frac{n\beta}{q}\frac{W_{TO}}{S} = \frac{9(0.8905)}{1188 \text{ lb/ft}^2}\left(100 \text{ lb/ft}^2\right) = 0.675$$

$$C_D = C_{D_0} + kC_L^2 = 0.0198 + 0.117(0.675)^2 = 0.073$$

and the turn fuel weight fraction is

$$\frac{W_f}{W_{TO}} = \beta\left(1 - \exp\left\{-\left[c_t \Delta t \Big/ \left(\frac{C_L}{C_D}\right)\right]\right\}\right)$$

$$= 0.7055\left(1 - \exp\left\{-\left[(0.799)(0.0122)\Big/\left(\frac{0.675}{0.073}\right)\right]\right\}\right) = 0.001$$

$$\beta_8 = \beta_7 - 0.001 = 0.7035 - 0.001 = 0.7025$$

The mission analysis continues with air-to-air weapons expenditure and then a return flight and loiter. The total fuel weight fraction for the mission is obtained by summing the fuel weight fractions for each leg where fuel was burned (not the weapons expenditure legs):

$$\frac{W_{f_{\text{mission}}}}{W_{\text{TO}}} = \frac{W_{f_{\text{takeoff}}}}{W_{\text{TO}}} + \frac{W_{f_{\text{accel}}}}{W_{\text{TO}}} + \frac{W_{f_{\text{climb}}}}{W_{\text{TO}}} + \frac{W_{f_{\text{cruise}}}}{W_{\text{TO}}} + \cdots\cdots + \frac{W_{f_{\text{loiter}}}}{W_{\text{TO}}}$$

$$\frac{W_{f_{\text{mission}}}}{W_{\text{TO}}} = 0.0022 + 0.0108 + 0.05 + 0.0419 + \cdots$$

$$\frac{W_{f_{\text{mission}}}}{W_{\text{TO}}} = 0.31$$

At this point, the airplane can be sized. First, the model of the F-16 built from simple shapes is used to determine the required areas for structural weight calculations. Once these reference areas are known,

$$\frac{W_{\text{wing}}}{S} = 0.04 \frac{n_{\text{max}}^{0.2} \, \mathcal{R}^{1.8} \, (1+\lambda)^{0.5}}{(t/c)^{0.7} \cos \Lambda_{\text{LE}}} = 0.04 \frac{9^{0.2} \, 3.0^{1.8} \, (1+0.212)^{0.5}}{(0.04)^{0.7} \cos 40 \deg}$$

$$= 6.13 \; \text{lb/ft}^2$$

$$\frac{W_{\text{structure}}}{S} = \frac{W_{\text{wing}}}{S} + \frac{W_{\text{horiz.tail}}}{S_t}\frac{S_t}{S} + \frac{W_{\text{vert.tail}}}{S_v}\frac{S_v}{S} + \frac{W_{\text{fuselage}}}{S_{\text{wet}_{\text{fuselage}}}}\frac{S_{\text{wet}_{\text{fuselage}}}}{S}$$

$$= 6.13 \; \text{lb/ft}^2 + 4.0 \; \text{lb/ft}^2 \frac{108 \; \text{ft}^2}{300 \; \text{ft}^2} + 6.0 \; \text{lb/ft}^2 \frac{51.25 \; \text{ft}^2}{300 \; \text{ft}^2} + 4.8 \; \text{lb/ft}^2 \frac{859 \; \text{ft}^2}{300 \; \text{ft}^2}$$

$$= 22.35 \; \text{lb/ft}^2$$

$$\frac{W_{\text{airframe}}}{S} = \frac{W_{\text{structures}}}{S} + \left(\frac{W_{\text{lndg gear}}}{W_{\text{TO}}} + \frac{W_{\text{misc}}}{W_{\text{TO}}}\right)\frac{W_{\text{TO}}}{S}$$

$$= 22.35 \; \text{lb/ft}^2 + (0.033 + 0.17)(100 \; \text{lb/ft}^2) = 42.65 \; \text{lb/ft}^2$$

$$\frac{W_{\text{engine}}}{S} = \frac{W_{\text{engine}}}{T_{\text{SL}}}\frac{T_{\text{SL}}}{W_{\text{TO}}}\frac{W_{\text{TO}}}{S} = 0.13(0.9)100 \; \text{lb/ft}^2 = 11.7 \; \text{lb/ft}^2$$

$$\frac{W_{f_{\text{mission}}}}{S} = \frac{W_{f_{\text{mission}}}}{W_{\text{TO}}}\frac{W_{\text{TO}}}{S} = 0.31(100 \; \text{lb/ft}^2) = 31 \; \text{lb/ft}^2$$

$$S = W_{\text{payload}} \bigg/ \frac{W_{\text{TO}}}{S} - \frac{W_{\text{airframe}}}{S} - \frac{W_{\text{engine}}}{S} - \frac{W_{f_{\text{mission}}}}{S}$$

$$= \frac{5000 \; \text{lb}}{(100 - 42.65 - 11.7 - 31) \; \text{lb/ft}^2}$$

$S = 1408 \; \text{ft}^2$! and $W_{\text{TO}} = S(W_{\text{TO}}/S) = 140,800$ lb, which is a Boeing 737-size airplane!!

Clearly, the F-16 in its present form is not well suited to this design mission. The aircraft designer has a wide range of tools to use to modify the design geometry, constraints, and mission to produce an airplane concept that can fly the design mission but be of a size that is more reasonable and affordable. Many design iterations and some long discussions with the customer might be needed to achieve an acceptable result.

8.10 AeroDYNAMIC

Mission analysis and sizing is tedious, and typically must be repeated many times during the design process. The AeroDYNAMIC software that accompanies this textbook allows the user to easily input the characteristics of several different design missions, quickly "fly" one or more design concepts through any or all of the missions, and make changes to the designs to improve their performance. The program gives the user instant feedback on the impact on mission performance of a particular design change, making it possible to rapidly optimize a design concept for best performance on a particular mission.

8.11 Cost

Methods for estimating cost are plentiful, but most of them are extremely complex. Many analysts are employed by industry and government agencies to attempt to estimate what a new aircraft will cost. A comprehensive method for aircraft cost estimation must include the effects of design and development hours, manufacturing methods and materials, added costs of new technology, labor and factory facility costs, the number of aircraft in the production run, marketing, flight test and certification, and a myriad of other considerations. To the buyer, operating costs are just as important as the initial purchase price of the aircraft. These costs depend on a different set of factors. For airlines, the bottom line in cost analysis is the *return on investment*, the total amount of money the airline can expect to make as profit for each aircraft it purchases, operates, and eventually retires from service.

8.11.1 Purchase Price

A simple first guess on purchase price can be made based on the fact that the cost is generally proportional to aircraft weight, and the cost per pound of most types of aircraft has been steadily increasing. Table 8.5 lists the 2001 purchase prices and price per pound of several aircraft types. These can be taken as a lower bound for the possible price per pound of future aircraft, unless new manufacturing technology reverses the upward price trend.

The most cursory survey of aircraft price trends suggests an alarming rate of price increases. Consider, for instance, the 52,000-lb, twin-engine F-4 Phantom, which sold in 1966 for $5 million, and its replacement, the 40,000-lb, single-engine F-16, which was considered a bargain in 1995 at $24 million! The causes for this rapid rise in aircraft prices include increased manpower, energy, and technology costs. There is good reason to believe that introduction of advanced engineering and manufacturing methods and aggressive cost controls can reverse or at least slow this trend.

Table 8.5   Purchase price per pound of several aircraft types[a]

Type	Purchase price, million 2001 dollars	W_{TO}, lb	Price/W_{TO} $/lb
Jet airliners/transports			
Boeing 777	176.5	600,000	294
Boeing 767	100	387,000	258
Boeing 737-700	38	149,000	255
Fokker 70	25	81,000	308
McDonnell–Douglas C-17	294	585,000	502
Jet bombers			
Northrop–Grumman B-2	2,230	376,000	5,930
Jet fighters			
Lockheed–Martin F-22	81	60,000	1,350
Boeing F-15E	65	80,000	812
Boeing F/A-18E	44	42,000	714
Grumman F-14	71	70,400	1,021
Lockheed–Martin F-16C	24	40,000	600
Lockheed–Martin F-35A	28	50,000	560
Lockheed–Martin F-35B	35	50,000	700
Lockheed–Martin F-35C	38	57,000	666
Lockheed–Martin F-117	43	52,500	811
Mitsubishi FS-X	80	49,000	1,632
Tornado IDS/ADV	33	61700	534
Eurofighter Typhoon	53	46,300	1,146
Dassault Rafale C	47	47,400	981
Dassault Rafale M	52	47,400	1,097
Saab Gripen	30	27,500	1,090
Business jets			
Learjet 45	6.2	19,500	318
Learjet 60	10	23,500	425
Cessna Citation V	5.4	16,000	337
Cessna Citation VII	9.8	22,450	437
Cessna Citation X	13.1	34,500	380
Bombardier Challenger	20	45,000	444
Gulfstream 4SP	25	74,600	335
Gulfstream 5	30	85,100	352
General aviation			
Cessna 172	0.1	2,500	40
Mooney M20R	0.19	3,200	59
Mooney M20M TLS	0.29	3,300	88

[a]Aircraft statistics and prices obtained from current published data.

8.11.2 Operating Costs

The primary costs for aircraft operations are the price of fuel, crew salaries, and maintenance expenses. Fuel costs can be estimated based on the expected fuel consumption for the design mission and expected fuel prices, adjusted for anticipated inflation. Aircrew salaries vary. Some are based on the number of flight hours flown, whereas others must be paid whether the aircraft flies or not. The customer will generally be the best source of information on crew salary costs. Maintenance costs can be estimated by historical trends and by modeling the required maintenance procedures to determine the number of work hours and the type and cost of replacement parts that are required to perform them. Computer simulations have recently significantly improved the accuracy of these predictions. More detailed information on cost-estimating methods can be found in Refs. 2 and 3.

8.12 Chapter Summary

8.12.1 Internal Layout

1) Place expendable payload near c.g.
2) Distribute fuel ahead and behind c.g. to balance there
3) Put heavy items close to main structure
4) Keep the guns away from sensitive avionics

8.12.2 Structures and Weight

1) Given as weight per unit area (wing loading)
2) Wing:

$$\frac{W_{\text{wing}}}{S} = 0.04 \frac{n_{\max}^{0.2} \, \mathcal{R}^{1.8} \, (1+\lambda)^{0.5}}{(t/C)^{0.7} \cos \Lambda_{\text{LE}}}$$

3) Other component weights are given as fixed number averaged from data
4) Fuselage weight based on wetted area rather than planform
5) Gear and all else given as fractions of W_{TO}
6) Engine(s):

$$\frac{W_{\text{engine}}}{S} = \frac{W_{\text{engine}}}{T_{\text{SL}}} \frac{T_{\text{SL}}}{W_{\text{TO}}} \frac{W_{\text{TO}}}{S}$$

8.12.3 Geometry Constraints

1) Volume: The outside has to be bigger than the inside.
2) Landing gear: 10–20% weight on nose gear, 80–90% on mains
   a) For 15% nose, 85% mains, c.g. is ahead of mains 15% of distance between nose and mains.

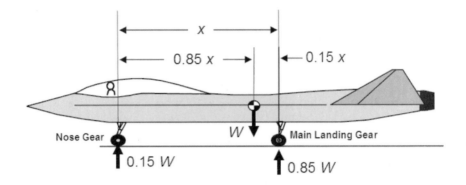

b) Tip-back angle must be less than angle of most aft c.g. ahead of main gear.

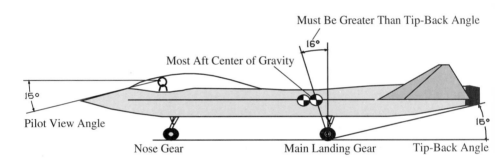

8.12.4 Mission Analysis

$$\beta_{\text{takeoff}} = 1 \quad (\text{because } W = W_{\text{TO}})$$

$$\beta_{\text{accel}} = \beta_{\text{takeoff}} - \frac{W_{f_{\text{takeoff}}}}{W_{\text{TO}}}$$

$$\beta_{\text{climb}} = \beta_{\text{accel}} - \frac{W_{f_{\text{accel}}}}{W_{\text{TO}}}$$

$$\beta_{\text{cruise}} = \beta_{\text{climb}} - \frac{W_{f_{\text{climb}}}}{W_{\text{TO}}}$$

$$\beta_{\text{loiter}} = \beta_{\text{cruise}} - \frac{W_{f_{\text{cruise}}}}{W_{\text{TO}}}$$

$$\frac{W_{f_{\text{mission}}}}{W_{\text{TO}}} = \frac{W_{f_{\text{takeoff}}}}{W_{\text{TO}}} + \frac{W_{f_{\text{accel}}}}{W_{\text{TO}}} + \frac{W_{f_{\text{climb}}}}{W_{\text{TO}}} + \frac{W_{f_{\text{cruise}}}}{W_{\text{TO}}} + \frac{W_{f_{\text{loiter}}}}{W_{\text{TO}}}$$

8.12.5 Sizing

$$S = W_{\text{payload}} \left/ \frac{W_{\text{TO}}}{S} - \frac{W_{\text{airframe}}}{S} - \frac{W_{\text{engine}}}{S} - \frac{W_{f_{\text{mission}}}}{S} \right.$$

8.12.6 Weight and Balance

1) Total weights and moments, divide to get total c.g. location
2) Center-of-gravity excursion diagram

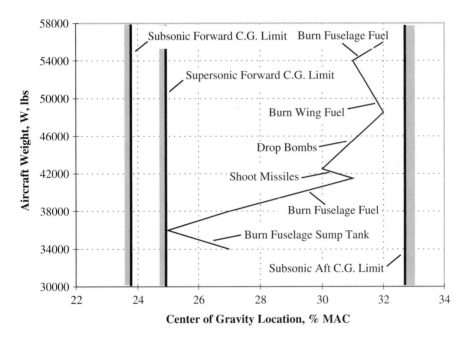

8.12.7 Cost

1) Purchase price
2) Operating costs
3) Very complicated, many factors
4) Dollars per pound (like hamburger) works pretty well for first guess for purchase price; can be way off if only a few aircraft purchased (B-2)

References

[1]Raymer, D. P., *Aircraft Design: A Conceptual Approach*, AIAA Education Series, AIAA, Washington, DC, 1989, Chap. 15.
[2]Roskam, J., *Airplane Design Part VIII*, Roskam Aviation and Engineering Corp., Ottawa, KS, 1990.

[3]Smith, H. W., and Burnham, R., "The Outside Has to Be Bigger Than the Inside," AIAA Paper 80-0726, May 1980.

[4]Mattingly, J. D., Heiser, W. H., and Daley, D. H., *Aircraft Engine Design*, AIAA, Education Series, AIAA, New York, 1987, Chap. 3.

Problems

Synthesis Problems

S-8.1 Brainstorm at least five different ways to reduce crew salary costs on a jet airliner.

S-8.2 A turbojet-powered long-distance cross-country racer is being designed. To be competitive, the jet must fly at speeds very near or above $M = 1.0$. Based on your knowledge of the relative magnitudes of parasite, induced, and wave drag at these speeds and the variation of wing structural weight with aspect ratio and sweep, what do you think the planform of an optimized racer will look like?

S-8.3 Drag of externally carried weapons severely reduces maximum speed and range of conventional fighter aircraft. Brainstorm at least five different aircraft configurations that overcome or at least reduce this problem.

Analysis Problems

A-8.1 A design concept for a multirole fighter aircraft has a design wing loading of 70 lb/ft^2, design thrust-to-weight ratio of 0.7, an engine thrust-to-weight ratio of 8, a structural wing loading portion of 20 lb/ft^2, a mission fuel fraction of 0.25, and a mission payload of 6000 lb. Assuming its configuration does not have to change when it is sized, what is its sized wing area?

A-8.2 A design concept for a 100-passenger regional airliner with a design wing loading of 100 lb/ft^2, a design thrust-to-weight ratio of 0.3, and a misson fuel fraction of 0.35 was initially sized with a wing area of 1500 ft^2 and a total volume of 20,000 ft^3. Allowing 1000 ft^3 for baggage and cargo and assuming a standard flight deck crew of two, does this concept have sufficient volume for the fuel it needs to fly the misson?

A-8.3 You are tasked to estimate the purchase price of a new business jet concept. The airplane is expected to have a design maximum takeoff weight of 40,000 lb. What is your initial purchase price estimate for this aircraft in 1996 dollars?

9
Putting It All Together: Conceptual Aircraft Design

"Man's mind and spirit grow with the space in which they are allowed to operate."
Kraft A. Ehricke, rocket pioneer

"One test result is worth one thousand expert opinions."
Dr. Wernher von Braun

9.1 Introduction

The very first sentence in Raymer's text on aircraft design[1] makes a significant and often misunderstood point: Design is a separate aeronautical engineering subdiscipline that has an equal place with aerodynamics, structures, stability and control, performance, and propulsion. Although design can serve as a "capstone" course that pulls together skills taught in these other subdisciplines, it is more than the sum of those subdisciplines or simply an exercise of those subdisciplines at the undergraduate level. Design is set apart from these for several reasons. First, it contains two essential activities not required for the others: *sizing* (and its interactions with *constraint* and *mission analysis*) and *optimization* of the whole airplane while considering the interaction and constraints of one aspect of the airplane with all other aspects. For example, airfoils and wings must achieve some acceptable level of aerodynamic effectiveness and performance *and* provide sufficient volume for fuel and landing gear *and* do so without imposing unacceptable structural weight penalties or unnecessarily complicated load paths. Second, design is iterative and cyclical in these activities.[2] Third, in many cases, students are expected to exercise synthesis and decision-making skills that are fundamentally different from the analysis skills they have used so far in their training. The success of a student's first attempt at design requires both understanding of each of the activities just mentioned in isolation and understanding of how they relate to each other. Merely understanding material presented in previous engineering courses is not enough: an effective course in conceptual aircraft design is an essential part of each student's training.

This chapter has two purposes. The first is to identify for other engineers and design instructors 12 aircraft conceptual design activities introduced in the preceding chapters of this book, explain how they work together in the conceptual design process, and show how they are taught and learned at one undergraduate institution. The second is to provide students with a "big-picture" overview of a typical aircraft design curriculum. The chapter is organized into three parts. The first describes the 12 conceptual design activities, the second shows how they are

379

studied, practiced, and applied in an undergraduate aircraft design curriculum, and the third part contains concluding remarks.

9.2   Overview of the 12 Activities

The 12 technologies or activities are essentially steps in the conceptual design process, though they are often repeated in different orders several times during the conceptual design phase. Design begins (and ends) with *customer focus*, or determining customer needs and translating them into design requirements and desired design characteristics. The second activity is *design synthesis*, the creative process of conceiving an original design concept. The concept usually is described in a sketch, which shows the general configuration, relative positions of major systems, and any unique features. In the beginning, more than one concept can be generated. Once suitable concepts have been generated, they are described by engineering drawing and *geometry modeling,* which includes creating sufficiently detailed two-dimensional and three-dimensional representations of an aircraft design concept. CAD and three-dimensional solid modeling are primary tools, but simpler parametric representations such as the method described in Chapter 4 are also useful.

Once a design concept has been synthesized and its geometry modeled, *aerodynamic analysis* (the methods in Chapter 4 are an example) creates an aerodynamic model of the aircraft. *Propulsion modeling* (Sec. 5.3) creates a corresponding representation of the chosen propulsion system's characteristics. The aerodynamic and propulsion models are used in *constraint analysis* (Sec. 5.16), which represents customer design requirements as boundaries on a plot of aircraft wing loading (weight per unit planform area) vs thrust-to-weight ratio. From the constraint diagram, a design point (a set of target values for wing loading and thrust-to-weight ratio) is selected so that the aircraft will meet all reasonable design requirements. *Mission analysis* (Sec. 8.5) uses results of aerodynamic, propulsion, and constraint analysis to predict how much fuel the aircraft will burn on missions which the customer requires it to fly. *Weight analysis* (Sec. 8.3) predicts the weights of airframe and engine system components. *Sizing* (Sec. 8.6) uses the design point from constraint analysis, mission fuel fraction from mission analysis, component weights from weight analysis, and the required payload specified by the customer to determine how large the aircraft must be to meet all the design requirements. Because cost is often a primary factor in design decisions, *cost analysis* (Sec. 8.9) uses the results of sizing to give decision makers the necessary cost feedback. The design synthesis and geometry modeling activities and aerodynamic, propulsion, constraint, mission, sizing, and cost analyses are then repeated in an optimization cycle, which identifies the smallest, lightest, or most cost-effective aircraft configuration that can meet all customer needs. Once the optimum configuration is defined, *performance reporting* determines and communicates estimates of the final aircraft concept's characteristics, capabilities, and cost.

These 12 activities—customer focus, design synthesis, geometry modeling, aerodynamic analysis, propulsion modeling, constraint analysis, mission analysis, component weight analysis, sizing, cost analysis, optimization, and performance reporting—are essential tools that the aircraft design student must learn in order to practice conceptual design. Each of these activities and methods used to teach them are described in more detail in subsequent sections and illustrated by examples of student work.

9.3   Customer Focus

All aircraft design begins with a customer (or customers) with needs that can be met by an aircraft. Needs are generally expressed as performance requirements or specific characteristics and capabilities: maximum speed, range, and payload; minimum cost and maintenance man hours per flying hour; adequate durability, versatility, lethality, survivability, maintainabilty, and manufacturability; etc. The aircraft designer identifies these requirements, prioritizes them, and determines which design features will allow the design to meet the requirements. The House of Quality[3] (HOQ) (Sec. 1.8) is introduced as one means of organizing and documenting this process. Design students should be given an abbreviated request for proposals (RFP) for an aircraft type needed by a particular customer. They should perform HOQ analysis through research and conversations with their customer (either an actual customer or one simulated by their instructor). Students are reminded that the customer must be consulted in all phases of the design process to ensure an optimized design that meets all important customer needs at minimum cost.

Figure 9.1 illustrates a typical HOQ diagram created by undergraduate students[4] for an uninhabited combat air vehicle (UCAV), which could perform a deep interdiction mission. Customer needs are listed in order of priority along the left-hand side of the grid. Candidate design features are listed across the top of the central matrix, below the "attic" grid. Symbols in the central matrix indicate a judgement made by the students of how each design feature would contribute to the aircraft design to meet the customer needs. As indicated in the key in the upper-left corner of the chart, strong positive influence or correlation is denoted by a filled circle and is arbitrarily given a weight of 9. A moderate correlation is denoted by an open circle and given a weight of 3. A weak correlation is indicated by a triangle and has a weight of 1, and negative correlations have no symbol and a weight of 0. By multiplying correlation weights under each design feature by the priority weight for each item listed on the right side of the central matrix, and then summing the products under each feature, an overall relative weight or priority for that design feature is calculated. The students have highlighted the four most important design features at the bottom of their diagram by circling them. These priorities are

1) low radar cross section (RCS),

2) maximum use of off-the-shelf parts,

3) engine large enough to meet all performance requirements without using afterburner, and

4) a good electronic countermeasures (ECM) system.

The attic of the HOQ diagram, the triangular grid above the main correlation matrix, identifies design features that either complement each other and provide synergy or interfere with each other to some degree. For instance, early attempts to produce an aircraft with low RCS interfered with choices that produced an efficient, low drag, aerodynamic shape, and vice versa. To illustrate this effect in the HOQ diagram, an asterisk (or an "x" if the negative correlation was deemed

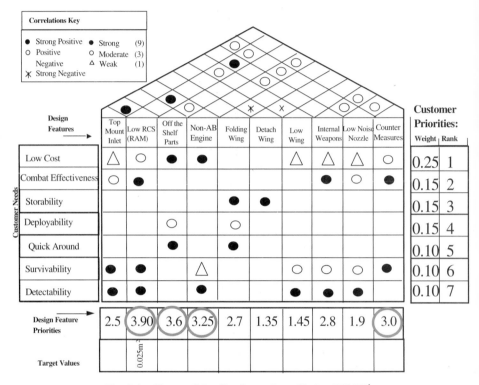

Fig. 9.1   House of Quality for an interdiction UCAV.[4]

moderate) would be placed in the attic grid at the intersection of the low RCS column and the efficient aerodynamic shape [or high-lift-to-drag ratio (L/D)] column. These symbols identify opportunities and potential problems that managers or team leaders should monitor carefully.

It is likely that experts would not agree with the students' choices of features, weights, correlations, and priorities on this HOQ diagram, but that is unimportant. The value of this exercise is not in the answers, but in understanding the process by which these results were obtained. Furthermore, the link has been established, in the students' minds, between customer needs and design features. Understanding this link is the key objective of the *customer focus* portion of the curriculum.

9.4   Design Synthesis

This consists of creating an aircraft design concept, either as a modification of an existing design or as a new concept created from scratch. Because this process is creative (right brain) rather than analytical (left brain, the type activity college students are most comfortable with), students typically have difficulty with design synthesis. They are often more comfortable modifying an existing design rather than creating one from scratch, but in doing so they frequently miss opportunities for designing a more capable aircraft and for gaining experience with creativity.

Students should be taught that creativity can be enhanced by proper preparation and effort. See Sec. 1.3.3 for more details.

9.5 Geometry Modeling and Engineering Drawing

Design synthesis usually results in some sort of sketch or parametric representation of a conceptual aircraft design concept. This concept must be defined in sufficient detail so that its aerodynamic, weight, and volume characteristics can be determined. This is frequently done by making an engineering drawing of the concept, then taking measurements from the drawing and using them to perform analysis.

Although making sketches of aircraft concepts should be highly encouraged, these sketches must be represented parametrically in sufficient detail to provide the information necessary for aerodynamic and weight analysis. The AeroDYNAMIC[5] software package provided with this book is a helpful tool for this process. Aircraft configurations are created by specifying parameters such as wing and tail span, sweep, chord and position; fuselage length, height, and width; landing gear and engine size and position; etc. The software automatically draws the aircraft configuration described by these parameters. Because concepts are defined parametrically, changes can be made quickly and easily, with the computer handling the tedious task of redrawing the modified configuration.

Very soon after they are created, parametric representations must be translated into engineering drawings or CAD models to help students visualize, optimize, and communicate their design concept's geometry. Engineering drawing is a communication task, whether using a simple sketch to communicate an initial concept for an aircraft configuration to other designers or using detailed construction drawings to communicate how a fully developed aircraft design is to be built. File translators in AeroDYNAMIC allow students to translate their parametrically described conceptual aircraft designs into drawing files which can be read by most popular CAD packages. Students then use a CAD program to add details, dimensions, and multiple views to their drawings.

9.6 Aerodynamic Analysis

Aerodynamic analysis methods used for the early conceptual phase of design are limited to simple, semi-empirical methods obtained from the U.S. Air Force Stability and Control DATCOM.[6] The only aerodynamic data needed in the early conceptual design phase are the aircraft drag polar and maximum lift coefficient as functions of configuration and Mach number.

Lift calculations begin by estimating the wing lift-curve slope C_{L_α}, using

$$C_{L_\alpha} = \frac{c_{l_\alpha}}{1 + 57.3 c_{l_\alpha}/\pi e \! R} \qquad (9.1)$$

where c_{l_α} is the lift-curve slope of an infinite wing with the same airfoil and e is the span efficiency factor, estimated empirically by

$$e = \frac{2}{2 - \! R + \sqrt{4 + \! R^2(1 + \tan^2 \Lambda_{t_{max}})}} \qquad (9.2)$$

In Eq. (9.1), $\mathcal{R} = b^2/S$ is the wing aspect ratio (b is wing span and S is aircraft reference planform area.) In Eq. (9.2), $\Lambda_{t_{\max}}$ is the sweep angle of the line of airfoil maximum thicknesses on the wing. Once C_{L_α} is known, $C_{L_{\max}}$ is estimated by assuming that $\alpha_{L=0}$ (the angle of attack at which the wing generates zero lift) is the same as the zero-lift angle of attack of the wing's airfoil, and that, at least for takeoff and landing, the maximum useable angle of attack is approximately 15 degs. With these assumptions $C_{L_{\max}}$ for a clean configuration is given by

$$C_{L_{\max}} = C_{L_\alpha} \cdot (\alpha_{\max} - \alpha_{L=0}) = C_{L_\alpha} \cdot (15 \deg - \alpha_{L=0}) \qquad (9.3)$$

To obtain $C_{L_{\max}}$ for takeoff and landing configurations, the effect of flaps is modeled as a shift in $\alpha_{L=0}$. As a first approximation, a 10-deg increment in $\alpha_{L=0}$ for takeoff flap settings and 15 deg for landing flaps is used. This must be scaled if the flaps do not extend across the entire span of the wing. $C_{L_{\max}}$ for takeoff or landing is then given by

$$C_{L_{\max}} \cong C_{L_{\max}(\text{no flap})} + C_{L_\alpha} \cdot \frac{S_f}{S} \cos \Lambda_{\text{h.l.}} \qquad (9.4)$$

where $\Lambda_{\text{h.l.}}$ is the sweep angle of the flap hinge line and S_f is the portion of the total wing planform that is affected by the flaps (or to which the flaps are attached).

A lift increment caused by a strake or leading-edge extension is modeled as a change in wing lift-curve slope after Lamar and Frink[7]

$$C_{L_{\alpha(\text{with strake})_s}} = C_{L_{\alpha(\text{without strake})}} \frac{S + S_{\text{strake}}}{S} \qquad (9.5)$$

where S_{strake} is the exposed planform area of the strake.

The aircraft drag polar is represented as

$$C_D = C_{D_0} + k_1 C_L^2 + k_2 C_L \qquad (9.6)$$

where C_{D_0} is the zero-lift or parasite drag coefficient, $k_1 = 1/(\pi e_O \mathcal{R})$ is the induced drag factor, and k_2 is chosen to represent the effect of camber. The drag polar calculation is accomplished in four steps. The subsonic zero-lift drag coefficient C_{D_0} is predicted using the equivalent skin-friction drag coefficient method:

$$C_{D_0} = C_{f_e} \frac{S_{\text{wet}}}{S} \qquad (9.7)$$

An appropriate value for C_{f_e} for a given aircraft is obtained from historical data such as those summarized in Table 9.1.

Once the subsonic zero-lift drag coefficient has been determined, the supersonic values are predicted by adding to the subsonic C_{D_0} a predicted supersonic wave-drag coefficient, which is predicted using an empirically derived modification to the theoretical minimum wave drag of Sears–Haack bodies[8] at each supersonic Mach number of interest:

$$C_{D_{\text{wave}}} = \frac{4.5\pi}{S} \left(\frac{A_{\max}}{l} \right)^2 E_{\text{WD}} (0.74 + 0.37 \cos \Lambda_{\text{LE}}) \left[1 - 0.3\sqrt{M - M_{C_{D_0} \max}} \right]$$

$$(9.8)$$

Table 9.1   Common C_{fe} values.

Type	C_{fe}
Jet bomber and civil transport	0.0030
Military jet transport	0.0035
Air Force jet fighter	0.0035
Carrier-based Navy jet fighter	0.0040
Supersonic cruise airplane	0.0025
Light single-propeller airplane	0.0055
Light twin-propeller airplane	0.0045
Radio-controlled model airplane	0.014

where Λ_{LE} is the wing leading-edge sweep angle and

$$M_{C_{D_0}\,\text{max}} = \frac{1}{\cos^{0.2}\Lambda_{LE}} \tag{9.9}$$

estimates the Mach number where the maximum value of C_{D_0} occurs. As a reasonably accurate simplifying approximation, variation of C_{fe} and hence C_{D_0} with Reynolds number is ignored.

To complete the estimation of the variation of C_{D_0} with Mach number, critical Mach number M_{crit} is estimated using

$$M_{\text{crit}} = 1.0 - 0.065 \left(100\frac{t_{\text{max}}}{c}\right)^{0.6} \cos^{0.6}\Lambda_{LE} \tag{9.10}$$

Then, C_{D_0} is assumed to be constant below M_{crit}, the wave-drag coefficient is added to the subsonic C_{D_0} to calculate total C_{D_0} at Mach numbers above $M_{C_{D_0}\,\text{max}}$, and a curve is drawn connecting the subsonic and supersonic C_{D_0} values to estimate transonic C_{D_0}. The simplest of these connecting curves, a straight line, is frequently used as an adequate approximation for transonic C_{D_0}, provided the aircraft being designed is not required to cruise in the transonic regime on its design mission. Figure 9.2 compares predicted vs actual variation of C_{D_0} with Mach number for the F-106.

With C_{D_0} calculated, the subsonic induced drag factor is predicted using

$$k_1 = 1/(\pi e_O \mathcal{R}) \tag{9.11}$$

where e_O is Oswald's efficiency factor, which is predicted empirically by

$$e_O = 4.61(1 - 0.045\,\mathcal{R}^{0.68})(\cos \Lambda_{LE})^{0.15} - 3.1 \tag{9.12}$$

Supersonic induced drag factor is estimated using

$$k_1 = \frac{\mathcal{R}(M^2 - 1)}{(4\,\mathcal{R}\sqrt{M^2 - 1}) - 2} \cos \Lambda_{LE} \tag{9.13}$$

and transonic k_1 values are approximated by fitting a smooth curve between the subsonic and supersonic values.

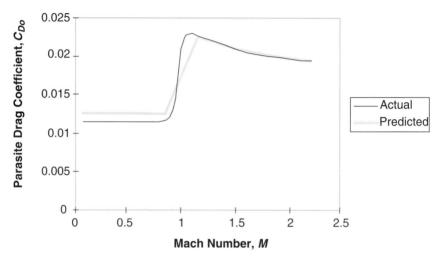

Fig. 9.2   Variation of actual and predicted C_{D_0} with Mach number for the F-106.

The final constant in the drag polar k_2 is chosen such that the minimum drag coefficient occurs at the angle of attack where the airfoil of the aircraft's wing generates minimum drag. This generally occurs near $\alpha = 0$, but for cambered airfoils the wing lift coefficient is nonzero. At this stage, effects of wing twist and incidence angle are assumed to cancel each other, so that the wing C_L for minimum drag is calculated using the already estimated wing lift-curve slope, and

$$C_{L_{\min D}} = C_{L_\alpha}\left(\frac{-\alpha_{L=0}}{2}\right) \qquad (9.14)$$

where $\alpha_{L=0}$ is the angle of attack where the wing's airfoil generates zero lift. Supersonic values of k_2 are assumed to equal zero, and transonic values are estimated by fitting a smooth curve between subsonic and supersonic values.

When $k_2 \neq 0$, C_{D_0} is no longer the minimum drag coefficient. For these situations, $C_{D_{\min}}$ is calculated using Eq. (9.7) and then C_{D_0} is calculated using

$$C_{D_0} = C_{D_{\min}} + k_1 C_{L_{\min D}}^2 \qquad (9.15)$$

Students are expected to learn how to perform all aerodynamic analysis calculations by hand and to program them in a spreadsheet before using the AeroDYNAMIC software. See Chapter 4 for more details on aerodynamic analysis.

9.7   Propulsion Analysis

Once aerodynamic analysis has created a model of the design concept's aerodynamic behavior, the designer uses propulsion modeling to create a similar representation of the aircraft's propulsion system's performance. Many simple yet elegant and accurate models are available,[9] but for an initial student conceptual design experience it is sufficient to use the following representation for nonafterburning performance of turbojets and turbofans

$$T_A = T_{SL} \left(\frac{\rho}{\rho_{SL}} \right) \tag{5.11}$$

where T_A is thrust available for a given flight condition density ρ and the subscript SL identifies similar quantities for standard sea-level conditions. An almost equally simple model for afterburning thrust of turbojets and low bypass ratio turbofans is

$$T_A = T_{SL} \left(\frac{\rho}{\rho_{SL}} \right) (1 + 0.7 M_\infty) \tag{5.12}$$

Likewise, variation of turbojet and turbofan engine thrust specific fuel consumption c_t is modeled by

$$c_t = c_{t_{SL}} \left(\frac{a}{a_{SL}} \right) \tag{5.16}$$

9.8 Constraint Analysis

Constraint analysis[9] is based on a modification of the equation for specific excess power.

$$P_s = \frac{(T - D)V}{W} = \frac{dh}{dt} + \frac{V}{g} \frac{dV}{dt} \tag{9.16}$$

where T is thrust available, D is drag, W is weight, V is true airspeed, h is altitude, and g is the acceleration of gravity. Dividing by velocity yields

$$\frac{T}{W} - \frac{D}{W} = \frac{1}{V} \frac{dh}{dt} + \frac{1}{g} \frac{dV}{dt} \tag{9.17}$$

Substituting the following relations into Eq. (9.17),

$$T = \alpha T_{SL}$$

where α, the thrust lapse ratio, is determined in thrust modeling;

$$W = \beta W_{TO}$$

where β is the weight fraction, the ratio of the specified weight for that design requirement to the maximum takeoff gross weight W_{TO};

$$q = \frac{1}{2} \rho V^2$$

is dynamic pressure;

$$D = C_D q S = \left(C_{D_0} + k_1 C_L^2 + k_2 C_L \right) q S$$

$$C_L = \frac{L}{q S} = \frac{n W}{q S}$$

produces the master equation for constraint analysis:

$$\frac{T_{SL}}{W_{TO}} = \frac{q}{\alpha}\left[\frac{C_{D_0}}{(W_{TO}/S)} + k_1\left(\frac{n\beta}{q}\right)^2\left(\frac{W_{TO}}{S}\right) + k_2\left(\frac{n\beta}{q}\right)\right] + \frac{\beta}{\alpha}\left[\frac{1}{V}\frac{dh}{dt} + \frac{1}{g}\frac{dV}{dt}\right]$$

$$(9.18)$$

Equation (9.18) is written in a form that expresses the minimum T_{SL}/W_{TO} that can achieve a given performance requirement at a particular value of W_{TO}/S. The aircraft drag polar determined from aerodynamic analysis is used to calculate the drag term in this equation, and all other variables in Eq. (9.18) are specified by each design requirement. A separate form of the master equation is written using values of velocity, load factor, acceleration, rate of climb, etc. for each design requirement. Lines representing these equations are plotted as boundaries on a thrust-to-weight ratio vs wing loading diagram. Combinations of design thrust-to-weight ratio and wing loading that fall outside the boundary specified by each constraint result in an aircraft design that will not meet that design requirement. When all of the constraint boundaries are plotted, they define a solution space, which, for all combinations of thrust-to-weight ratio and wing loading within the space, allow the aircraft to meet all of the design requirements. See Sec. 5.16 for more details on constraint analysis. Figure 9.3 shows a constraint diagram created by a student using the AeroDYNAMIC software.

The constraint diagram can identify one or more performance requirements that are much more difficult to achieve than the others. This might be caused by the aircraft configuration being analyzed (for instance, a low-speed turn requirement for an aircraft configuration that has a wing with a very low aspect ratio). It

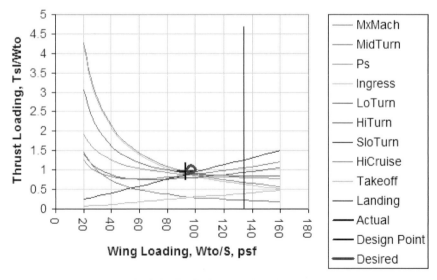

Fig. 9.3   Typical constraint diagram.

might also be that two requirements are so diametrically opposed to each other (e.g., a long loiter and a high maximum speed requirement) that no configuration can adequately meet both or that one requirement is unrealistic (e.g., requiring Mach 3.0 capability for an F-15 replacement). It is a good idea to put at least one such unreasonable requirement in the request for proposal (RFP), so that students must identify it and determine how to handle it. Students are encouraged to talk to the customer and determine how important this requirement is. They must also perform trade studies to show the customer how much they could save or how much additional capability they would lose or gain for various amounts of relaxation of a difficult design requirement.

If a difficult performance requirement is identified by the customer as being a high priority, it becomes a *design driver*. The characteristics of the design will be predominantly influenced by this requirement and relatively insensitive to changes in other requirements. In most recent military aircraft procurement programs, cost, and not any particular performance requirement, has been the most important design driver. Cost is not normally shown directly on the constraint diagram, but is calculated at the end of the sizing process. However, it can usually be assumed that a design point having higher values of thrust-to-weight ratio and lower wing loading (and therefore a larger engine and larger wing) on the constraint diagram will result in a more expensive aircraft. See Ref. 10 for a more general formulation of this analysis.

9.9 Mission Analysis

Mission analysis uses the aircraft's drag polar and the design point determined from aerodynamic and constraint analysis, plus the appropriate model for engine thrust and specific fuel consumption, to predict the amount of fuel the aircraft will burn on its design mission(s). The master equation for mission analysis is essentially the time integral form of Eq. (5.75), substituting actual thrust for thrust available and multiplying by c_t to yield fuel flow rate $c_t T = -W_f$. The minus sign in front of fuel flow rate indicates that aircraft weight decreases as fuel is burned. The integral of fuel flow rate yields fuel burn, which, divided by weight, is the fuel fraction

$$\int_{t_0}^{t_1} \frac{c_t T}{W_{\text{TO}}} \, dt = \int_{t_0}^{t_1} \frac{-W_f}{W_{\text{TO}}} \, dt$$

$$= \int_{t_0}^{t_1} c_t q \left[\frac{C_{D_0}}{(W_{\text{TO}}/S)} + k_1 \left(\frac{n\beta}{q} \right)^2 \left(\frac{W_{\text{TO}}}{S} \right) + k_2 \left(\frac{n\beta}{q} \right) \right] dt$$

$$+ \int_{t_0}^{t_1} c_t \beta \left[\frac{1}{V} \frac{dh}{dt} + \frac{1}{g} \frac{dV}{dt} \right] dt \tag{9.19}$$

which is easily integrated for the most common cases of constant velocity climbs and cruise, constant altitude accelerations, and for relatively short mission legs where it is acceptable to assume the aircraft's drag polar and β are approximately constant:

$$\frac{W_0 - W_1}{W_{TO}} = \frac{\Delta W_f}{W_{TO}}$$

$$= c_t q \left[\frac{C_{D_0}}{(W_{TO}/S)} + k_1 \left(\frac{n\beta}{q}\right)^2 \left(\frac{W_{TO}}{S}\right) + k_2 \left(\frac{n\beta}{q}\right) \right] (t_1 - t_0)$$

$$+ c_t \beta \left[\frac{1}{V}(h_1 - h_0) + \frac{1}{g}(V - V_{0_1}) \right] \qquad (9.20)$$

Note that W_{TO}/S in Eq. (9.20) is for this analysis a constant because it was chosen as a result of constraint analysis. The assumption of constant β is not acceptable for long cruise or loiter legs, and the assumption of constant drag polar is not acceptable for transonic accelerations, but these are easily accommodated in computer implementations of the method by breaking long legs into many similar short legs. See Sec. 5.16 for more details and Ref. 10 for a more general formulation of this analysis.

When this master equation is applied to each leg of the design mission, the fuel fractions from all preceding legs of the mission are used to determine the weight fraction for each subsequent leg. When the fuel fractions for each leg are summed, the total fuel fraction for the mission is obtained:

$$\beta_{\text{takeoff}} = 1 \qquad (\text{because } W = W_{TO})$$

$$\beta_{\text{accel}} = \beta_{\text{takeoff}} - \frac{W_{f_{\text{takeoff}}}}{W_{TO}}$$

$$\beta_{\text{climb}} = \beta_{\text{accel}} - \frac{W_{f_{\text{accel}}}}{W_{TO}}$$

$$\beta_{\text{cruise}} = \beta_{\text{climb}} - \frac{W_{f_{\text{climb}}}}{W_{TO}}$$

$$\beta_{\text{loiter}} = \beta_{\text{cruise}} - \frac{W_{f_{\text{cruise}}}}{W_{TO}}$$

$$\frac{W_{f_{\text{mission}}}}{W_{TO}} = \frac{W_{f_{\text{takeoff}}}}{W_{TO}} + \frac{W_{f_{\text{accel}}}}{W_{TO}} + \frac{W_{f_{\text{climb}}}}{W_{TO}} + \frac{W_{f_{\text{cruise}}}}{W_{TO}} + \frac{W_{f_{\text{loiter}}}}{W_{TO}} \qquad (9.21)$$

9.10   Weight Analysis

Once mission analysis has predicted the mission fuel burn and fuel fraction, weight-estimating methods are used to predict the structural and propulsion system weight fractions. These are commonly expressed as fractions of the aircraft gross weight, or as portions of the total aircraft wing loading. The second approach is more convenient, because, as described in Tables 8.1 and 8.2 in Sec. 8.3, some of the simplest structural weight estimates are expressed as weight per unit area of a given component, or in other words, as wing loadings. The wing loading portion of the entire airframe is expressed as the sum of the wing loadings of its various components, scaled by their appropriate reference areas. See Sec. 8.3 for more details.

Calculating the wing loading portion of the propulsion system begins with an estimate of the propulsion system's own thrust-to-weight ratio. Inverting this ratio and multiplying by thrust loading and wing loading chosen in constraint analysis then yields the required wing loading portion:

$$\frac{W_{\text{engine}}}{S} = \frac{W_{\text{engine}}}{T_{\text{SL}}} \frac{T_{\text{SL}}}{W_{\text{TO}}} \frac{W_{\text{TO}}}{S} \tag{8.22}$$

9.11 Sizing

Once mission fuel fraction is determined, the aircraft is sized. Sizing determines how large and expensive an aircraft concept must be so that it can fly all required missions carrying all required payloads. The sizing equation is derived from the fact that when properly sized the sum of the portions of the aircraft's wing loading as a result of its structure, systems, payload, engine, and fuel must equal the total wing loading chosen in constraint analysis

$$\frac{W_{\text{TO}}}{S} = \frac{W_{\text{airframe}}}{S} + \frac{W_{\text{payload}}}{S} + \frac{W_{\text{engine}}}{S} + \frac{W_{f\text{mission}}}{S} \tag{9.22}$$

The payload (including crew and passengers and their luggage and/or mission-specific systems) is specified by the RFP or the designer, the airframe and engine wing loading portions are obtained from weight analysis with corrections for the impact of new technology, and wing loading portion for mission fuel burn is obtained from the mission fuel fraction calculated in mission analysis:

$$\frac{W_{f\text{mission}}}{S} = \frac{W_{f\text{mission}}}{W_{\text{TO}}} \frac{W_{\text{TO}}}{S} \tag{9.23}$$

Equation Eq. (9.22) is solved for the only variable not specified, the reference planform area in the payload wing loading portion term:

$$S = W_{\text{payload}} \Big/ \frac{W_{\text{TO}}}{S} - \frac{W_{\text{airframe}}}{S} - \frac{W_{\text{engine}}}{S} - \frac{W_{f\text{mission}}}{S} \tag{9.24}$$

$$W_{\text{TO}} = S \left(\frac{W_{\text{TO}}}{S} \right) \tag{9.25}$$

The result of the calculation in Eqs. (9.24) and (9.25) is final, provided the aircraft configuration does not need to change after it is sized. Volume occupied by payload, passengers, crew, etc. does not change when the airplane is resized, which might require that the relative sizes of the fuselage and wings, for instance, change when the airplane size changes. At best, this effect will be small, and results of an initial sizing calculation will be sufficiently accurate. At worst, the newly sized aircraft will have to be redrawn and new structural estimates made. A new drag polar must be calculated for the new configuration, and a new mission fuel fraction must be calculated. These new results are used in another sizing calculation, and the process is repeated until it converges to an aircraft wing size, which satisfies the sizing equation and internal volume requirements.

9.12   Cost Analysis

Once the sized wing area is determined, it is multiplied by the wing loading chosen in constraint analysis to determine the sized aircraft weight. Engine thrust is determined by multiplying the aircraft weight by the thrust-to-weight ratio chosen in constraint analysis. Once weight and engine thrust are known, acquisition and life-cycle costs are estimated. Initial estimates are based on price-per-pound data for similar aircraft, as in Sec. 8.9. The cost-estimating methodology presented in Ref. 1 is more complicated and might not give more accurate estimates. However, exposure to this methodology is a good idea because it gives a better feel for all of the factors that influence cost.

9.13   Optimization

Students naturally will assume that once they complete the first 10 conceptual design activities for their design concept they are done. Not so! They must optimize their design by repeating the geometry modeling/aerodynamic analysis/propulsion modeling/constraint analysis/mission analysis/weight analysis/sizing/cost analysis cycle, varying aircraft characteristics such as wing sweep and aspect ratio, etc., and determining the combination of these parameters, which results in the smallest (or cheapest) aircraft that still meets all of the design requirements. It is often convenient to summarize the results of these parametric studies in carpet plots, which makes it easier to pick out the optimum configuration. Figure 9.4 illustrates a carpet plot for wing optimization created by students for the deep interdiction UCAV.

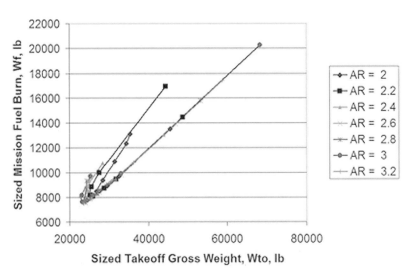

Fig. 9.4   Wing optimization carpet plot.[4]

Specific Excess Power Contours

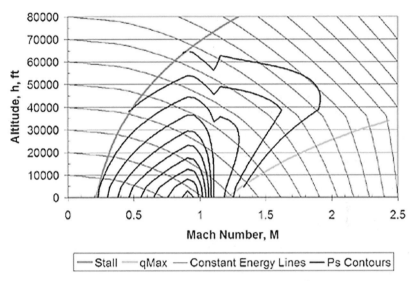

Fig. 9.5   Specific excess power diagram.[9]

9.14   Performance Reporting

Performance reporting is another communication task, like drawing. Indeed, two fundamental products of conceptual design are aircraft drawings and performance and cost estimates. The most common methods of reporting military fighter aircraft performance are specific excess power diagrams (at several values of load factor) and maneuverability diagrams (at several altitudes). Figure 9.5 illustrates a specific excess power diagram created by a student using the AeroDYNAMIC software. The diagram describes the 1-g climb and acceleration capabilities of a student-designed replacement for the F-111 and F-15E. Figure 9.6 illustrates a maneuverability diagram for the same aircraft.

9.15   Putting It All Together

Ideally the undergraduate aircraft conceptual design experience is presented as a two-semester sequence of three-semester-hour courses taken during the students' senior year. The experience should build on exposure to as many of the 12 activities as possible in previous courses. The 12 design activities are taught and practiced during the Fall semester. Then, in the spring, students are challenged with a new RFP, or with one that is substantially changed from the Fall RFP. This allows students to practice the activities until they understand them, then cement this understanding by applying their skills to a new problem.

In the Fall, students are novices. Each design activity is explained to them, and they are required to practice each method and demonstrate that they understand it.

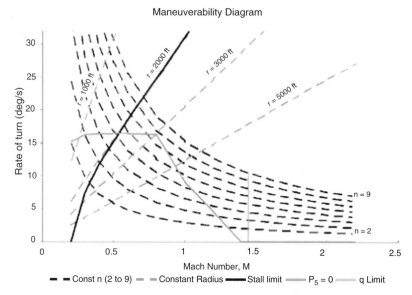

Fig. 9.6   Maneuverability diagram.[11]

Students can begin by researching and modeling an existing aircraft, which has a design mission similar to the one specified in the RFP, and then applying each of the analysis methods to their aircraft. They should also research new technology, which could be useful for their new aircraft design. This exercise increases their facility with the newly learned activities, and at the same time familiarizes students with the RFP and with past and potential future solutions to similar design problems.

Once students have become familiar with the 12 design technologies, two-person teams can be tasked to evaluate a notional design proposal. This exercise can be used to expose students to both good and (intentionally) bad examples of design decision-making and reporting, and provide practice at critical thinking in conjunction with the design analysis tools and methods. When students are frustrated by being unable to determine some important characteristic of the aircraft described in the notional proposal, they learn the importance of accuracy, completeness, and attention to detail in their own design proposals. Care must be taken that the notional aircraft does not adequately meet the design requirements of the RFP so that students do not conclude that no new design is needed. After completing this exercise, students should be ready to generate their own aircraft concept.

Initial instruction and student practice of the conceptual design process can be expected to consume 75% of the Fall semester and include development, sizing, and optimization of a unique airframe concept, selection of jet-engine cycle parameters, and airframe/propulsion system integration. Plan to devote over half of class time to work sessions to facilitate instructor–student interaction during the design process.

During the final 25% of the course, students can work as a design team of 10–14 students. From the airframe concepts and jet-engine cycles developed by the two-person teams, they down-select a single airframe and engine combination which they continue to define, refine, and optimize. This larger group allows students to specialize in such areas as aerodynamics, stability and control, structures, propulsion, and design integration. The semester's work should be concluded with a large formal report and hour-long design briefing to instructors and representatives of industry and military agencies with an interest in the particular RFP. This interaction with actual practicing engineers is extremely valuable and should be fostered whenever possible.

The ideal second-semester design course is much less rigidly structured. Students are given a new or substantially changed RFP (changes require resizing) and told to use the technologies they learned in the previous semester to create a new conceptual design. They are expected to accomplish this task as a 10–14 person team in the first 25% of the course, so that they can devote the remainder of the course to designing, building, and flight testing a radio-controlled subscale flying model of their design concept. The students are responsible for organizing their activities and developing a business plan. Examples of the results of the design-build-fly process can be found in Ref. 9. The process of using the 12 activities to solve a new problem significantly increases students' understanding of aircraft conceptual design. Solving all of the problems required to successfully fly a model of their concept gives them an appreciation for, and some facility with, preliminary and detail design.

References

[1]Raymer, D. P., *Aircraft Design: A Conceptual Approach,* 2nd ed., AIAA Education Series, Washington, DC, 1992.

[2]Nicolai, L., "Designing the Engineer," *Aerospace America*, April 1992, pp. 72–78.

[3]King, R., "Listening to the Voice of the Customer: Using the Quality Function Deployment System," *National Productivity Review*, Summer 1987, pp. 277–281.

[4]Houk, A. L. et al., *UB-40 Flying Burrito*, Dept. of Aeronautics, USAF Academy, CO, 1998.

[5]Brandt, S. A, Nielsen, M. L., and Greeder, M. M., "Software Simulations for Problem-Based Aeronautical Engineering Education," World Aviation Congress, Anaheim, CA, Oct. 1997.

[6]Finck, R. D., "USAF Stability and Control DATCOM," Wright-Patterson AFB, OH, 1975.

[7]Lamar, J. E., and Frink, N. T., "Aerodynamic Features of Designed Strake-Wing Configurations," *Journal of Aircraft*, Vol. 19, No. 8, 1982, pp. 639–646.

[8]Sears, W. R., "On Projectiles of Minimum Wave Drag," *Quarterly of Applied Mathematics,* Vol. 4, No. 4, 1947.

[9]Mouch, T. N., et al., "Design-Build-Fly and Real Customers as Motivators in Aircraft Design Courses," World Aviation Congress, Los Angeles, CA, 1996.

[10]Mattingly, J. D., Heiser, W. H., and Daley, D. H., *Aircraft Engine Design*, AIAA Education Series, AIAA, New York, 1987.

[11]Roy, K. R., et al., *YF-98 OTSINE*, Dept. of Aeronautics, USAF Academy, CO, 1998.

10
Case Studies and Future Aircraft Designs

"[Concerning] engines of war, the invention of which has long since reached its limit, and for the improvement of which I see no further hope in the applied arts ..."

Sextus Julius Frontius, Roman engineer

"In the development of air power, one has to look ahead and not backward and figure out what is going to happen, not too much of what has happened."

Billy Mitchell

10.1 Introduction

In this final chapter, you are invited to consider several examples of the design method as it has been applied to several interesting and important aircraft and to thoughts of future aircraft. The knowledge and skills in aeronautics and the design method that you have gained by working through the preceding chapters should have given you a greater ability to appreciate the material that follows. These design examples or case studies include descriptions of the influence of technology and customer needs on each design and of the many decisions and compromises that had to be made in order to bring each aircraft into existence. Each case study also shows that each aircraft was shaped by many iterations through the design cycle, often involving progressive improvement of earlier designs into the final form of the aircraft.

Chapter 1 gave you an overview of the design method with an example, a description of engineering design in an aeronautical context, and a brief introduction to the history of aircraft design. Figures 1.3 and 1.7 should have served to inform you that design can be thought of as a cyclical process that requires creative synthesis, that it is heavily dependent on detailed analysis, and that, in practice, it must converge on a solution. Engineering design is not only a matter of focusing on the right questions; decisions and corresponding solutions are required!

Chapters 2–9 were created to introduce you to some of the terminology, procedures, and issues of aeronautics in the context of aircraft design. In this last chapter, we want to leave you with some background on design successes of the past as well as some of the challenges and opportunities that exist in aircraft design today. Recall the following quote from Theodore von Kármán: "A scientist discovers that which exists. An engineer creates that which never was." The central theme of this book is that design, and the creative work that it entails, is the essence of engineering.

10.2   Case Study 1: 1903 Wright Flyer

The body of written material on the Wright brothers is immense. Their epic achievement, which culminated in the first sustained flight of a powered aircraft carrying an adult human, is truly one of the turning points in human history. The next few paragraphs will attempt only to identify how the Wright brothers used the scientific (design) method described in Chapter 1 to make the many breakthroughs that ushered in the aviation age. For some extremely interesting and inspiring reading in more detail on the Wrights and their work, see *Miracle at Kitty Hawk*[1] and *The Wright Brothers*[2] by Fred C. Kelly, *Wilbur and Orville*[3] by Fred Howard, *The Bishop's Boys*[4] by Tom D. Crouch, or any of the many other books on this subject. The comments that follow are distilled from the four sources just listed.

Though their education did not extend beyond secondary school and they earned their living by running a bicycle shop, the Wright brothers were true scientists and engineers. They followed the steps of the design method faithfully and methodically, making many iterations through the design cycle, and successively solving each of the problems they encountered. The paragraphs that follow will list some of the activities and accomplishments of the Wrights in each step of the design process. As a result, this case study will follow a topical rather than a chronological organizational pattern. It is hoped that this arrangement of the material will make it easier for you to appreciate how faithfully the Wrights followed the design method and how remarkable an achievement they produced as a result.

10.2.1   Defining the Problem(s)

The Wrights, especially Wilbur, had been interested in the idea of human flight since their boyhood. They initially framed the problem by stating the fact that various types of animals and insects were able to fly with apparent ease and that they believed that with the right combination of knowledge, skill, and technology, man could be enabled to do the same. The problem was to determine how this could be done. As they studied the problem further, they broke it down into many subproblems, the first being how to shape wings so that they generated lift when moved through the air with minimum drag penalty. They needed to determine what camber, location of maximum camber, and thickness to use for their airfoils, and what aspect ratio, planform shape, and structure to use for their wings. They wondered whether a monoplane, biplane, or triplane was a more efficient design concept for their airplane. They also identified the need for an engine/propeller propulsion system that could generate sufficient power while not weighing too much. These considerations were in the thinking of most aeronautical researchers of the day. One unique aspect of the Wrights' work was in their recognition of the need for adequate control of their aircraft about all three axes. In fact they stated on several occasions that they believed that solving the problems of stability and control was the key to successful manned flight. Perhaps their extensive experience with designing, building, and riding bicycles gave them a better appreciation for the interaction between humans and machines and for the role of control and stability in the design of a successful man-operated machine.

As their work progressed, the Wrights identified and solved countless other problems: where to conduct their flying experiments, what kind of control surfaces

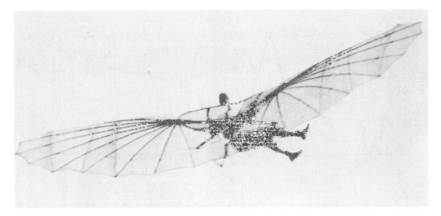

Fig. 10.1   Otto Lilienthal flying one of his gliders in 1894 (courtesy National Air and Space Museum, SI A-627-B, Smithsonian Institution).

and mechanisms to use, how to transmit the engine power to the propellers, even how to transport their airplane from Dayton, Ohio, to Kitty Hawk, North Carolina. In each case, they tackled the problem or question deliberately and methodically, in close adherence to the steps of the design method.

10.2.2   Gathering Information

The spark of interest in flight which the Wrights had held since their boyhood was kindled into a flame when they read in a newspaper in 1896 of the death after a flying accident of Otto Lilienthal, a German scientist and flight pioneer, and at the time the world's most experienced glider builder and pilot (see Fig. 10.1). They gradually came to the opinion that sustained manned flight was possible and began pursuing the problem in earnest. Their experience in the bicycle shop gave the brothers a solid engineering background and good mechanical skills, but they had very little background in flight-related topics. They began reading encyclopedias and library books about flight in nature and anything else flight-related that they could obtain. As they read and discussed, the many ideas stirred by their reading, the brothers' interest and enthusiasm grew. In 1899, Wilbur wrote a letter to the Smithsonian Institution seeking a list of good books on manned flight. The list of authors and written works that he received in return read like a who's who of recent aeronautical research. Included in the list were a pamphlet by Lilienthal, a pamphlet and a book by Samuel P. Langley, secretary of the Smithsonian (who had built and successfully flown a small uncrewed steam-powered aircraft in 1896), and a comprehensive review of the aeronautical state of the art by Octave Chanute, a French-born American civil engineer who was the foremost promoter of aviation in the United States at the time (see Fig. 10.2).

The book by Chanute gave the Wrights a great deal of information, and they were apparently impressed by Chanute's broad knowledge and experience, and so in 1900 Wilbur wrote to Chanute asking for his recommendation for a good place to conduct their flying experiments. Chanute suggested several good sites in reply and over the ensuing years provided the Wrights with considerable

Fig. 10.2   Biplane glider designed, built, and tested by Octave Chanute in 1896 (courtesy Library of Congress).

encouragement, some equipment, and a great deal of (often unwanted) advice. The Wrights also wrote to the U.S. Weather Service, and based on the information they received selected the beaches and dunes near Kitty Hawk, North Carolina, as a site with reliable winds and good weather during late autumn and early winter. Chanute, at the Wrights' invitation, visited Kitty Hawk several times while they were conducting their experiments.

The Wrights' careful reading and studying of the aeronautical literature of the day brought their knowledge of the basic theories and technologies of flight to a level of understanding that was equal to that of anyone in the world at the time. As they read and discussed (frequently argued) the various issues raised by these books and pamphlets, the Wrights identified gaps in their knowledge and doggedly sought out the missing information. They paid particular attention to the writings of Otto Lilienthal and came to agree with him that actual flight experience was essential to their mastery of the problem. Perhaps, once again, their experiences designing and building bicycles had taught them how important it was for a vehicle designer to be an experienced operator of such vehicles, in order to appreciate all of the factors that contributed to good performance and handling qualities.

As the Wrights began using the wealth of information they had obtained, they discovered errors, even in the data supplied by Lilienthal. As a consequence, they were forced to make their own tests of airfoils and wings. They built a wind tunnel with a delicate and intricate force balance to measure lift and drag (Fig. 10.3). They also performed ground tests of engines and propellers and flight tests of a variety of kites and gliders to obtain the information they required. In all cases

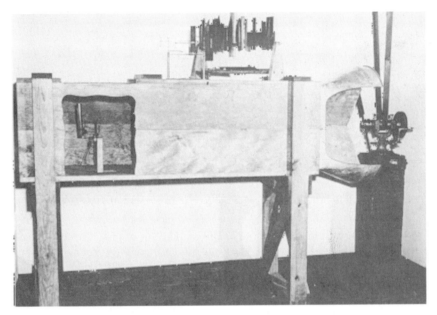

Fig. 10.3   Mockup of the Wright brothers' wind tunnel (courtesy National Air and Space Museum, NASM A-41899-D, Smithsonian Institution).

their research was extremely methodical, meticulous, and systematic, so that the data they obtained were the most accurate of its type at the time.

10.2.3   Creating Design Concepts

The overall configurations of the Wrights' airplanes were initially quite similar to those of other researchers and in fact resembled several biplane gliders that had been built and flown by Chanute. However, they made many original inventions both on their airplanes and in the equipment they built to obtain test data. Their wind tunnel and force balance, so essential to obtaining accurate aerodynamic data, were all of their own design. The Wrights also designed and built their own engine, propellers, and drive system.

The fact of Lilienthal's fatal crash, in which his ability to control his glider's attitude by shifting his weight was insufficient to recover from the affects of a wind gust, haunted the Wrights' research and discussions and challenged them to create a more effective control system. The control system design that resulted, especially wing warping for roll control, was so completely new that the Wrights were able to obtain a patent on it. The roll control system in their 1902 and 1903 aircraft (see Figs. 10.4 and 10.5) included a rudder that moved in concert with the wing warping to help coordinate the turn. This control system, which gave the Wrights the ability to maneuver like a bird, was the feature unique to the Wright aircraft that made the planes not only possible but practical.

Unfortunately, in the area of pitch control the brothers overcorrected for the inadequacies of Lilienthal's machine. They placed their pitch control surface

Fig. 10.4   Wright brothers' 1902 glider at launch (courtesy Library of Congress).

forward of the wing to give it more control authority, but they also made their aircraft statically unstable in pitch. The result was an aircraft that required constant pilot control inputs to remain in level flight and that was prone to sudden pitch-ups and dives, which caused numerous crashes. So, though their bank control system and coordinated rudder made their aircraft the first really practical flying machine, their pitch control system made it relatively difficult and dangerous to fly (see Fig. 10.6).

Unable to find a willing manufacturer, the Wrights designed and built their own engine. It was extremely simple and did not put out much power for its weight, but it was adequate. The brothers also designed and built their own propellers and the system to transmit engine power to them. Not too surprisingly, their power transmission system used bicycle chains and sprockets.

10.2.4 Analysis

The books by Lilienthal, Langley, and Chanute taught the brothers how to predict the lift and drag that would be generated by a wing and what amount of engine

Fig. 10.5   Wright 1903 Flyer, ready for a flight attempt on 14 December 1903 (courtesy Library of Congress).

Fig. 10.6   Wright 1903 Flyer, damaged after its fourth flight on 17 December 1903, showing clearly the engine, chain drives, and hand-carved propellers (courtesy Library of Congress).

power would be needed to give an aircraft a specified performance. However, as they used these equations in designing their first aircraft, they discovered to their dismay that the predictions they made were incorrect. By carefully testing many wing shapes in their wind tunnel and by building a series of kites and gliders for flight test, they determined more accurate values for the coefficients in the equations they were using. When they finally built their first powered aircraft, they knew exactly how much power they needed from their engine, how fast their aircraft would be able to fly, and how much range it would have on the tiny amount of fuel they carried.

10.2.5 Decisions

Hesitantly at first, but with increasing confidence, the Wright brothers made a series of design decisions that shaped their aircraft. The initial choice of a configuration similar to Chanute's was a conservative one. The biplane arrangement offered the ability to build a very light truss-type structure with bracing wires between the wings. The tail in the rear gave Chanute's and Lilienthal's gliders natural stability, but the Wrights, concerned about the lack of control which caused Lilienthal's death, put a moveable control surface in front of the wings, reducing stability but vastly increasing control. As the design and construction process continued through wind-tunnel tests, concept demonstrator and prototype construction, and flight tests, the Wrights made countless design decisions. Each decision was based on careful analysis and hours of discussion. The result was one of the most incredible achievements in history.

10.3   Case Study 2: Douglas DC-3

In the entire history of commercial air transportation, one aircraft stands out above all others, the DC-3, famous third model in the Commercial series of the Douglas Aircraft Company (see Fig. 10.7). The DC-3 has been used for every

Fig. 10.7   Douglas DC-3 (courtesy National Air and Space Museum, SI 95-8328, Smithsonian Institution).

conceivable purpose to which an aircraft can be put, but above all it surely must be considered as the most successful airliner ever built.

10.3.1   Air Transport in the 1920s

Great strides were being made towards improving the world economy at this time. An important factor was the increasingly intensive operation of transport aircraft. Immediately after World War I the highest aircraft utilization rates attained in practice were usually in the range of 500–800 h per year. Often they were a great deal less than this. In the late 1920s, however, rates rose towards 1000 h and, in the 1930s still further to between 1000 and 2000 h. The actual values varied between operators.

From about 1922, it began to be realized that a primary technical requirement for future transport aircraft was that they should be freed, as soon as possible, of the need to land immediately in the event of engine failure. None of the early twins had acceptable single-engine performance. Furthermore, there was no prospect, at the stage then reached in aircraft development, of producing a twin that would fly satisfactorily on one engine while carrying a commercial load. The obvious solution was a three-engined aircraft.

10.3.2   Technical Developments That Led to the Modern Airliner

As with many innovative, record-breaking aircraft, the DC-3 was made possible and successful by advances in aviation technology that gave it cost, safety, and performance advantages. Each of these advances made significant improvements in the capabilities of all aircraft that made use of them, but they came together most effectively in the DC-3. The synergy of these technological advances being brought together in a single new design produced a quantum jump in air transport capability, which made the DC-3 one of the most prolific and long-lived aircraft ever built.

Table 10.1   Engine characteristics[5]

Type and date	Aircraft	Takeoff power, bhp[a]	Cruising SFC, lb/bhp/h	Weight, lb/bhp	Fuel octane number	Initial overhaul life, h
P&W Wasp R-1340C (1930)	Ford 5-ATC Northrop Alpha	450	0.55	1.7	73	300
Wright Whirlwind J-6 (1930)	Fokker F-7 Ford 4-ATC	300	0.55	1.8	73	250
Bristol Jupiter XIF (1931)	Handley-Page HP.42	525	0.53	1.84	73	200
P&W Wasp R-1340 SIDI (1933)	Boeing 247	550	0.526	1.79	80	N/A
Wright Cyclone R-1820-F3 (1934)	Douglas DC-2	710	0.57	1.45	87	N/A
P&W Twin Wasp R-1830B (1936)	Douglas DC-3	1,000	0.46	1.67	87	500

[a]Brake horsepower.

10.3.3   Engine Developments

All through the 1920s and into the 1930s, steady progress was made improving the reliability and economy of airline operations. Apart from the switch from single- and twin- to three-engine types, much was being done to improve the powerplants themselves. Some indication of this is given by the gradual lengthening of the running times authorized between overhauls, which stood at between 15 and 50 h during World War I. In the early 1920s liquid-cooled engines, exclusively used at that time, had overhaul lives of 120–150 h. These were gradually superseded by the air-cooled radial engine of much improved reliability and economy but not of greatly increased power. By the middle to late 1920s, when air-cooled radials were coming into general use, overhaul lives had increased to 200–300 h, as shown in Table 10.1.

American engine makers were beginning to get substantial civil sales by 1930, and they began to modify development procedures to give the airlines what they wanted. In 1931 Pratt and Whitney (P&W) decided to add 50 h of running at high temperature to the 150-h bench test that the government required for new engines to prove durability. The next year, engines for the Boeing 247 were put through even more stringent tests to prove their reliability, including 100 h of full-throttle running and 500 h at cruising power. This made the tests three to four times more severe than for military engines. Maintenance expenses for engines fall almost in direct relation to the time between overhauls. Long engine overhaul lives and high reliability were (and still are) vitally important to the economy of airline operations. This is indicated by the fact that no less than 40–50% of the variable operating costs at that time were attributed to engine maintenance and overhaul.

With the emphasis almost entirely on improved reliability and durability, there was no marked increase in power output. Indeed, until the modern-type monoplane transports began to appear, in the early 1930s, increases in aircraft performance remained very modest. Cruising speeds continued to be around 100 mph, the fastest being 125 mph. To get a Certificate of Airworthiness, all civil aircraft were supposed to be capable of reaching a height of 66 ft (20 m) on takeoff within a distance of 656 ft (200 m) at full load in still air conditions. This was the so-called ICAN screen requirement. However, Junkers and other European monoplane designers soon started exploiting the advantages of higher wing loadings. This meant that airfield performance of their aircraft tended to be significantly inferior to that of the more traditional biplanes. Even so, major airports did not need to have runs of more than about 3000 ft, and grass runways were still in common use.

10.3.4 Developments in Structural Design

For the first few years after 1919, the biplane transport aircraft were of the traditional stick-and-string form of construction, which had been used since the earliest days. Airframes consisted essentially of wooden, wire-braced, frameworks that were covered by fabric. The only really important structural advance in biplane designs was the substitution, in the 1920s, of metal for wood in their framework structures. These normally had a practical life of about 5000 flying hours.

Two technical developments, both of which first appeared during World War I, were to lead to the first revolution in transport aircraft design. These were the Junkers all-metal cantilever low-wing monoplane, and the Fokker high cantilever wing of all-wood construction mounted on a fabric-covered fuselage of welded metal tubes. Junkers produced his first low-wing monoplane transport, the F.13, in 1919, and this started an evolutionary process that was to prove outstandingly successful between the wars. The F.13 and a number of other designs were built in quantity and widely used in many countries. They used a new type of all-metal construction with a corrugated aluminum alloy skin. The metal skin contributed a little to the strength of the structure, by helping the internal spars to carry the shear loads, but it was not stressed to the extent of the modern-type airliner of the 1930s. Airframe lives increased to 20,000 h or more. The corrugations imposed a drag penalty so that the advantages of the robust and durable metal structure were to some extent offset by their aerodynamic "dirtiness."

It was not until 1924 that Junkers produced the first of his three-engine monoplanes, which, in a later (1932) version, the Ju52/3M, persisted in widespread use until 1945. Junkers was followed by Fokker, who flew his famous Trimotor in 1925. Then, in 1926, Henry Ford in the United States used a derivative of the Junkers all-metal corrugated structure for a high-wing monoplane of broadly Fokker configuration. The Ford 5-AT was put into intensive production, with over four per week being produced during 1929. It was the first successful airliner built in America, but its real significance was that Henry Ford was marketing it. When there were virtually no passenger services in the United States, he at first used the aircraft to operate a freight service, but encouraged airlines to think of carrying passengers. The Ford was faster than the Fokker or Junkers, having more power, and it might have been slightly better streamlined.

The trimotor monoplanes were a parallel development to biplanes and had similar characteristics, except that they required rather bigger airfields, were a bit faster (cruising speeds up to 125 mph), and proved to have more durable structures. These were the first transport aircraft to be built in quantity: about 200 each of the Fokker and Ford and eventually about 4000 of the Junkers, mostly for the Luftwaffe.

10.3.5 Stressed-Skin Construction

A further advance in metal usage, originated by the German designer Adolf Rohrbach at the end of World War I, brought about the second transport aircraft revolution. He used all-metal cantilever structures in which the box-like member, which made up the greater part of the wing, used the skin to carry a major share of the load. Heavier metal structures became feasible by making aircraft smaller and more compact in relation to their weight.

Whereas the Junkers and the Ford metal structures had corrugated metal skins that added little strength, Rohrbach used a fully stressed smooth metal skin for a number of types in the 1920s. These inspired American designers, notably John Northrop, who independently originated the extremely durable multicellular wing structure. His simple method of building was cheap. Having been conceived by Northrop for that reason, he found by testing it that the strength was greater than expected, for the sheet skin did not fail after it began to buckle. Even after it had developed permanent buckling and so failed by normal standards, it still carried around 90% of its maximum load. This finding was crucial to the progress of metal aircraft. It had first been discovered in 1925 by H. Wagner, who was then working for Rohrbach, but his findings were not published until 1928, and Northrop's work was done independently.

What nobody fully appreciated at the time was how enormously long an aircraft built this way could last, for the many load-bearing members meant that the stress in any one member was low and fatigue deterioration was therefore slight. Recognizing the financial value of the durability it offered, Northrop applied it to other designs. For these the structural form was modified and applied to the DC-1-2-3 series in a form where the wing skin and the spars were formed from separate sheets, so that the metal could be of different thickness. The result was a light, durable structure that gave the DC-3 a better payload fraction and a much longer service life.

10.3.6 Variable-Pitch Propellers

The previous generation of aircraft had takeoff, climbing, and cruising speeds sufficiently close together for fixed-pitch propellers to be reasonably efficient in all three regimes. This was no longer the case for the streamlined aircraft that appeared in the 1930s. A fixed-pitch propeller designed to give satisfactory takeoff was inefficient while cruising at height, and the aircraft became uneconomical. Alternatively, if designed for good cruise performance it gave an unacceptably poor takeoff. Variable pitch was the only way of achieving a wider range of operating speeds and altitudes, and at the same time it increased the propeller's efficiency in cruising flight by around 5%. The first variable-pitch propellers had been introduced at the end of World War I, though they were extremely heavy as well as being mechanically unsatisfactory. In the 1920s practical experimental variable-pitch propellers appeared, and it was not until 1928 that a design was flown and proved satisfactory. This was the British Hele-Shaw Beacham, which was particularly notable because it had the important constant-speed feature that was to be adopted by Hamilton Standard and all other types from the mid-1930s. But it was slightly ahead of its time, and only the advent of the faster monoplanes was to justify its weight and complexity. More than any other feature of the new monoplanes, variable-pitch propellers made possible the great improvement in

Fig. 10.8   Boeing 247 (courtesy National Air and Space Museum, SI 99-42679, Smithsonian Institution).

performance. The wider speed range and greater operating heights were made possible by supercharged engines. On the Boeing 247 (see Fig. 10.8), the use of variable-pitch propellers meant that the takeoff run was reduced by 20%, rate of climb increased by 22%, and cruising speed by 5%.

10.3.7   Modern Airliners

The DC-3 did not suddenly appear on the aviation scene, but evolved from a series of less-capable progenitors and in competition with other very capable aircraft that were also making use of new technology. A discussion of these other aircraft points out the contribution of each of the aeronautical engineering disciplines in the success of an aircraft.

10.3.8   Boeing 247

The Boeing Company Model 200, known as the Monomail of 1930, was a single-engine, six-passenger transport. Only two were built, the type not going into production because its single engine was unattractive for passenger carrying. In addition, the large number of airliners like the Ford 5-AT that had been built in 1928–1929 meant that there was little demand for more until 1932–1933. Though the full performance of the Monomail's low-drag design could not be realized, because of the limitations of its fixed-pitch propeller its cruising speed of 135 mph was significantly better than the 100–120 mph of airliners then in service. As a follow-up of their work on the Monomail, the Boeing Company developed the B-9 twin-engined bomber for the U.S. Army Air Corps. This aircraft made use of the same structural design techniques as the Monomail and showed that a twin could be more efficient than a trimotor.

By 1932 Boeing was giving serious thought (in conjunction with their associate company United Air Lines, just formed out of an amalgamation of a number of earlier operators, including Boeing Air Transport) to a smaller civil outgrowth of the B-9 configuration. The result was the famous Boeing 247, which made its first flight in February 1933 and started in United Air Lines' service less than two months later. Though the 247 carried only 10 passengers, fewer than the Ford, it cruised 30 mph faster at 155 mph, which gave it an immediate competitive advantage. The scale of U.S. aircraft production, even at this early date, is shown by the fact that 30 aircraft of this type had been delivered to United by the end of June 1933. United had ordered it off the drawing board at a cost of $50,000 each and eventually had a fleet of 60 (Ref. 5).

The original Boeing 247 weighed 12,650 lb loaded, and its passengers, each with individual windows, sat in single seats arranged along each side of a central aisle in a soundproofed fuselage. The fuselage had a rounded cross section 5×5 ft, though the two spars of its thick wing caused an obstruction by projecting up and across the floor of the passenger cabin.

Despite the 247 being an aircraft generally accepted as representing a significant advance over the Fokkers, Fords, and others, it had a weakness. This was the retention of a low-wing loading. This was necessary because of the lack of wing flaps and because the 247 had low power in relation to its weight. So, the potential of the 247's basic design was not realized. Limiting capacity to 10 passengers was in itself bound to give relatively high costs per seat mile and was a surprising choice when the Ford carried 13.

The two P&W Wasp engines were supercharged to 5000 ft and gave 550 hp on takeoff. For the first 247s the engine nacelles were not aerodynamically very clean. They had short-chord ring cowlings around the cylinders of the type that had been developed by Townend at the National Physical Laboratory in England. The aircraft was designed to the U.S. airworthiness requirements of the period, which required the ability to maintain at least 2000 ft with full load after losing an engine. At first the aircraft had fixed-pitch propellers, but its takeoff performance, particularly from higher elevation airports, was inadequate. However, Hamilton Standard were producing their first production variable-pitch propellers at about this time, and these were introduced on the 247 and on the improved 247D of 1934.

The 247 was a sleek aircraft that could cross the continent in 19 h, including refueling stops. Of its introduction *Aviation Magazine* later said: "The year 1933 marked a milestone in air transportation when United put its revolutionary Boeing 247s into service. They immediately outmoded all passenger airplanes then in service."[6]

10.3.9 Douglas DC-1

A few months after Boeing finalized their discussions with United Air Lines on the original 247 project the Douglas Company 1000 miles away in Los Angeles was setting out on a path that was soon to establish it as the world's premier transport aircraft manufacturer. United Air Lines, with its large order of 60 of the new 247s, had ensured that no spare aircraft for any airline competitors would be available for some time from Boeing. Civil air transport was on the verge of realizing its potential as a major form of long distance transportation within the United States and the airline with the best equipment stood to reap huge profits.

Transcontinental and Western Air (TWA) at that time was interested in exploring the possibilities of an aircraft for its transcontinental routes in place of the Ford and Fokker Trimotors it then used. By August 1932, the month in which Ford stopped making his Trimotor, TWA had begun to worry about United's monopoly situation. Although not destined to fly with United until 1933, it was clear to TWA that the Boeing model would be popular, and they did not want to be left behind. TWA had asked Boeing if it could buy some of its aircraft before the order for United was completed, but Boeing refused.

Jack Frye, vice president for operations at TWA, had anticipated Boeing's response and started soliciting the other aircraft manufacturers, including Douglas, on the design that his airline needed to compete with United's 247s. The requirement was for a three-engine aircraft that was to weigh about 14,000 lb and carry 12 passengers for 1000 miles at 150 mph and have acceptable performance after one engine failure.

At this time rumors of Boeing's promising work on its 247 were no doubt reaching Santa Monica, and Douglas, with little experience of civil aircraft, had become involved in the design of stressed-skinned metal aircraft. This was a result of acquiring Northrop Aircraft as a subsidiary in January 1932. After a brief meeting with his designers, Donald Douglas decided to enter the competition. Five days later his representatives were on a train with the initial drawings and proposal, working out final details en route to the TWA offices in Kansas City.

What Douglas proposed to TWA was a twin-engine aircraft of similar general configuration to the Boeing 247 but incorporating several Northrop features (TWA already operated the Northrop Alpha) including long-chord NACA engine cowlings. TWA had asked for an aircraft with unusually good performance on one engine, so that in going for a twin design Douglas had set themselves a more difficult task. It was this goal that led Douglas to incorporate more of the devices available to make the aircraft more efficient than did the 247, especially wing flaps. They won the contract on 20 Sept. 1932, and the DC-1 was born (see Fig. 10.9).

Fig. 10.9   Douglas DC-1 (courtesy National Air and Space Museum, SI 00038742, Smithsonian Institution).

"It was the challenge of the 247 that put us into the transport business," Donald Douglas later said.[7]

The original design of the DC-1 (Douglas Commercial-1) did not have variable-pitch propellers. These were added in time for the first flight in July 1933, 11 months after Frye's letter. The weight of the aircraft structure turned out to be 30% more than estimated, and so variable-pitch propellers plus higher power engines were therefore necessary if the specification was to be met at the higher weight. The use of a higher wing loading than the 247 was thus largely fortuitous and not a conscious decision to give higher efficiency, but it made the decision to use wing flaps all the more important. Aerodynamically the DC-1 was slightly more efficient than the 247, for its engine cowlings were of more refined design.

A distinctive feature of the new Douglas airliner was a degree of sweepback on the wings. It was used to position the aerodynamic center of the aircraft in proper relationship to the c.g. The original design did not employ sweepback but had a slightly tapered straight wing. As design of the aircraft progressed, however, it became evident that the c.g. was further aft than had been anticipated. Giving the outer panels some sweepback offered a simple means for moving the aerodynamic center rearward into the correct position. The wing section of the DC-1 was the NACA 2215 at the root and 2209 at the tip (i.e., 15 and 9% thick, respectively). These were thin sections for the time made practicable by the multicell form of construction that enabled the wing center section to pass beneath the cabin floor. However, the passenger appeal of the Boeing 247 in service with United was already winning much of the traffic. The aircraft was showing clear signs of lower operating costs that later proved to be 20% lower per seat mile than those of the Ford Trimotor.[5]

10.3.10   Douglas DC-2

The possibility of improving the DC-1 was obvious by the time the one and only copy was built. Even before it had completed its flight trials, the Wright Company had produced its more powerful R-1820-F3 engine to replace the original -F used on the DC-1. With the new powerplant and the strength of an off-the-drawing-board order from TWA for 26 (later increased to 41), Douglas developed the DC-2, derived directly from the DC-1 (see Fig. 10.10).

The DC-1 was appreciably larger than the 247, even more so the DC-2 with passenger accommodation up from 12 to 14 in a cabin 5.5 ft wide and 6.25 ft high. The DC-2's fuselage was 2 ft longer than the DC-1, which together with some rearrangement of the cabin layout, made possible the addition of another row of seats. The cabin was not only 6 in. wider and 3 in. higher than the 247, but it was completely unobstructed because the wing spars passed through the fuselage beneath the floor. The seats in the DC-2 were 19 in. wide, and there was an aisle of 16 in. width between them.

The first DC-2 flew on 11 May 1934. It was certificated and put into TWA service in July, 16 months after the first Boeing 247 entered service with United. On the Newark–Pittsburgh–Chicago run, TWA's DC-2s made the trip in 5 h rather than the 5.5 h taken by the 247. In an effort to regain the upper hand, United, in 1934, undertook conversion of its fleet to 247Ds, an improved version with new and more powerful engines and several interior refinements. The DC-2 had a cruising speed at 8000 ft of 170 mph on 50% takeoff power that made it about 10 mph faster than

Fig. 10.10   Douglas C-33 military transport version of the DC-2 (U.S. Air Force photograph).

the Boeing 247D. It was also longer ranged, offering a maximum practical stage length of about 1000 miles as against 750 miles for the Boeing. The DC-2 could carry its maximum payload for about 500 miles, which was the only parameter where it showed no improvement over the 247.

As soon as the DC-2 was established in service, its outstanding qualities became apparent. The type sold quickly, and American Airlines and five other U.S. domestic operators followed TWA's lead, along with KLM and Swissair. By the end of 1934, Douglas was producing 10 DC-2s a month, and the 100th was delivered to American Airlines in June 1935. A total of 220 of these aircraft was built (160 for airlines) before the type was succeeded in production by the larger DC-3 in 1936. The initial development of the DC-1 and DC-2 series is said to have cost just over $300,000 (Ref. 5). Production aircraft were sold at $65,000 each, and the break-even point is believed to have been reached when 75 had been sold. Low as the cost of the DC-2 seems today, it was an expensive aircraft compared with the Fokker and Ford trimotors of the previous era that had cost between $35,000 and $50,000 each.

The engines of the DC-2 were supercharged and geared Wright R-1820 Cyclone air-cooled radials of 710 hp each on takeoff. Their NACA cowlings were of very clean design and had a new type of cylinder baffling within the cowling (first developed by United Aircraft and Transport Corporation), which greatly improved engine cooling. Hamilton Standard variable-pitch propellers became standardized for the DC-2. They gave engine-out performance much superior to that of any previous twin.

In addition, the sudden jump in the quality of fuel available accounts for the rise in engine power and drop in fuel consumption that occurred between 1928 and 1933 and did much to make the DC-2 successful. For the first time, more expensive fuel of higher octane rating, than was used for cruising, was fed from a separate tank for takeoff. By providing as much power from two engines as was available from three, the better fuels allowed the aircraft designer to produce an aircraft as big as the 247/DC-2 that used the more efficient twin-engine layout.

Compared with the Boeing 247, the most important aerodynamic innovation in the Douglas aircraft was the addition of wing flaps. They were of the simple split trailing-edge type and helped reduce landing speed and steepen the glide angle. Improving the streamlining of aircraft caused them to land less steeply, which made life difficult for the pilot. The stalling speed (flaps down) was 60 mph, which was comparable with that of the Boeing 247 in spite of the fact that, at 19 lb/ft^2, the wing loading was 15% greater. The higher wing loading was undoubtedly an important factor in the improved performance and economy of the Douglas design. Although flaps were not new, the use of them was crucial for the success of the DC-2.

10.3.11 Boeing 247D

As a later design, the DC-1 and its DC-2 production derivative had a number of important advantages over the 247. It was soon apparent that the Boeing design had become uncompetitive. Its improved 247D was an attempt by Boeing to match the DC-2. The 247D's loaded weight was increased to 13,650 lb, though with no increase in passenger capacity, and it entered service with United in 1934. It had more highly supercharged Wasp engines, cowled within the very much cleaner NACA long-chord cowlings that had been developed at Langley Field in 1927. These had already been used with great success on the DC-1 and DC-2. With its uprated engines the 247D could maintain 6000 ft on one engine at full gross weight. Nevertheless, it still lacked flaps, which imposed the relatively low wing loading of 16.3 lb/ft^2. Its coast-to-coast schedule of 20 h with seven stops was improved, but its economic characteristics remained inferior. The Boeing design was eclipsed: where Boeing built 75 247s, Douglas sold 220 DC-2s, many of them abroad.

Table 10.2 summarizes the characteristics of these aircraft. The zero-lift drag coefficient of the DC-3 is about 17% higher than that of the Boeing 247D; however, the values of the parameter C_{fe} are quite close together. The larger C_{D_o} of the DC-3 results from the larger ratio of wetted area to wing area caused by the larger fuselage of the DC-3, which was designed to carry three abreast compared with two abreast in the Boeing 247. The higher lift-to-drag ratio (L/D) of the DC-3 is accounted for by its higher aspect ratio. The higher wing loading of the DC-3 (compared with the Boeing 247) is not reflected in its stall speed because of the use of trailing-edge flaps.

If it is assumed that the technical efficiency of an airliner can be measured in three ways, namely, its drag, its lift/drag ratio, and the proportion of the weight at takeoff that forms its useful load (payload plus fuel and crew), we can compare aircraft. If the DC-2 and DC-3 are compared with their predecessors of the early 1920s, the biggest improvement was in aerodynamic drag, which of course influences the other two criteria. The drag was roughly half that of biplanes and poorly streamlined monoplanes like the Ford Trimotor. The reduction in drag explains how the DC-2 and DC-3 were able to cruise 50% faster with no more power per passenger than the Ford whose engines gave 97 bhp per passenger and those of the DC-3 gave 95 hp per passenger. The lower drag also accounts for the improvement in lift/drag ratio. For the DC-2 and DC-3 the L/D was around 14, whereas the 1920s' aircraft had ratios of around 7.5–8.5. There appears to have been a reduction in the ratio of useful load to takeoff weight, which for the DC-3 was around 33%, whereas for the Ford and Fokker Trimotors it was 40–45%. The effect on costs of the heavier

Table 10.2   Characteristics of the first modern transport aircraft[8]

Aircraft	Ford 5-AT Trimotor (1926)	Boeing 247 (1933)	Boeing 247D (1934)	Douglas DC-1 (1933)	Douglas DC-2 (1934)	Douglas DC-3 (1935)
Engines	3 P&W R-1340	2 P&W R-1340	2 P&W R-1340	2 Wright R-1820	2 Wright R-1820	2 Wright R-1820
Takeoff power, bhp	3×420	2×550 (525?)	2×600	2×710	2×710	2×1000
MTOW, lb	13,500	12,650	13,650	17,500	18,080	24,000
Power, weight, bhp/lb	0.0933	0.087	0.088	0.081	0.078	0.083
Wing area, ft^2	835	836	836	942	942	987
Wing span, ft	77.8	74	74	85	85	95
Length, ft	50.3	54.3	54.3	60	62	64.5
Wing loading, lb/ft^2	16.2	15.1	16.3	18.6	19.2	24.3
Aspect ratio	7.24	6.55	6.55	7.67	7.67	9.14
C_{D_0}	0.0471		0.0212			0.0249
C_{f_e}	0.0142		0.0057			0.0062
$C_{L_{max}}$	1.6		1.7			2.1
$(L/D)_{max}$	9.5		13.5			14.7
Cruising speed at 50% takeoff power, mph	100	155	160 at 7500 ft	190 at 10,000 ft	200 at 10,000 ft	192 at 10,000 ft
Range, miles				1000	1050	1500
Type of flaps	None	None	None	Split	Split	Split
Stalling speed, mph	64	59	60	59	61	65
Passengers	13–15	10	10	12	14	21

structure was reduced by the 50% increase in cruise speed, which increased the output productivity of the aircraft in relation to its weight and size, just as a reduction in structure weight would have done. Though the increased weight was partly the result of better equipment and more comfortable accommodation for passengers, the Douglas's were simply stronger and heavier than they needed to be.

10.3.12 Douglas DC-3

Early in 1935 discussions started between Douglas and one of its leading customers, American Airlines, about a scaled-up version of the DC-2 to be used as an overnight coast-to-coast sleeper transport in succession to the Curtiss Condor biplanes then in service. Although the sleeper idea was popular with passengers, the Condor was not; it was slow. The new aircraft for American was to provide sleeping accommodation for 14 passengers. Work started in mid-1935 on the first Douglas Sleeper Transport (DST), later to become famous as the DC-3, and the prototype flew for the first time on 22 December the same year.

The widened and lengthened fuselage could accommodate 21 day passengers arranged two abreast on one side of a 19-in. aisle and in single seats on the other with seats 20 in. wide. Even with two-and-one seating the passengers had more space than in the DC-2, the internal section of the cabin being 7 ft 8 in. wide and 6 ft 6 in. high. Widening the fuselage to produce the DC-3 gave another fortuitous improvement in efficiency, for the fatter shape was better streamlined, and the 50% increased capacity caused little increase in drag. The result of these changes was an aircraft that had no contemporary rival and that established new standards of operation wherever it was used. With a seating capacity now 110% higher than its Boeing rival, its operating cost per seat mile represented a quantum improvement of 25% on that of the Boeing 247D, making it the most economical passenger carrier in existence. Compared even with the DC-2, the DC-3 usually carrying 50% more seats, was only 10–12% more expensive per mile flown.

The important feature of the Douglas type of wing construction was that it incorporated a number of spanwise plate webs, the DC-1, DC-2, and DC-3 each had three, which took the vertical shear loads and, in conjunction with the stiffened metal surface of the wing, acted as a main structural member. Closely similar to the DC-2, being built of the same materials and employing basically the same structural design, the DC-3 retained the multispar wing and tail. This form of construction played an important role in the airframe's outstanding fatigue-resistant and fail-safe qualities. Individual DC-3's flew as many as 70,000 h without major rebuild and thus give evidence of a durability that more recent designs have found hard to rival. The wing area was increased about 5% over that of the DC-2 and span from 85 to 95 ft by the addition of new and pointed wing tips. Wing loading went up from 19 to 24 lb/ft^2.

The first DC-3 (a DST version) was delivered to American Airlines in July 1936. It covered the 2600 miles from New York to Los Angeles overnight with three or four stops in 16 flying hours and made the return trip in 14 h. The total elapsed time for the one-way journey was usually about 24 h. This compares with the 8-h nonstop schedule achieved later by those descendants of the DC-3, the DC-7, and Super Constellation and the 6 h of today's jet airliners.

Whereas the sales success of the DC-2 had been outstanding, that of the DC-3 was phenomenal. Orders poured in from all over the world, 150 had been ordered, and most delivered by the end of 1937. The type soon came to be accepted as the universal standard both abroad and on U.S. domestic routes. The latter were now expanding so vigorously that they gained a proportion of traffic that was equal to that of the rest of the world's airlines put together. Douglas now had a monopoly position in the supply of airliners up to the outbreak of war, apart from Lockheed's smaller and faster aircraft and a few bigger ones from Boeing. The success of the DC-2 and DC-3 allowed Douglas to grow from sales of $2,294,000 in 1933 to $20,950,000 in 1937, 60% of which came from civil sales. Their military sales were based on the DC series. Replacing the 700-hp Wright Cyclones of the DC-2 with Wright Cyclones of 1000 hp (or later on, P&W Twin Wasps of 1200 hp), provided the extra power the DC-3 needed and enabled the loaded weight to be raised initially to 24,000 lb. In military service takeoff weight rose to much higher values.

Table 10.3 serves to show the range of improvement over less than a decade and measures the direct gain in efficiency provided by the innovations in aircraft design that were introduced between 1927 and 1936. It can be seen that the DC-3s costs were less than half that of the Trimotor and 60% of those of the Boeing 247.

Reducing the drag and increasing size (via the number of seats) did most to cut operating costs between the Ford and the DC-3. The cost of fuel per seat mile was 40% less for the DC-3 than the Ford, despite the increase in speed. Improved engine design (with the more expensive higher octane fuels used) helped by the use of variable-pitch propellers, probably accounted for about a third of this reduction in fuel costs, but reduced drag did the rest. By making possible the higher cruise speed the reduction in drag also helped to reduce crew, maintenance, and depreciation costs per seat mile, by increasing the amount of revenue earned in a given time. The greater seating capacity meant more passengers per crew (rather like the improvement in the Boeing 747-400 where the flight deck crew was cut from three to two). The maintenance costs of the DC-3 were over 60% lower than those of the Ford; other features reducing costs were the use of two engines rather than three, more durable engines and the superior design of the airframe. Just how durable the airframe would prove to be was unknown to the designers and operators of the

Table 10.3   Comparative operating costs (cents [1937] per available seat-mile)[5]

Aircraft	Introduction	Passengers	Flying operations	Direct maintenance	Depreciation	Total
Ford Trimotor	1928	13	1.34	0.67	0.62	2.63
Lockheed Vega	1929	6	1.56	0.58	0.37	2.52
Boeing 247	1933	10	1.19	0.43	0.49	2.11
Douglas DC-3	1936	21	0.69	0.24	0.24	1.27

DC-2 and DC-3. As the Boeing 247 and DC-2 had rendered the Ford obsolete, so then the DC-3 rendered them obsolete in turn. Though the DC-3 was to have so long a service life, because it was not rendered obsolete by more modern designs depreciation charges on it were unnecessarily high, even though they were roughly half those of the Ford.

10.3.13   World War II Service of the DC-3

When America entered World War II on 7 Dec. 1941, there were approximately 360 DC-3s in U.S. airline service. At the time, the DC-3 was the most thoroughly proven transport aircraft available in America. It was immediately pressed into service under a number of designations, the most common were the C-47 Skytrain (cargo carrier), C-53 Skytrooper (troop transport), and R4D (U.S. Navy) in America and Dakota in the United Kingdom. There seems to be no end to stories of the amazing feats of the DC-3 during World War II. Its ruggedness and versatility were unequalled. It was consistently flown overloaded. For example, a 21 seater belonging to China is known to have taken off with 74 passengers. Despite having large sections of wing and tail missing, as well as being riddled with bullet holes, it was still controllable. Another lost its starboard wing, which was replaced with a wing from a DC-2 (5 ft shorter) and then flew 900 miles back to base for repairs! This was the only DC-$2\frac{1}{2}$!

Douglas is said[6] to have spent about $1,200,000 on the initial development of the DC-3. The purchase price of the production aircraft was between $90,000 and $115,000 of which about $20,000 was for the engines. Douglas's break-even point for the aircraft was about 50. A total of 10,926 DC-3 type aircraft were built in the United States between 1936 and 1945. Of this total about 10,000 were procured for military use. This represents a remarkable return on investment.

Not all of the aircraft survived World War II of course, but at the end of hostilities there were about 4000 military DC-3s to be disposed of by the U.S. Government. Many former service personnel started cargo operations or nonscheduled airlines, and for them the low-cost wartime aircraft were ideal; not only were the aircraft cheap and plentiful, they were the type with which so many had a wealth of experience. A large part of its sterling performance during the war was as a result of the exceptional durability of the airframe. It had been designed in 1933 when much less was known about many aspects of aeronautical engineering. As a consequence, the DC-3 was overdesigned in many respects and certainly more so than modern airliners. This was partly fortuitous because it gave the DC-3 great ruggedness that was to prove particularly appropriate both in terms of its service in World War II, the Berlin Airlift (see Fig. 10.11), Korean and Vietnam wars, and also as an aircraft for smaller airlines and executive transport use.

10.3.14   Conclusions

1) The DC-3 was basically an improved and stretched version of the DC-2 that likewise was an outgrowth of the DC-1. It was not a startling innovation; it was the logical evolution of an earlier model. The impetus for its creation, however, was broadly similar to that behind the DC-1, namely, competition among domestic

Fig. 10.11   C-47 military transport versions of the DC-3 during the Berlin Airlift, an action in which they changed the lives of thousands of Berliners (U.S. Air Force photograph).

airlines. Just as the DC-1 grew out of TWA's need to replace its aging trimotor fleet with something better than United's Boeing 247s, so the DC-3 allowed American Airlines to offer the public the ultimate in transcontinental sleeper service.

2) The DC-3 owes much of its spectacularly economical combination of payload capabilities and range characteristics to its combination of robust structure, good aerodynamics, available engines, variable-pitch propellers, and appropriate fuels that could do the job. Some of this might have been a result of luck. Nevertheless, the design was a brilliant achievement and a magnificent balancing of numerous desirable qualities, but it presupposed the powerplants capable of delivering the requisite power.

3) The DC-3 was a thoroughly proven and enthusiastically accepted airliner by the late 1930s. It was designed early enough to be relatively simple and thus easy to maintain and repair, while incorporating enough of the essential technology to keep it competitive after years of use. Regardless of the influence that wartime service might have had on its later and continuing success, it was already firmly established on its own merits at the beginning of the U.S. involvement in World War II.

4) There is an inherent quality that further helps to explain its longevity, namely, its size. It was an aircraft specifically designed to be able to fly into and out of small airfields and to be economical over relatively short distances. This has turned out

to be an asset in countries with less developed air transport networks than those in the United States.

5) The cantilever low-winged monoplane, with riveted light-alloy stressed skins, retractable landing gear, and with two carefully cowled supercharged radial air-cooled engines of sufficient power driving variable-pitch propellers, remained fundamentally unchanged into the 1950s. The Boeing and Douglas companies set a pattern in 1933 that has persisted in remarkable fashion. Only in the 1950s with swept-wing turbojet-powered aircraft was there a major shift from the long-established aerodynamic/engine formula. It is difficult to imagine that any other aircraft will ever grow up with such a unique combination of factors in its favor as the DC-3.

10.4 Case Study 3: Evolution of the F-16

The Lockheed-Martin F-16 Fighting Falcon multirole tactical fighter must be regarded as one of the must successful and prolific (and certainly one of the most esthetically pleasing) jet fighter aircraft of the post-Vietnam War period. Its design evolution is an interesting one because it was developed as a lightweight air-to-air fighter (a replacement for the F-5) but went into service as a medium-weight multirole fighter (a replacement for the F-4). Its further development into the F-16XL deep-interdiction aircraft (a replacement for the F-111) is also discussed.

As stated by Bradley,[9]

> The design of tactical military aircraft presents quite a challenge to the aerodynamicist because of the vast spectrum of operational requirements encompassed by today's military scenario. The designer is faced with a multitude of design points throughout the subsonic-supersonic flow regimes plus many off-design constraints that call for imaginative approaches and compromises. Transonic design objectives are often made more difficult by restraints imposed by subsonic and supersonic requirements. For example, wings designed for efficient transonic cruise and maneuver must also have the capability to accelerate rapidly to supersonic speeds and exhibit efficient performance in that regime.
>
> The design problem is further complicated by the fact that the weapon systems of today are required to fill multiple roles. For example, an aircraft designed to fill the basic air superiority role is often used for air-to-ground support, strike penetration, or intercept missions. Thus, carriage and delivery of ordinance and carriage of external fuel present additional key considerations for the aerodynamicist. The resulting aircraft flowfield environment encompasses a complex mixture of interacting flows.
>
> For example, the need for rapid acceleration to supersonic flight and efficient supersonic cruise calls for thin wing sections with relatively high sweep and with camber that is designed to trim out the aft ac movement at supersonic flight. However, these requirements are contrary to those requirements for efficient transonic maneuver, where the designer would prefer to have thicker wing sections designed with camber for high C_L operation and a high-aspect-ratio planform to provide a good transonic drag polar.

Harry Hillaker, who was the chief project engineer on the YF-16 and vice president and deputy program director of the F-16, began the plane's design process by

Configuration Evolution

- Preliminary Configuration Definition
- Experimental Data (Wind Tunnel) Based
- 78 Combinations of Variables
  - Wing Planform
    - Airfoil Section – Fixed and Variable
    - Wing-Body Relationship
    - Inlet Location and Type
    - Single vs Twin Vertical Tall
    - Forebody Stakes

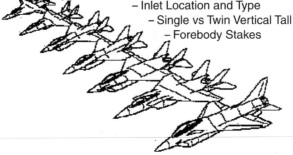

Fig. 10.12   F-16 evolution (courtesy Lockheed Martin).

clearly defining the problem and collecting data (Hillaker, H. J., private communication, Jan. 1997). Examination of results from air-to-air combat over Southeast Asia/Vietnam revealed that aircraft in the U. S. inventory had only marginal success over their opponents. From 1965 to 1968, engineers at General Dynamics in Fort Worth examined the data from the Southeast Asian conflict to determine what parameters provided an edge in air-to-air combat. Wing loadings, thrust loadings, control issues, g tolerances for the pilot, and empty-weight fraction were identified as key parameters in air-to-air combat. Because these studies were focused on technology, there were no configuration designs.

Hillaker (private communication, Jan. 1997) continued: "From 1969–1971, an extensive wind-tunnel program was conducted in which data were obtained on a wide variety of configurations over a range of free-stream test conditions and angles of attack." As shown in the sketches of Fig.10.12, configuration variables included wing planforms, airfoil sections, wing-body relationships, inlet locations, horizontal and vertical tail configurations, and forebody strakes. Buckner et al.[10] reported,

> The 1969–1970 period produced technology studies in four areas very important to air combat maneuvering fighter design:
>
> 1. Project Tailormate (1969, 1970)—an experimental study of a wide variety of inlet types and locations on typical fighter designs with the goal of maintaining low distortion and high-pressure recovery over wide ranges of angle of attack and sideslip. The YF-16 inlet location is largely a result of the experience gained in this work.
>
> 2. Wing Mounted Roll Control Devices for Transonic, High-Light (sic) Conditions (1969)—a study of a variety of leading- and trailing-edge devices for roll control in the combat Mach number and high-angle-of-attack range.

3. Aerodynamic Contouring of a Wing-Body Design for an Advanced Air-Superiority Fighter (1970)—an add-on to the roll-control study, which produced analytical information on a blended wing-body design. This experience was helpful in the later development of the YF-16 overall planform and blended cross-section concept.

4. Buffet Studies (1969–1970)—a series of efforts producing new knowledge and methodology on the phenomenon of increased buffet with increasing angle of attack. The methodology, in turn, allowed the YF-16 to be designed for buffet intensity to be mild enough to permit tracking at essentially any angle of attack the pilot can command in the combat arena.

The above studies produced knowledge that was brought to bear on the configuration studies, specifically for the development of a lightweight low-cost fighter, accomplished in 1970, 71.

The lightweight, low-cost fighter configuration studies intensified when a new set of guidelines were defined for application to combat scenarios in Europe. Hillaker (private communication, Jan. 1997) noted,

These mission rules included a 525 nautical mile (nm) radius, four turns at 0.9M at 30,000 feet, accelerate from 0.8M to 1.6M, three turns at 1.2M at 30,000 feet, and pull a four-g sustained turn at 0.8M at 40,000 feet. Furthermore, the aircraft had to demonstrate operability both for the U. S. Air Force and its NATO allies.

As noted by Buckner et al.,[10]

The YF-16 really got its start, then, as a result of the 'in-house' studies initiated in late 1970 in response to the new mission rules. . . . Major emphasis was placed on achieving flight and configuration characteristics that would contribute directly to the air-to-air kill potential in the combat arena—specifically maximizing maneuver/energy potential and eliminating aerodynamic anomalies up to the maneuver angle-of-attack limits.

Buckner and Webb[11] reported,

Examples of the "design to cost" in the case of the YF-16 aerodynamic features are (1) a single engine, eliminating the complex question of what to do between the nozzles, (2) an empennage/nozzle integrated design, devoid of adverse interference, (3) a single vertical tail tucked in safely between the forebody vortices, (4) a simple underslung, open-nosed inlet with no complex moving parts, (5) a thin-wing airfoil with only slight camber, minimizing the question of Reynolds number effects on transonic shock locations, and (6) simple trailing-edge ailerons. All of these features reduced the cost through virtual elimination of design changes in the refinement stage after contract go-ahead and through simplification of the task required to fully define the vehicle aerodynamics.

In 1972, aerodynamic data were obtained over a wide range of Mach number, of Reynolds number, and of angle of attack. The ranges of these variables, as taken from Ref. 9, are reproduced in Fig. 10.13.

Buckner et al.[10] reported,

From these data it is obvious that the taper ratio (λ) should be as low as practical, consistent with reasonable tip-chord structural thickness and early tip-stall

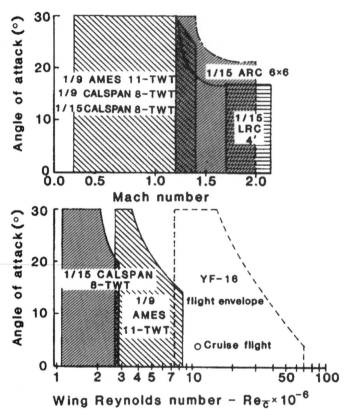

Fig. 10.13   Aerodynamic data used in YF-16 development (courtesy Lockheed Martin).

configurations. The final selection was a taper ratio of 0.227. The start combat weight is lowest at a wing aspect ratio of 3.0, the final selected value, and is relatively independent of wing sweep in the range of 35 to 40 degrees. The greater wing sweep is beneficial to the supersonic performance, giving reduced acceleration time and increased turn rate. Some penalty in aircraft size is noted as the wing sweep is increased to 45 degrees, and the higher-sweep wings are more prone to have aileron reversal because of aeroelastic effects. As a result, wind tunnel tests were limited to sweep angles between 35 and 45 degrees. Perturbation of wing thickness ratio (t/c) indicates a lighter weight airplane with a thicker wing, but supersonic maneuverability improve with thinner wings. The desire to achieve a balance in subsonic and supersonic maneuver capability dictated selection of as thin a wing as practical (t/c = 0.04), consistent with flutter and aileron reversal considerations.

Bradley[9] notes,

The tactical military aircraft design problem is made more difficult by the supersonic acceleration requirement. The wings must be as thin as structurally feasible to reduce drag, but fixed camber suitable for optimum transonic maneuver is not practical because of supersonic camber drag. An obvious solution is a smoothly varying wing camber design. However, structural and actuation system weights prove to be prohibitive for thin wings. Simple leading edge and trailing edge flaps often prove to be the most practical compromise for high-performance, multiple design point configurations.

The extreme possibilities are apparent. One may design a wing with optimized transonic maneuver camber and twist and attempt to decamber the wing with simple flaps for supersonic flight. On the other hand, one may design the wing with no camber or with a mild supersonic camber and attempt to obtain transonic maneuver with simple flaps.

Even though the designer has worked carefully to design a wing for attached flow and to optimize its performance, there are points in the sustained and instantaneous maneuver regimes where the flowfield contains extensive regions of separated flow. The manner in which the flow separates will strongly affect the vehicle's drag and its controllability at the higher values of the lift coefficient. For many designs, strakes or leading-edge extension devices are used to provide a controlled separated flow. Controlled vortex flow can then be integrated with the variable camber devices on the wing surface to provide satisfactory high-lift, stability and control, and buffet characteristics. The resulting flowfield is a complex one, combining attached flows over portions of the wing with the vortex flow from the strake (see Fig. 10.14).

One of the questions addressed by the designers of the Lightweight Fighter was single vertical tail vs twin-tailed configurations (see Fig. 10.10). NASA data available at this time indicated advantages of the twin-tailed configurations. However, as noted by Hillaker (private communication, Jan. 1997) the NASA data were

Fig. 10.14   Complex flowfield over a wing with leading-edge vortices (courtesy Lockheed Martin).

limited to angles of attack of 15 deg or less. Buckner et al.[10] noted:

> To the dismay of the design team, however, the directional stability character-
> istics of the twin-vertical-tailed 401F-0 configuration were not as expected. In
> fact, a severe loss of directional stability occurred at moderate-to-high angles of
> attack.... Analysis of oil flow visualization photographgraphs led to the belief
> that forebody flow separations and the interaction of the resulting vortices with
> the wing and vertical tail flow fields were major causes of the stability problem.

Modifications were made to delay forebody separation to higher angles of attack.
Buckner et al.[10] reported:

> It was more difficult to make the twin-tail configurations satisfactory compared to
> the single vertical tail configurations (in addition, some combinations of angle of
> attack and sideslip produced visible buffeting of the twin tails). Beneficial effects
> on the directional stability derivatives were noted when relatively small highly
> swept 'vortex generators' (strakes) were located on the maximum half-breadth of
> the forebody.
>   At this point, NASA Langley Research Center aerodynamicists were consulted
> and they suggested that the lift of the wide forebody could be increased by sharp-
> ening the leading edge to strengthen the vortices rather than weaken them as our
> earlier attempts had done. The point was that forebody separation is inevitable
> at very high angle of attack; therefore, the lift advantages offered by sharp lead-
> ing edges should be exploited. This also would allow the forebody vortices to
> dominate and stabilize the high-angle-of-attack flow field over the entire aircraft,
> improving, even, the flow over the outboard wing panels.

Once the YF-16 was in the flight-test program with the YF-17, it was no longer
called the Lightweight Fighter. It was called the Air Combat Fighter (ACF). Orders
for the ACF, the F-16, came in 1975. The first operational units were formed in
1978. In the late 1970s, production versions of the F-16 were quickly modified to
add a full radar capability, to add hard points to accommodate the ability to handle
air-to-ground capability, and to increase the combat radius to 725 n miles. Thus,
the aircraft had become a true multirole fighter.

Bradley[9] notes,

> Perhaps the greatest irony for the tactical aircraft designer results from the fact
> that an aircraft designed to be the ultimate in aerodynamic efficiency throughout
> a performance spectrum is often used as a 'truck' to deliver armaments. Aircraft
> that are designed in a clean configuration are often used operationally to carry an
> assortment of pylons, racks, missiles, fuel tanks, bombs, designator seeker pods,
> launchers, dispensers, and antennas that are attached to the configuration at any
> conceivable location.
>   The carriage drag of the stores is often of the same order of magnitude as
> the total minimum drag of the aircraft itself. For example, the minimum drag of
> the F-16 aircraft is compared in [Fig. 10.15] with and without the air-to-ground
> weapons load. It is readily seen that the store drags themselves present as large
> a problem to the aircraft designer as the drag of the clean configuration. Store
> carriage on modern tactical aircraft is extremely important, particularly as one

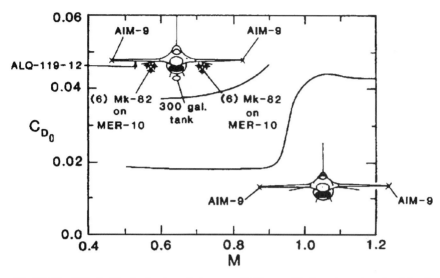

Fig. 10.15   Effect of typical external stores on F-16 parasite drag (courtesy Lockheed Martin).

approaches the transonic regime, where the interference effects of the stores and pylons are highest and most detrimental to performance.

Not only must external stores be carried efficiently by tactical aircraft, but they must release cleanly and follow a predictable trajectory through the vehicle's flowfield. The store trajectory is governed by the highly unsteady forces and moments acting on the store produced by the nonuniform flowfield about the configuration and the aerodynamic characteristics and motions of the store itself. The problem is complicated by realistic combat requirements for jettison or launch at maneuver conditions and multiple release conditions where the weapons must not 'fly' into one another.

Military aircraft of the future will have added emphasis on store carriage and release early in the design process. Some possible concepts for weapons carriage are contrasted in Table 10.4.

Bradley[9] presented wind-tunnel measurements that compared the drag for conformal carriage vs conventional installations of MK-82 bombs. As shown in Fig. 10.16, conformal carriage permits one to carry 14 MK-82 bombs at substantially less drag than a 12 MK-82 pylon/multiple bomb rack mounting. Bradley noted,[9] "Significant benefits for the conformal carriage approach, in addition to increased range, are realized: increased number of weapons and carriage flexibility; increased penetration speed; higher maneuver limits; and improved supersonic persistence. Lateral directional stability is actually improved with the weapons on."

Hillaker[12] noted that in 1974 General Dynamics embarked on a Supersonic Cruise and Maneuver Program (SCAMP) to develop a supersonic cruise derivative of the F-16. The supercruiser concept envisioned optimization at supersonic cruise lift conditions so that sustained cruise speeds on dry power (nonafterburner power) could be achieved in the Mach 1.2 to 1.3 speed range. Tradeoffs to the

Table 10.4   Weapon carriage concepts[9]

Store carriage concepts	Advantages	Disadvantages
Wing pylon carriage	Most flexible carriage mode: large payloads, inefficient store shapes	High drag
		High RCS
Internal carriage	Low drag	Limited weapon flexibility
	Low RCS	Increased fuselage volume
Semisubmerged carriage	Low drag	Holes must be covered up after weapons drop
	Low RCS	
		Severely restricted payload
Conformal carriage	Most flexible of low drag carriage concepts	Size restrained

aerodynamics required for supersonic cruise, subsonic cruise, and maneuvering flight were explored. The goal was to arrive at a design that would offer at least a 50% increase in the supersonic L/D, that would retain a high subsonic L/D ratio, and that would provide the level of maneuverability of a fighter.

Hillaker (private communication, Jan. 1997) noted that, by 1977 to 1978, it was clear that the F-16 was to serve both the air-to-air mission and the air-to-ground mission (with the air-to-ground mission dominating). Thus, engineers sought to develop a design that was a straightforward modification of the F-16. The approach was to build on the technology developed for the original SCAMP configuration adding a maneuver requirement to the supersonic cruise capability. From 1974 to 1982, the SCAMP/F-16XL configurations underwent significant refinements. The configuration evolution, as taken from a paper by Hillaker,[12] is reproduced in Fig. 10.17. Hillaker reported:

The planform requirements included a forebody blend (strake area) for high angle-of-attack stability, an inboard trailing-edge extension for pitching moment

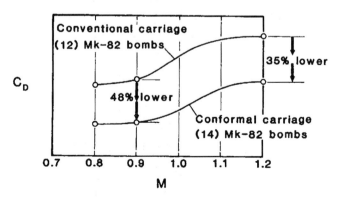

Fig. 10.16   Drag benefits of conformal weapons carriage (courtesy Lockheed Martin).

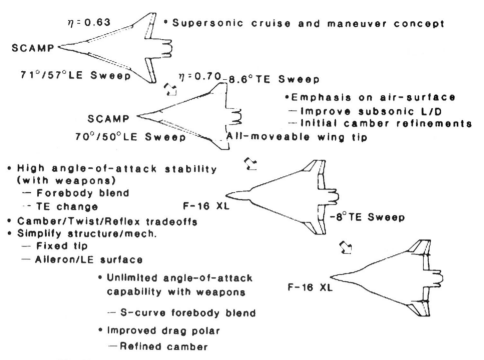

Fig. 10.17   F-16XL configuration evolution (courtesy Lockheed Martin).

improvement, and fixed wing tips with ailerons and leading-edge device to enhance the flow over the aileron at high angles of attack. The combination of
ailerons and a leading-edge device (flaps for flow control, not lift) was developed
to resolve the mechanical and structural complexities of the all-movable wing
tip on the previous configuration. All-movable wing tips provided high roll rates
and adverse yaw for resistance to yaw divergence at high angles of attack. The
aileron-leading-edge device combination provided the same aerodynamic advantages and simplified wing structure, reduced weight, increase fuel volume, and
allowed tip missile carriage.

The benefits of the F-16XL configuration include 40 to 85% lower drag with
the integrated weapons carriage, 50% better supersonic lift-to-drag ratio with no
subsonic penalty, 17% lower wave drag with 83% more internal fuel, and increased
lift with expanded maneuver angle of attack.

The first of two demonstrator aircraft (modified from two F-16A aircraft) took
to the air at Carswell Air Force Base in Fort Worth, Texas, on 3 July 1982. They are
still being used by NASA for flight research (Nipper, M. J., private communication,
Nov. 1996).

10.5   Opportunities in Aircraft Design

The aircraft concepts that are presented in this section come from a study by the
U. S. Air Force Scientific Advisory Board.[13] Entitled *New World Vistas*, this study
was done in recognition of the 50th anniversary of a technology forecasting effort

led by one of the most famous of all aeronautical engineers, Theodore von Kármán. Following the spirit of von Kármán's work, the challenge for the authors of the *New World Vistas* report was to search the world for the most advanced aerospace ideas and project them into the future. The complete report, published in 1996, contained more than 2000 pages of text in 12 volumes including one specifically dedicated to aircraft and propulsion systems.[14]

As you learned in Chapter 1, new design concepts come from identification of mission needs. In the following paragraphs we describe four different air vehicle concepts that might be developed to satisfy future mission requirements. Each design idea is briefly presented in terms of its purpose (mission) and the enabling technologies that are required to make the concept a reality.

10.5.1 Uninhabited Combat Air Vehicle (UCAV)

It was not long after the invention of the airplane that it was first used in war. World War I fighters formed the beginning of a long line of aircraft designs that have led today to the F-22. Fighters and fighter pilots are indeed legendary. However, in the future, it is likely that we will see many more uninhabited aircraft used in armed conflict. Improved antiaircraft missile systems, cockpit information overload, and new developments in sensors and information technology are all contributing to this change. Removing humans from the air vehicle also relieves constraints in *g* limits, agility, and flying/handling qualities. Lower weight and cost (no pilot or displays or life support systems), improved stealth (no need for cockpit area) and endurance, and significant reductions in in-theater support (no need to transport and assemble shelters, workstations and environmental control units) can all be derived through the development of uninhabited vehicles. The cockpit of a traditional aircraft is simply an information and control center. The question being asked today is: "instead of bringing information to the pilot in a high-stress, high-threat environment, why don't we bring the pilot to a safe information center and operate the aircraft remotely?" Information from many sources, including the UCAV itself, will be brought to an Execution Control Center over high-speed, massively redundant fiber or satellite communication routes where well-rested pilot/technicians will control the battle (Fig. 10.18).

Missions.   "Uninhabited aircraft are those air vehicles that do not have on-board presence of a pilot or aircrew. All control functions normally associated with piloting are performed by on-board controllers. Information used by those controllers may be effected by humans off-board, providing input via data links."[14] Today's uninhabited vehicles have limited capabilities—they are used primarily as cruise missiles or expendable reconnaissance probes. However, in the future a mix of inhabited and uninhabited vehicles will constitute the warfighting system. and some of the uninhabited aircraft will be fully combat capable. Current operating envelopes will be expanded and a whole range of mission options are possible. For example, in addition to the UCAV concept described above ($+/-20$-g super high agility fighter), a very high-altitude and long-endurance reconnaissance aircraft (see next section), or a miniaturized stealth attack aircraft might also be developed.

Enabling technologies.   Replacing the human brain at the center and sight of aircraft operations will be neither easy nor inexpensive. Cognitive engineering is

Fig. 10.18   UCAV concept.

the term being used today to describe the work that has to be done to allow off-board human controllers to effectively interact with onboard systems and make UCAVs mission capable. Cognition refers to functions of the mind like thinking, knowing, remembering, problem solving, and decision making. Situation awareness is the ability of an individual to receive and integrate information and then to understand its significance for the task at hand. The primary challenge in cognitive engineering for UCAVs is the design of human/machine interfaces that provide pilots at a remote location a high level of situation awareness, cognitive support to deal with vast amounts of information, and reliable communication/control systems that allow the pilots to fly the aircraft. Intelligent signal/data processing (using artificial intelligence, neural networks, fuzzy logic hardware and software), secure and redundant data links, and advanced control systems all will be required. Also, in the case of miniaturized vehicles, additional developments in aerodynamics for low-Reynolds-number flows, low-cost and small propulsion systems, low-cost plastic structures, and low-cost control/sensor/avionics packages are required.

10.5.2   High-Altitude, Long-Endurance (HALE) Aircraft

Development of Earth-orbiting satellites has had a tremendous impact on our ability to perform communication, reconnaissance, and environmental-monitoring missions. However, satellites are very expensive to build and place in orbit, and they are also difficult to maneuver once there. A high-altitude (> 80,000 ft), long-endurance (HALE) autonomous aircraft provides a very attractive alternative, and a variety of new technologies are making it possible to turn this concept into a reality. AeroVironment, an engineering firm headed by Paul McCready who is famous for his work on human-powered aircraft, has built a prototype of this concept called Pathfinder.[15] The aircraft is a flying wing with a span of 100 ft and a chord of 8 ft. It is powered by eight electric powered propellers and has 26 individual elevators on the trailing edge of the wing to control pitch and dampen structural deformation caused by turbulence.

Missions.   In addition to supporting wireless communication, reconnaissance, and remote sensing, a high-altitude aircraft might also be used as a weapons platform. Specific HALE requirements include altitude greater than 80,000 ft, low subsonic speed, indefinite endurance, and a payload of 2000 lb. These requirements obviously demand a very high-aspect-ratio, very low wing loading aircraft. A number of operational problems will exist with this concept. The aircraft will be vulnerable to gusts and will have to avoid storms and ice buildup during ascent and descent. Also, ground effect will make landing difficult.

Enabling technologies.   Because this aircraft will be required to cruise indefinitely at a very high altitude, it is obvious that high *L/D* (large aspect ratio) and low Reynolds number, laminar flow airfoils must be built into the aerodynamic design. Also, in order to minimize wetted area, and therefore power required, it is likely that it will be a flying wing design. The altitude and endurance requirements also suggest that electric-motor-driven propellers will be used for propulsion. Solar cells will no doubt be the power source because batteries would be too heavy, but either batteries, or fuel cells, or flywheels will be required for power storage at night. Highly efficient, compact electric motors have already been developed, but more work needs to be done on solar cells. It is estimated that a 20% increase in solar-cell conversion efficiency and an 80% reduction in cell weight with respect to current cell technology is required. Electrical power will be required not only

Fig. 10.19   HALE concept.

for propulsion but also for sensors, processors, and transmitters. The aircraft will probably be built from composite materials to provide necessary strength and stiffness. Even with high stiffness materials, the very high aspect ratio will probably require active control of structural response to atmospheric disturbances.

10.5.3 Large, Long-Range Aircraft

The Boeing 747 was designed in the 1960s, and, along with the C-5, it is still one of the world's largest aircraft. Its wing span is approximately 200 ft, it is more than 60 ft high, and its maximum takeoff weight is in excess of 750,000 lb. Imagine an aircraft twice the size of a 747—that is what we mean by large. Such an aircraft would obviously provide greatly enhanced payload capability. If new performance technologies are added to this concept, it could also have an expanded operating envelope including improved unrefueled range and increased speed. Also, operational enhancements such as reduced maintenance, fuel burn, logistics and crew requirements would be important design considerations. Very large aircraft will also require new operational concepts for passenger gates, runway design, cargo loading, etc.

Missions.   Large aircraft serve as both transports and bombers in today's aircraft inventory. Advanced bombers are likely to be transonic and/or supersonic and will obviously need to be stealthy. We restrict our discussion here to transonic transports. Defining performance parameters for a large transport are a range of 12,000 n miles, a payload of up to 500,000 lb, and short field takeoff capability.

Enabling technologies.   Transports have a mission that is mostly high-altitude cruise. Therefore, the aerodynamic goal is to provide high L/D. Through viscous drag-reduction techniques such as laminar flow airfoils and riblets and the use of very high aspect ratio, strut braced wings, L/D might be extended from 20 to nearly 40. Advanced high-lift systems employing microvortex generators and circulation control could be used to provide short field length. The Air Force's Integrated High Performance Turbine Engine Technology program has already demonstrated high thrust-to-weight, low thrust-specific fuel-consumption engines that will add to the range capability of large aircraft. Range and payload capability can also be

Fig. 10.20   Very large cargo aircraft concept.

achieved through reduced aircraft weight. Composites will be used throughout the structural design. Also, the high-aspect-ratio wings will require active load control and flutter suppression. Flight systems will include advanced external vision for all weather operation, fly-by-light/power-by-wire controls, global positioning system (GPS) navigation and guidance, and a highly automated flight deck to reduce crew requirements to a minimum.

10.5.4 Hypersonic Vehicles

Since 1903 when the Wright brothers flew at Kitty Hawk, the ability to fly faster and higher has been a driving force in the development of aircraft technology. Today, our space launch vehicles are able to achieve escape velocity (36,000 ft/s), and we have accomplished manned flights as high as the moon. However, hypersonic flight (greater than $M = 4$) *within the atmosphere* still presents aircraft designers with extraordinary technical challenges. Therefore, the design of hypersonic vehicles remains one of the areas rich in potential for future generations of aerospace engineers.

Missions.   One can imagine a number of missions for hypersonic aircraft. For example, consider an airbreathing global-reach aircraft capable of taking off from a runway, flying to any point on earth at a Mach number in excess of 15, and then returning to home base—in less than 2 h! Such an aircraft would have an enormous impact on our thinking about both military and civilian transportation. If this design were transatmoshperic, it might also serve as a reusable launch system for placing space payloads in orbit. Other examples of hypersonic vehicles, especially important for the military, include high-speed missiles and maneuverable reentry vehicles. The hypersonic speed capability of these vehicles would serve to reduce their vulnerability and, at the same time, increase their lethality.

Enabling technologies.   Because of the large number of possible hypersonic missions and associated vehicle concepts, we restrict our discussion here to an

Fig. 10.21 Hypersonic vehicle concept.

airbreathing global reach aircraft with maximum Mach number of approximately 15. The global reach mission requires an aircraft with high cruise efficiency and thus low drag and high L/D at the design Mach number. At the high cruise Mach number in this case, aerodynamic heating will be very severe (orders of magnitude worse than the space shuttle),[16] and so the aircraft will require a sophisticated cooling system. Boundary-layer behavior in the complex, chemically reacting flowfield around the aircraft must be modeled and understood by the designer. Also, the design must account for control effectiveness in this very high dynamic pressure environment. The propulsion system will no doubt require a combination of different engine types (i.e., conventional turbojet/fan, ramjet, scramjet, rocket) to operate effectively over such a wide range of Mach numbers.[17] Aeropropulsion integration will be especially important because a large portion of the airframe will also serve as the engine when operating in the scramjet mode. The fuel must have very high energy content to provide the necessary thrust and high cooling capacity to keep the aircraft structure within temperature limits. Liquid hydrogen is a likely fuel candidate but presents problems with its low density and cryogenic nature. The low density results in high fuel tank volume and thus larger size and weight for the overall vehicle and cryogenic fuels require special tankage and handling. Materials for such an aircraft are also a particularly difficult design challenge. The structure must be lightweight and strong at very high temperatures. Ceramic matrix composites and carbon/carbon composites have shown some promise in materials research to date.

10.6 Design Thinking in Other Contexts

We conclude with the idea that the way of thinking we have developed in this book, design thinking, can be transferred to other problems you encounter in your life. Recall Koehn's definition of the engineering method: "The engineering method (design) is a strategy for causing the best change in a poorly understood or uncertain situation within available resources."

This broad definition of the design process tells us that action (change) is required and suggests that change can be caused by human intervention. It also recognizes the uncertainty and less than perfect knowledge characteristic of most real design situations and reminds us that solutions must be found within existing constraints on time, money, and human energy. Keep this definition in mind, but, for a moment, extend your thinking beyond engineering work and the design of physical systems.

David Perkins,[18] a cognitive psychologist, offers another interesting perspective on the idea of design. He argues that while we often think of knowledge as simply a collection of information, in fact, knowledge is constructed as a result of human design activity. He says, "Knowledge as information purveys a passive view of knowledge, one that highlights knowledge in storage rather than knowledge as an implement of action. Knowledge as design might be our best bet for a first principle in building a theory of knowledge for teaching and learning." He also suggests that it is helpful to use four key design questions when thinking about any human design activity, for example, the organization of the U.S. Senate, the organization of a paragraph, the form of an Italian sonnet, or a deceptive practice in advertising. The four questions are: What is the design's purpose? What is its structure (its general characteristics)? What are model cases of it? What are the arguments that explain and evaluate it? The reader might take a moment to review one of the four future design concepts in light of these four questions. Also, take a moment to think how these same questions would be useful in analyzing an English composition.

As a second example of design thinking in other than an engineering context, we shift our focus to a contemporary description of organizations. Peter Senge[19] has written extensively about *learning organizations*, "organizations where people continually expand their capacity to create the results they truly desire, where collective aspiration is set free, and where people are continually learning how to learn together." In speaking about the new role of leaders and managers in such organizations, he refers to the idea of *manager as researcher and designer*. "What does she or he research? Understanding the organization as a system and understanding the internal and external forces driving change. What does she or he design? The learning processes whereby managers throughout the organization come to understand these trends and forces."

So, our parting message is this. The experience you have gained by bringing a design perspective to an introduction to aeronautical engineering is transferable! As you encounter other ill-defined problems in your professional life, think about the purpose to be achieved, the structure of the issue, existing examples, and relevant arguments. Keep in mind that human knowledge is the construction that results from design thinking. In reflecting on the problem at hand, consider constraints that might affect your decisions as well as the sensitivity of the outcome to variables in your control. And finally, if you are in a leadership position, remember that you are not only responsible for the product of your organization, but also the design of the learning processes that are integral to it.

References

[1]Kelly, F. C., *Miracle at Kitty Hawk*, Farrar, Straus, and Young, New York, 1951.

[2]Kelly, F. C., *The Wright Brothers*, Harcourt, Brace and Company, New York, 1943.

[3]Howard, F., *Wilbur and Orville*, Knopf, New York, 1987.

[4]Crouch, T. D., *The Bishop's Boys,* W. W. Norton, New York, 1989.

[5]Miller, R., and Sawers, D., *The Technical Development of Modern Aviation*, Routledge & Kegan Paul, London, 1968.

[6]Brooks, P. W., *The Modern Airliner,* Putnam, London, 1961.

[7]Yenne, B., *McDonnell & Douglas: The Tale of Two Giants*, Arms and Armour Press, London, 1985.

[8]Loftin, L. K. Jr., *Subsonic Aircraft: Evolution and Matching of Size to Performance*, NASA R P 1060, 1980.

[9]Bradley, R. G., "Practical Aerodynamic Problems—Military Aircraft," *Transonic Aerodynamics*, edited by David Nixon, *Progress in Aeronautics and Astronautics*, Vol. 81, AIAA, Washington, DC, 1982.

[10]Buckner, J. K., Hill, P. W., and Benepe, D., "Aerodynamic Design Evolution of the YF-16," AIAA Paper 74-935, Aug. 1974.

[11]Buckner, J. K., and Webb, J. B., "Selected Results from the YF-16 Wind-Tunnel Test Program," AIAA Paper 74-619, July 1974.

[12]Hillaker, H. J., "A Supersonic Cruise Fighter Evolution," Paper 14, 1982.

[13]McCall, Gene H., *New World Vistas*, Summary Volume, USAF Scientific Advisory Board, Pentagon, Washington, DC, 1995.

[14]Bradley, R. G., *New World Vistas*, Aircraft and Propulsion Volume, USAF Scientific Advisory, Board, Pentagon, Washington, DC, 1995.

[15]Brown, S. F., "The Eternal Airplane," *Popular Science*, April 1994, pp. 70–76.

[16]Bertin, J. J., *Hypersonic Aerothermodynamics*, AIAA Education Series, AIAA, Washington, DC, 1993.

[17]Heiser, W. H., and Pratt, D., *Hypersonic Airbreathing Propulsion*, AIAA Education Series, AIAA, Washington DC, 1993.

[18]Perkins, David, *Knowledge as Design*, Lawrence Erlbaum, New Jersey, 1986.

[19]Senge, Peter, *The Fifth Discipline*, Doubleday/Currency, New York, 1990.

Appendix A
Glossary

aerodynamic center a point on an airfoil about which pitching moment is independent of angle of attack.

aerodynamic twist a spanwise change in airfoil camber on a wing.

aerodynamics the study of phenomena associated with flowing gases.

aeronautics the science of flight.

afterburner a device attached to the exhaust of a turbojet or turbofan engine in which additional fuel is added to the exhaust gases and burned to increase engine thrust.

aileron control surfaces mounted on wings used to control roll.

air force military organization that primarily uses aircraft for its mission.

aircraft vehicle that is capable of moving through the air.

aircraft reference line an arbitrarily chosen line on an aircraft from which angle of attack is measured.

airfoil a streamwise cross section of a wing. Also, a wing that has the same cross-sectional shape across its span and that spans the test section of a wind tunnel, so that the flow around it is two dimensional.

airplane aircraft that is heavier than air and uses wings to generate lift in order to remain aloft.

airship lighter-than-air powered aircraft. A dirigible.

airspeed an aircraft's speed relative to the air mass.

  calibrated indicated airspeed corrected for position error.

  equivalent calibrated airspeed corrected for nonstandard pressure. At sea level on a standard day, true airspeed is equal to equivalent airspeed.

  indicated the airspeed displayed on the cockpit pitot-static airspeed indicator.

  indicator a differential pressure gauge connected to a pitot tube and a static port and calibrated so that pressure differences can be read as airspeed.

  true equivalent airspeed corrected for nonstandard density.

altitude height above a reference surface, geometric altitude.

  density that altitude in the standard atmosphere that has a particular air density of interest.

  pressure that altitude in the standard atmosphere that has a particular air pressure of interest.

  temperature that altitude in the standard atmosphere that has a particular air temperature of interest.

amplitude the maximum magnitude of displacement achieved by an oscillating system.

analysis process that uses calculations, computer simulations, laboratory tests, etc., to determine the effectiveness of a design concept.

angle of attack angle between a body's velocity vector and its reference line or axis.

absolute   angle between an airfoil's or wing's zero lift line and the relative wind.

area   a quantitative measure of a surface or region, the area of a rectangle is its length multiplied by its width.

  planform   the area of the shadow of a level object over a flat surface in sunlight at noon, the reference area of a lifting surface.

  rule   a mathematical model for the ideal variation of a body's cross-sectional area with axial distance for minimum wave drag.

  wetted   the amount of surface area of an object that would get wet if it were immersed in water.

aspect ratio   the ratio of a wing's span to its mean chord, $R = b^2/S$.

axis   one of the orthogonal rays of which a coordinate system is composed.

  lateral   an aircraft coordinate system axis that runs spanwise.

  longitudinal   an aircraft coordinate system axis that runs lengthwise or streamwise.

  vertical   an aircraft coordinate system axis that is orthogonal to the lateral and longitudinal axes, positive downward.

bank angle   the tilt of an aircraft's wings and lift vector in a level turn.

barometer   a tube with one end sealed and the other exposed to the atmosphere and filled with a liquid, used to measure atmospheric pressure.

Bernoulli, Daniel   Swiss mathematician and physicist, studied pressure in flowing fluids.

Bernoulli's equation   a relationship between velocity and static pressure in a flowing fluid.

Boeing   aircraft manufacturer. Builders of the 7-series airliners.

boundary layer   the region next to a body where flow velocities are less than the freestream velocity.

  laminar   a boundary layer in which the flow is in orderly layers, so that streamlines remain parallel to the body surface.

  separation   the appearance of zero velocity in the boundary layer, causing the flow to move out away from the body surface.

  thickness   distance perpendicular to the surface of a body to a point in its boundary layer where flow velocity equals 99% of freestream.

  turbulent   a boundary-layer flow characterized by large swirls or eddies, making it very disorderly and unsteady.

brainstorming   the creative process of generating ideas. Ideation.

bulkhead   structural frame generally perpendicular to the longitudinal axis of a fuselage.

bypass ratio   the ratio of mass flow rate of air that flows around the core of a turbofan engine to mass flow rate of air flowing through the core.

camber   curvature of an airfoil's mean camber line such that the airfoil's top and bottom surfaces have different shapes. Also, the maximum distance from an airfoil's mean camber line to its chord line.

canard   a pitch control surface placed forward of an airplane's wing.

ceiling   maximum altitude at which an aircraft is expected to operate.

absolute altitude at which maximum rate of climb is zero.

combat altitude at which maximum rate of climb is 500 ft/min.

service altitude at which maximum rate of climb is 100 ft/min.

center of gravity the point on a body at which its total weight force effectively acts. The point at which a body will balance.

center of pressure the point on a body at which the total aerodynamic force effectively acts.

chord a straight line drawn from the leading edge of an airfoil to its trailing edge.

climb a maneuver which increases an aircraft's altitude.

angle the angle between an aircraft's velocity vector and the horizon.

rate the rate of change of an aircraft's altitude in a climb.

compressibility effects changes to pressure forces generated by a moving gas due to changes in the gas density at high speeds.

compressor the portion of a turbine engine that increases the static pressure of the inlet air prior to it reaching the engine's burners.

conceptual design the first phase or stage of the design process. In this phase, the design problem is defined and studied, and a range of design concepts are generated. The goal of conceptual design is to select a workable design concept and optimize it as much as is feasible.

constraint analysis a method for analyzing a conceptual aircraft design to determine what characteristics it must have to meet all design requirements.

continuity equation a mathematical statement of the law of conservation of mass for a flowing fluid.

control authority the amount of change that a control surface is able to make on an aircraft's trim condition.

control surface a movable surface that, when deflected, generates moments to change an aircraft's trim condition.

core the compressor, burner, and turbine of a turbine engine.

damping ratio ratio by which the amplitude of an oscillation is reduced per cycle.

density the amount of mass per unit volume.

design the process of planning the physical characteristics and construction methods of a product.

cycle synthesis, analysis, decision making.

Mach number the Mach number at which an aircraft configuration or component, especially an engine inlet, is designed to operate most efficiently.

method a strategy for causing the best change in a poorly understood or uncertain situation within the available resources.

phases stages in design process: conceptual, preliminary, detail.

process define the problem, collect data, create a design concept, select and perform analysis, make decisions.

requirement a statement of a particular capability that the customer needs the aircraft to have.

spiral a concept for understanding design that emphasizes the fact that with each design cycle the designer gains more knowledge but a narrower range of feasible choices.

detail design   the third phase of the design process, which prepares the design for production. Detailed parts and manufacturing processes are designed in this phase.

downwash   a downward flow velocity component caused by wing tip vortices.

drag   component of aerodynamic force parallel to the freestream velocity.

  coefficient   a nondimensional measure of a body's drag, obtained by dividing the drag by dynamic pressure and a reference area.

  curves due to lift   all drag on an aircraft that varies with lift coefficient; the sum of induced drag and that part of profile drag that varies with lift.

  induced   drag resulting from tilting of the lift vector by downwash created by wing tip vortices on finite wings.

  parasite   all drag on an aircraft not due to lift.

  polar   the variation of drag coefficient with lift coefficient.

  pressure   drag arising from flow separation and loss of total pressure, so that the aft end of a body has lower static pressures on it than the front.

  profile   the sum of skin-friction drag and pressure drag.

  skin friction   drag arising from transfer of momentum from gas molecules moving parallel to and impacting the microscopically rough surface of a body.

  wave   drag arising from shock waves, which cause loss of total pressure in the flow around a body, so that the aft end has lower static pressures on it than the front.

drawing   a means of communicating information visually.

elevator   a control surface mounted on the horizontal stabilizer used to control pitch.

endurance   the amount of time an aircraft is able to fly.

engine   a device for converting chemical energy into thrust or power.

engineer   a person who follows engineering as a profession.

engineering   science by which matter and energy are made useful to man.

engineering method   *see* design method.

equation of state   the relationship between the density, temperature, and pressure of a gas, or a mixture of gases such as air.

Euler, Leonhard   Swiss scientist and mathematician.

Euler's equation   a mathematical statement (in differential form) of Newton's second law as it applies to a flowing fluid.

fighter   a military airplane used primarily to attack other military aircraft, secondarily to attack ground and naval forces.

flap   a movable portion of a wing which, when deflected, increases its camber and its lift.

flight regime   a range of Mach numbers for which aerodynamic phenomena are essentially of the same type for a given body.

flight-path angle   the angle between an aircraft's velocity vector and true horizontal.

flowfield   a region or field of flowing fluid.

Fowler flap   a trailing-edge flap that extends aft as well as down when deployed, increasing the lifting surface area of the wing.

freestream   term used to identify conditions at a point in the flowfield where the effects of the body are negligible.

friction   transfer of momentum and production of heat between two objects moving relative to each other.

   coefficient   a nondimensional parameter that describes the ratio of frictional force experienced by a body moving relative to another to the magnitude of the normal force pressing the two bodies together.

   rolling   the force that resists the motion of rolling wheels, caused by the making and breaking of microscopic bonds or welds between the wheel and the surface, as well as between the wheel and its axle.

   skin   transfer of momentum between a body and a viscous fluid flowing over it, *see* skin-friction drag.

fuel   chemical that, when oxidized, releases heat energy that can be converted into thrust or power by an engine.

   consumption   the rate at which fuel is burned by an engine.

   fraction   the ratio of an aircraft's fuel weight to its total weight.

fuselage   the central payload-carrying body of an aircraft.

geometric twist   construction of a wing in such a way that the airfoils along the span are not all at the same orientation or angle of attack.

glide   descending flight without power, using a component of weight to overcome drag instead of thrust.

   angle   the flight-path angle of an aircraft in a glide.

   ratio   the ratio of horizontal distance traveled to altitude lost in a glide.

gradient layer   a portion of the atmosphere where temperature varies with altitude.

ground speed   speed relative to the ground. It is true airspeed corrected for wind.

high-lift devices   devices attached primarily to the wings of aircraft that increase lift.

Hillaker, Harry   designer of the F-16 and F-16XL.

house of quality   a chart used to prioritize design features based on customer needs.

hydrostatic equation   a mathematical description of the vertical force balance on a static particle of fluid.

hypersonic   associated with flight Mach numbers above 5.

ideation   the creative process of generating ideas. Brainstorming.

ill-defined problems   problems for which only partial information is available and/or more than one solution may be acceptable.

incidence   the angle of a lifting surface chord line relative to the aircraft reference line.

incompressible flow   a flow in which the density remains constant.

inlet   the portion of a wind tunnel, jet engine, etc., where air flows into the system or device.

inviscid flow   a flow that is frictionless.

isothermal region   an altitude band in the atmosphere in which temperature does not vary with altitude.

Johnson, Clarence L. "Kelly"   founder and head of the Lockheed Advanced Aerospace company, the Skunk Works, and designer of (among others) the P-38, P-80, U-2, YF-12, and SR-71.

landing gear   the wheeled structures an aircraft rests on when on the ground.
leading edge   the farthest forward point of an airfoil or wing.
    extension   *see* strake.
    flap   a movable portion of a wing leading edge that increases camber.
lift   the component of the aerodynamic force that is perpendicular to the free-stream velocity vector.
    coefficient   a nondimensional measure of an airfoil's or wing's lift, obtained by dividing the lift by dynamic pressure and airfoil or wing area.
    curve   a plot of lift coefficient vs angle of attack.
    curve slope   the ratio of change in lift coefficient to change in angle of attack.
load factor   ratio of aircraft lift to weight. The "gs" the aircraft is generating.
Lockheed   aircraft manufacturing company, builders of P-38, F-104, C-5, P-3, C-141, U-2, F-117, SR-71, etc. Now part of Lockheed Martin.
longeron   structural skin stiffener running lengthwise on a fuselage.

Mach, Ernst   an Austrian scientist who first identified the Mach number.
Mach angle   the angle at which a Mach wave trails back from a moving body.
Mach number   the ratio of the flow velocity to the speed of sound.
    critical   the freestream Mach number at which the flow at some point in the flowfield around a body first reaches $M = 1$.
    drag divergence   the freestream Mach number at which drag on a body begins to rise rapidly.
Mach wave   a pressure disturbance generated by an infinitesimally small body moving through the air at or above the speed of sound.
maneuverability   the ability of an aircraft to change its velocity vector.
    diagram   a plot of instantaneous and sustainable turn radius and rate.
manometer   U-shaped tube containing fluid used to measure pressure differences.
manometry equation   A mathematical expression of the relationship between pressure differences and fluid column height differences in a manometer.
mass   a quantity of matter.
    flow rate   the rate at which matter is flowing through a system.
mean aerodynamic chord   an aerodynamically weighted average chord length of a wing.
mean camber line   a line drawn equidistant between an airfoil's upper and lower surfaces.
minimum drag condition   operating conditions for which drag is a minimum, $(L/D)_{\max}$.
mission analysis   mathematical simulation that predicts the fuel used to fly a mission.
moment   torque or twisting force; force multiplied by lever arm.

NACA   National Advisory Committee for Aeronautics, a U.S. government agency created by Congress on 3 March 1915, to promote development of the aeronautical sciences in the United States.

NASA National Aeronautics and Space Administration, successor to NACA, created by Congress on July 29, 1958, to promote development of aeronautical and space sciences in the United States. NASA took over the personnel and programs of the NACA when it was created and became the managing agency for the U.S. space exploration program.

neutral point the location of an aircraft's center of gravity that causes it to have neutral static stability.

nozzle the rear portion of a jet engine that accelerates the exhaust gases.

one-dimensional flow a flow in which the properties in a plane perpendicular to the flow velocity vector are constant.

Oswald's efficiency factor a correction factor that allows the equations for the lift and drag of elliptical wings to be used for the lift and drag of airplanes.

perfect gas law *see* equation of state.

pitot tube a tube placed end on into a flowing fluid and connected to a pressure-measuring device to measure total pressure of a flow.

Pitot, Henri an 18th-century French scientist, inventor of the pitot tube.

pitot-static tube a device with pressure ports for measuring both total pressure and static pressure of a flow.

planform area the shadow area of a wing, including the portion inside the fuselage.

power thrust multiplied by velocity.

    available thrust available multiplied by velocity.

    excess power required subtracted from power available.

    required drag multiplied by velocity.

    specific excess excess power divided by weight.

position error obtained from flight test, a correction between indicated and calibrated airspeed, used to account for error in static port placement and instrument errors.

Prandtl–Glauert correction a correction made to lift coefficients to account for compressibility effects.

preliminary design the second stage or phase in the design process, during which specific characteristics, dimensions, materials, structures, functions, etc., of a conceptual design are worked out. Computer and physical models of the design are also built and tested in this phase.

pressure a force exerted by a liquid or gas per unit area.

    altitude that altitude in the standard atmosphere that has a particular pressure. Measured by setting an altimeter to standard sea level reference pressure.

    dynamic pressure due to the transfer of momentum to a surface arising from the directed motion of fluid molecules in a flow.

    static the pressure due to the transfer of momentum to a surface from randomly moving gas molecules.

    total the pressure that exists at a stagnation point, or would exist at any point in the flow, if the flow were slowed isentropically to zero velocity. A point property that is the sum of static and dynamic pressures.

    vessel a closed container or reservoir that allows a gas inside it to have a pressure different from that on the outside of the container.

procurement process   a formal procedure for initiating and managing the design, construction, and purchasing of new military aircraft.

range   the distance an aircraft is able to fly.
rate of climb   the rate of change of altitude with respect to time.
reticulated foam   a spongelike material used to fill fuel tanks to suppress fire.
return on investment   a cost analysis that predicts the total profit that can be expected over the life of an aircraft.
rib   airfoil-shaped structural frame running chordwise in a wing or control surface.
rudder   a control surface mounted on the vertical tail used to control yaw.
ruddervator   a control surface mounted on a V-tail used to control both pitch and yaw.

sea level   altitude representing the average elevation of the ocean's surface.
Sears–Haack bodies   Bodies of revolution whose cross-sectional areas follow the area rule, so that they have the minimum wave drag for their volumes.
selection matrix   a chart for displaying the advantages and disadvantages of several design concepts, assigning numerical values and weights to these characteristics, and choosing the best concept based on these calculations.
separation   a condition of airflow around a body in which a stagnation point develops in the boundary layer, causing the flow to move out away from the body, causing pressure drag.
shock wave   a concentration of pressure disturbances into a single region of abrupt changes in flow velocity, pressure, and temperature caused by bodies moving faster than the speed of sound.
   bow   a shock wave in front of a blunt supersonic body. It is normal to the flow directly in front of the body and oblique off to the sides.
   normal   a shock wave that is perpendicular to the flow direction.
   oblique   A shock wave that is not perpendicular to the flow direction.
shock-induced separation   boundary-layer separation caused by the extreme adverse pressure gradient associated with a shock wave.
sink rate   the rate of change of altitude of a glider.
slat   a leading-edge flap that has a slot when deployed.
slot   a fixed opening slightly behind the leading edge of a wing, which admits high-energy air to delay boundary-layer separation.
sound   pressure waves propagating through a fluid.
   barrier   a rapid rise in drag at speeds just below the speed of sound that was once believed to be an impenetrable barrier to faster speeds.
   speed of   the velocity at which pressure waves move through a fluid.
span   distance from one wing tip of a wing to the other.
span efficiency factor   a correction factor that allows the equations for the lift and drag of elliptical wings to be used for other types of wings.
spanwise   the direction from one wing tip of a wing to the other.
spar   structural beam running spanwise in a wing or control surface.
stability   the tendency to return to equilibrium.
   dynamic   a system's tendency over time, when disturbed, to return to and remain at the equilibrium condition.
   longitudinal   the tendency of an aircraft, when disturbed, to return to trim.

static   a system's initial tendency, when disturbed from equilibrium, to move back toward the equilibrium condition.

stabilizer   a fixed aerodynamic surface other than the wing that aids in keeping an airplane pointed parallel to its velocity vector.

  horizontal   a stabilizing surface that forms the horizontal portion of an airplane's empenage and that adds to pitch stability.

  vertical   a stabilizing surface that forms the vertical portion of an airplane's empenage and that adds to directional stability.

stagnation point   a point in the flow where the velocity equals zero. The pressure at this point is always equal to the total pressure in the flow.

stagnation streamline   a streamline that leads to a stagnation point.

stall   a condition experienced by lifting surfaces and bodies at high angles of attack where further increases in angle of attack result in a decrease in lift coefficient instead of an increase.

  angle of attack   the angle of attack at which a lifting surface or body produces its maximum lift coefficient.

standard atmosphere   a mathematical model for the variation of temperature, pressure, density, etc., with altitude in the Earth's atmosphere.

static margin   nondimensional distance between an aircraft's neutral point and its center of gravity.

static port   an orifice oriented parallel to a flowing fluid so that it senses only the static pressure of the fluid.

steady flow   a flow in which the properties at any point are constant with respect to time.

strake   a highly swept surface ahead of the wing root that creates a leading-edge vortex to increase wing lift.

stratosphere   the layer of the Earth's atmosphere that is above the tropopause, where temperature is approximately the same at all altitudes.

stream tube   a tube composed of streamlines.

streamline   an imaginary line such that, at every point along the line, the flow velocity vector is tangent to the line. For a steady flow, this line coincides with the path of the fluid particles.

streamwise   in the direction of the freestream velocity vector.

stringer   structural skin stiffener running lengthwise on a fuselage and spanwise in wings and control surfaces.

structure   the material comprising the aircraft itself and giving it shape and strength as opposed to systems, engines, payload, etc.

subsonic   Mach numbers below M_{crit}.

supersonic   Mach numbers for which all flow around a body stays at $M > 1$.

sweep   the amount to which a wing's leading edge or other reference line is angled back from being perpendicular to the freestream velocity.

synthesis   the act of creating ideas and design concepts.

tail incidence angle   the angle between the horizontal tail chord line and the aircraft reference line.

taper ratio   the ratio of a wing's tip chord to its root chord.

temperature   a measure of the average kinetic energy of gas molecules as they move and collide with each other.

test section   the portion of a wind tunnel with the smallest cross-sectional area and highest velocities where measurements of aerodynamic forces and pressures on models are made.

thickness-to-chord ratio   the ratio of an airfoil's maximum thickness to its chord.

thrust   propulsive force produced by an aircraft engine.

    angle   the angle between the thrust vector and the velocity vector.

    dry   turbojet and turbofan engine thrust without afterburner.

    required   the amount of thrust required to overcome drag and sustain steady, level, unaccelerated flight.

    sea level   thrust in standard day sea level conditions.

    static   thrust when the aircraft has zero forward velocity.

    to-weight ratio   the ratio of maximum thrust to either aircraft or engine weight.

    wet   turbojet and turbofan engine thrust with afterburner.

trailing edge   the farthest aft point of an airfoil or wing.

trailing vortex   a tornadolike swirling flow trailing from a lifting wing's tips caused by the pressure imbalance between the wing's upper and lower surfaces.

transition   the changing of a boundary layer from laminar to tubulent flow. Also, the maneuver made at the end of a landing approach to level the aircraft off for a gentle touchdown. Also called roundout or flare.

trim   a condition in which the sum of the moments on an aircraft is zero.

    angle of attack   the angle of attack for which the sum of an airplane's pitching moments is zero.

    longitudinal   pitch trim.

    pitch   a condition in which the sum of the pitching moments on an aircraft is zero.

tropopause   boundary between troposphere and stratosphere, occurs at 36.152 ft.

troposphere   layer in the atmosphere that is closest to the Earth's surface.

turbofan   a turbine engine that has a fan attached to increase efficiency.

turbojet   a jet engine with a compressor, burner, and turbine.

turbine   a windmill-like device that extracts work from exhaust gases.

turn   to change the direction of a vehicle's velocity vector.

    radius   the distance from a turning vehicle to the centre of the circle described by its trajectory.

    rate   the rapidity with which a vehicle changes the direction of its velocity vector.

twist   *see* aerodynamic twist, geometric twist.

two dimensional   having variation of characteristics in only two rather than three orthogonal directions.

upwash   an upward component of flow velocity caused by wing tip vortices.

velocity   the net rate of motion of a fluid. Also, the rate of motion of an object relative to a frame of reference.

viscosity   resistance of a fluid to velocity discontinuities within it.

wake   the disturbed airflow behind a body moving through a fluid.

washin   geometric wing twist such that the tip is at a higher angle of attack than the wing root.

washout   geometric wing twist such that the tip is at a lower angle of attack than the wing root.

weight   mass multiplied by the acceleration of gravity.

   analysis   analysis that predicts the weight of aircraft components.

   fraction   the ratio of the weight of aircraft components, payload, fuel, etc., to the aircraft's maximum takeoff weight.

   fuel   the weight of fuel required to fly a mission.

   payload   the weight of payload required by the design mission.

   seats   the weight of seats predicted by weight analysis.

   takeoff   maximum takeoff gross weight.

wind tunnel   a device that creates a flow of air that can be used to measure lift, drag, etc., on models placed in its test section.

wing   a horizontal surface used to generate lift to support an airplane.

   loading   the weight of an aircraft divided by the reference planform area.

   root   the portion of the wing closest to the aircraft centerline.

   tip   the portion of the wing farthest from the aircraft centerline.

winglet   a nearly vertical surface attached to the tip of a wing to increase its effective aspect ratio and reduce induced drag.

wing tip vortex   a tornadolike swirling flow trailing from a lifting wing's tips caused by the pressure imbalance between the wing's upper and lower surfaces.

Wright, Orville and Wilbur   Designers, builders, and pilots of the first practical powered human-carrying airplane.

yaw   rotation about an aircraft's vertical axis.

Appendix B
Supplemental Data

Table B.1 Characteristics of the standard atmosphere in English
engineering units

Altitude (h), ft	Temperature (T), °R	Pressure (P), lb/ft²	Density (ρ), slug/ft³	Speed of sound (a), ft/s	Viscosity (μ) slug/ft s
0	518.69	2116.2	0.002377	1116.4	3.737E−07
1,000	515.1	2041	0.002308	1112.6	3.717E−07
2,000	511.6	1963	0.002241	1108.7	3.697E−07
3,000	508.0	1897	0.002175	1104.9	3.677E−07
4,000	504.4	1828	0.002111	1101.0	3.657E−07
5,000	500.9	1761	0.002048	1097.1	3.637E−07
6,000	497.3	1696	0.001987	1093.2	3.616E−07
7,000	493.7	1633	0.001927	1089.3	3.596E−07
8,000	490.2	1572	0.001869	1085.3	3.576E−07
9,000	486.6	1513	0.001811	1081.4	3.555E−07
10,000	483.1	1456	0.001756	1077.4	3.534E−07
11,000	479.5	1400	0.001701	1073.4	3.514E−07
12,000	475.9	1346	0.001648	1069.4	3.493E−07
13,000	472.4	1294	0.001596	1065.4	3.472E−08
14,000	468.8	1244	0.001546	1061.4	3.451E−07
15,000	465.2	1195	0.001496	1057.4	3.430E−07
16,000	461.7	1148	0.001448	1053.3	3.409E−07
17,000	458.1	1102	0.001401	1049.2	3.388E−07
18,000	454.6	1058	0.001355	1045.1	3.367E−07
19,000	451.0	1015	0.001311	1041.0	3.346E−07
20,000	447.4	973.3	0.001267	1036.9	3.325E−07
21,000	443.9	933.3	0.001225	1032.8	3.303E−07
22,000	440.3	894.6	0.001184	1028.6	3.282E−07
23,000	436.8	857.2	0.001144	1024.5	3.260E−07
24,000	433.2	821.2	0.001104	1020.3	3.238E−07
25,000	429.6	786.3	0.001066	1016.1	3.217E−07
26,000	426.1	752.7	0.001029	1011.9	3.195E−07
27,000	422.5	720.3	0.000993	1007.7	3.173E−07
28,000	419.0	689	0.000958	1003.4	3.151E−07
29,000	415.4	658.8	0.000924	999.1	3.129E−07

(Cont.)

Table B.1   Characteristics of the standard atmosphere in English
engineering units (continued)

Altitude (h), ft	Temperature (T), °R	Pressure (P), lb/ft^2	Density (ρ), slug/ft^3	Speed of sound (a), ft/s	Viscosity (μ) slug/ft s
30,000	411.9	629.7	0.000891	994.8	3.107E−07
31,000	408.3	601.6	0.000858	990.5	3.085E−07
32,000	404.8	574.6	0.000827	986.2	3.063E−07
33,000	401.2	548.5	0.000797	981.9	3.040E−07
34,000	397.6	523.5	0.000767	977.5	3.018E−07
35,000	394.1	499.3	0.000738	973.1	2.995E−07
36,000	390.5	476.1	0.000710	968.7	2.973E−07
36,152	390.0	472.7	0.000706	968.1	2.969E−07
37,000	390.0	453.9	0.000678	968.1	2.969E−07
38,000	390.0	432.6	0.000646	968.1	2.969E−07
39,000	390.0	412.4	0.000616	968.1	2.969E−07
40,000	390.0	393.1	0.000587	968.1	2.969E−07
41,000	390.0	374.8	0.00056	968.1	2.969E−07
42,000	390.0	357.2	0.000534	968.1	2.969E−07
43,000	390.0	340.5	0.000509	968.1	2.969E−07
44,000	390.0	324.6	0.000485	968.1	2.969E−07
45,000	390.0	309.5	0.000462	968.1	2.969E−07
46,000	390.0	295	0.000441	968.1	2.969E−07
47,000	390.0	281.2	0.00042	968.1	2.969E−07
48,000	390.0	268.1	0.000401	968.1	2.969E−07
49,000	390.0	255.6	0.000382	968.1	2.969E−07
50,000	390.0	243.6	0.000364	968.1	2.969E−07
51,000	390.0	232.2	0.000347	968.1	2.969E−07
52,000	390.0	221.4	0.000331	968.1	2.969E−07
53,000	390.0	211.1	0.000315	968.1	2.969E−07
54,000	390.0	201.2	0.000301	968.1	2.969E−07
55,000	390.0	191.8	0.000287	968.1	2.969E−07
56,000	390.0	182.8	0.000273	968.1	2.969E−07
57,000	390.0	174.3	0.00026	968.1	2.969E−07
58,000	390.0	166.2	0.000248	968.1	2.969E−07
59,000	390.0	158.4	0.000237	968.1	2.969E−07
60,000	390.0	151.0	0.000226	968.1	2.969E−07
61,000	390.0	144.0	0.000215	968.1	2.969E−07
62,000	390.0	137.3	0.000205	968.1	2.969E−07
63,000	390.0	130.9	0.000195	968.1	2.969E−07
64,000	390.0	124.8	0.000186	968.1	2.969E−07
65,000	390.0	118.9	0.000178	968.1	2.969E−07
66,000	390.0	113.4	0.000169	968.1	2.969E−07
67,000	390.0	108.1	0.000161	968.1	2.969E−07

(Cont.)

Table B.1   Characteristics of the standard atmosphere in English engineering units (continued)

Altitude (h), ft	Temperature (T), °R	Pressure (P), lb/ft²	Density (ρ), slug/ft³	Speed of sound (a), ft/s	Viscosity (μ) slug/ft s
68,000	390.0	103.1	0.000154	968.1	2.969E−07
69,000	390.0	98.25	0.001468	968.1	2.969E−07
70,000	390.0	93.67	0.00014	968.1	2.969E−07
71,000	390.0	89.31	0.000133	968.1	2.969E−07
72,000	390.0	85.14	0.000127	968.1	2.969E−07
73,000	390.0	81.17	0.000121	968.1	2.969E−07
74,000	390.0	77.4	0.000116	968.1	2.969E−07
75,000	390.0	73.78	0.00011	968.1	2.969E−07
80,000	390.0	58.13	8.683E−05	968.1	2.969E−07
82,346	390.0	51.93	7.764E−05	968.1	2.969E−07
85,000	394.3	45.83	6.77E−05	973.4	2.997E−07
90,000	402.5	36.29	5.25E−05	983.5	3.048E−07
95,000	410.6	28.88	4.1E−05	993.4	3.048E−07
100,000	418.8	23.09	3.21E−05	1003.2	3.048E−07

Table B.2   Characteristics of the standard atmosphere in SI units

Altitude (h), km	Temperature (T), K	Pressure (P), N/m²	Density (ρ), kg/m³	Speed of sound (a), m/s	Viscosity (μ), kg/m s
0.0	288.16	101325	1.225	340.3	1.79E−05
0.5	284.91	95461	1.1673	338.4	1.77E−05
1.0	281.66	89876	1.1117	336.4	1.76E−05
1.5	278.41	84560	1.0581	334.5	1.74E−05
2.0	275.16	79501	1.0066	332.5	1.73E−05
2.5	271.92	74692	0.95696	330.6	1.71E−05
3.0	268.67	70121	0.90926	328.6	1.69E−05
3.5	265.42	65780	0.86341	326.6	1.68E−05
4.0	262.18	61660	0.81935	324.6	1.66E−05
4.5	258.93	57752	0.77704	322.6	1.65E−05
5.0	255.69	54048	0.73643	320.5	1.63E−05
5.5	252.44	50539	0.69747	318.5	1.61E−05
6.0	249.2	47217	0.66011	316.5	1.6E−05
6.5	245.95	44075	0.62431	314.4	1.58E−05
7.0	242.71	41105	0.59002	312.3	1.56E−05
7.5	239.47	38299	0.55719	310.2	1.54E−05
8.0	236.23	35651	0.52578	308.1	1.53E−05
8.5	232.98	33154	0.49575	306.0	1.51E−05

(*Cont.*)

Table B.2   Characteristics of the standard atmosphere in SI units (continued)

Altitude (h), km	Temperature (T), K	Pressure (P), N/m²	Density (ρ), kg/m³	Speed of sound (a), m/s	Viscosity (μ), kg/m s
9.0	229.74	30800	0.46706	303.9	1.49E−05
9.5	226.5	28584	0.43966	301.7	1.48E−05
10.0	223.26	26500	0.41351	299.6	1.46E−05
10.5	220.02	24540	0.38857	297.4	1.44E−05
11.0	216.78	22700	0.3648	295.2	1.42E−05
11.5	216.66	20985	0.33743	295.1	1.42E−05
12.0	216.66	19399	0.31194	295.1	1.42E−05
12.5	216.66	17934	0.28837	295.1	1.42E−05
13.0	216.66	16579	0.26659	295.1	1.42E−05
13.5	216.66	15327	0.24646	295.1	1.42E−05
14.0	216.66	14170	0.22785	295.1	1.42E−05
14.5	216.66	13101	0.21065	295.1	1.42E−05
15.0	216.66	12112	0.19475	295.1	1.42E−05
16.0	216.66	10353	0.16647	295.1	1.42E−05
17.0	216.66	8849.6	0.1423	295.1	1.42E−05
18.0	216.66	7565.2	0.12165	295.1	1.42E−05
19.0	216.66	6467.4	0.10399	295.1	1.42E−05
20.0	216.66	5529.3	0.08891	295.1	1.42E−05
21.0	216.66	4728.9	0.07572	295.1	1.42E−05
22.0	216.66	4047.5	0.06451	295.1	1.42E−05
23.0	216.66	3466.9	0.05558	295.1	1.42E−05
24.0	216.66	2955.4	0.04752	295.1	1.42E−05
25.0	216.66	2527.3	0.04064	295.1	1.42E−05
30.0	231.24	1185.5	0.01786	295.1	1.49E−05

Table B.3   Compressibility correction f factors

Pressure altitude, ft	Calibrated Airspeed, kn								
	100	125	150	175	200	225	250	275	300
5000	0.999	0.999	0.999	0.998	0.998	0.997	0.997	0.996	0.995
10000	0.999	0.998	0.997	0.996	0.995	0.994	0.992	0.991	0.989
15000	0.998	0.997	0.995	0.994	0.992	0.990	0.987	0.985	0.982
20000	0.997	0.995	0.993	0.990	0.987	0.984	0.981	0.977	0.973
25000	0.995	0.993	0.990	0.986	0.982	0.978	0.973	0.968	0.963
30000	0.993	0.990	0.986	0.981	0.975	0.970	0.963	0.957	0.950
35000	0.991	0.986	0.981	0.974	0.967	0.959	0.951	0.943	0.934
40000	0.988	0.982	0.974	0.966	0.957	0.947	0.937	0.926	0.916
45000	0.984	0.976	0.966	0.956	0.944	0.932	0.920	0.907	0.895
50000	0.979	0.969	0.957	0.944	0.930	0.915	0.901	0.886	0.871

Conversion Factors

Distance

1 n mile = 6076 ft = 1.85 km
1 mile = 5280 ft = 1.61 km
1 ft = 0.3048 m

Force

1 lb = 4.4482 N

Mass

1 slug = 14.594 kg

Velocity

1 mph = 1.467 ft/s = 0.447 m/s
1 kn = 1.152 mph = 1.69 ft/s = 0.515 m/s

Pressure

1 atm = 29.92 in. Hg = 760 mm Hg = 2116.2 lb/ft^2
1 lb/in.^{2} = 144 lb/ft^2
1 lb/ft^2 = 47.88 N/m^2

Temperature

$^\circ$R = $^\circ$F + 460
K = $^\circ$C + 273
1.0 K = 1.80°R

Constants

Density of Water

ρ_{H_2O} = 1.938 slug/ft^3 = 1000 kg/m^3

Gravitational Acceleration

g = 32.2 ft/s^2
g = 9.80 m/s^2

Ratio of Specific Heats for Air

γ = 1.4

Specific Gas Constant for Air

R = 1716 ft · lb/slug · $^\circ$R
R = 287 J/kg · K

Airfoil Data

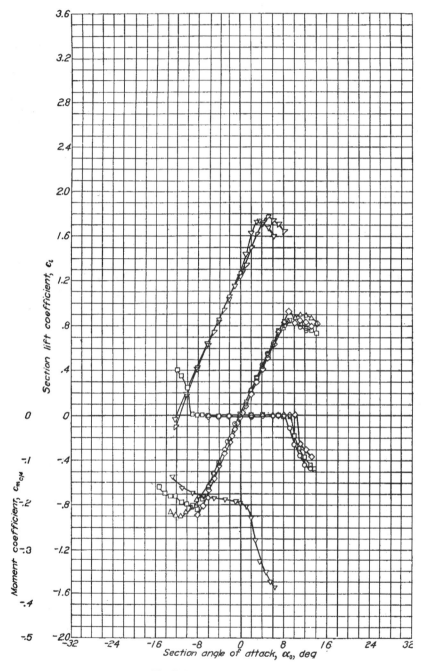

Fig. B.1a   NACA 0006.

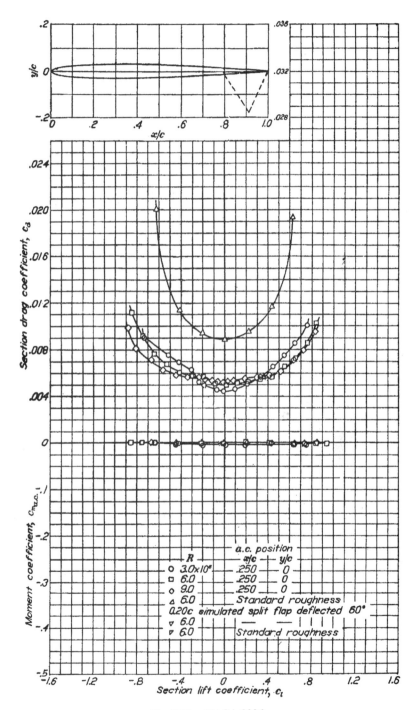

Fig. B.1b   NACA 0006.

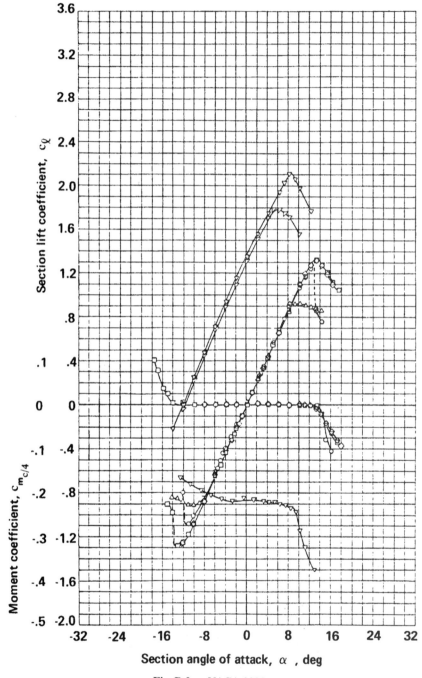

Fig. B.2a   NACA 0009.

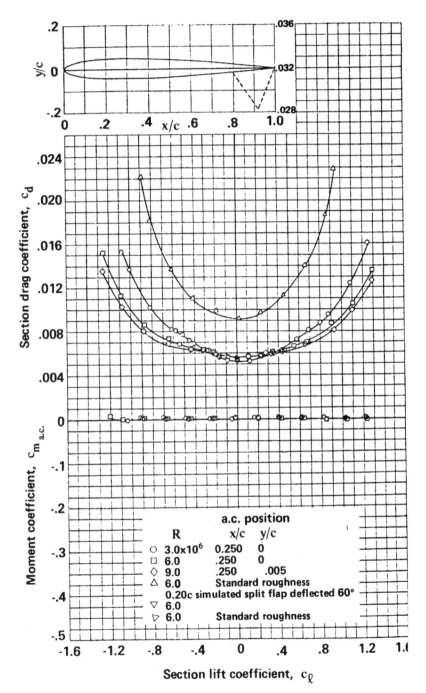

Fig. B.2b　NACA 0009.

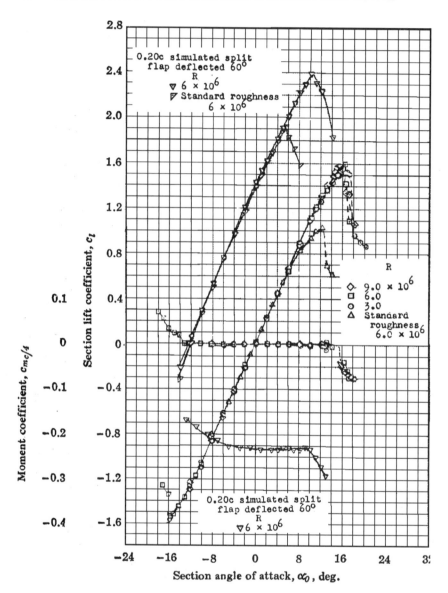

Fig. B.3a   NACA 0012.

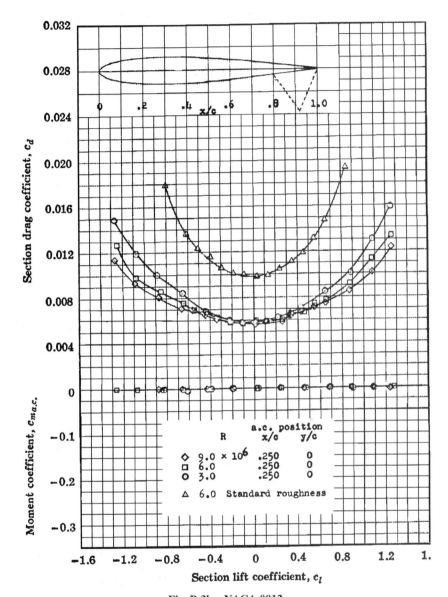

Fig. B.3b   NACA 0012.

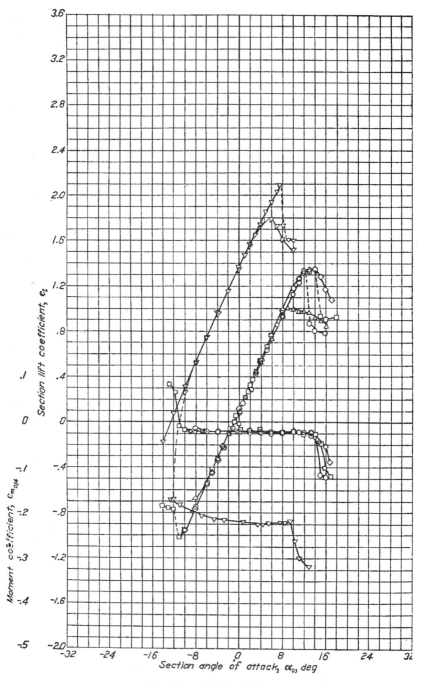

Fig. B.4a   NACA 1408.

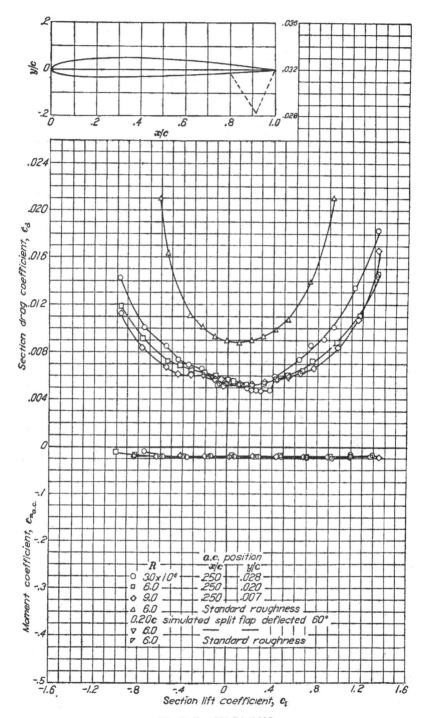

Fig. B.4b NACA 1408.

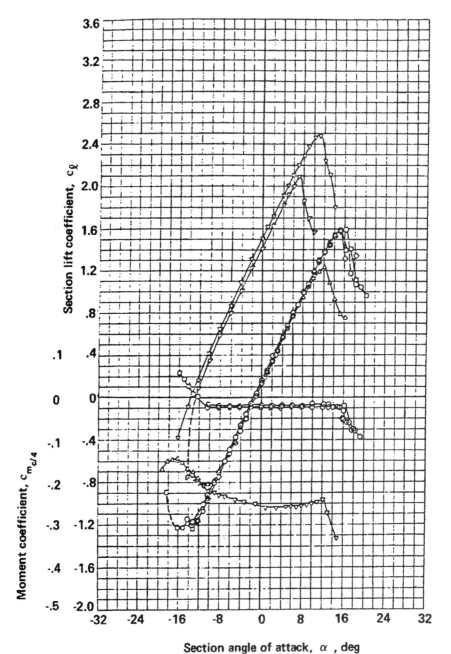

Fig. B.5a   NACA 1412.

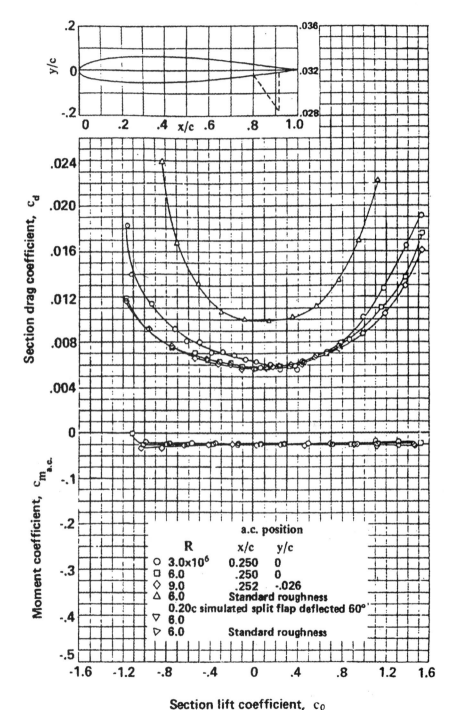

Fig. B.5b   NACA 1412.

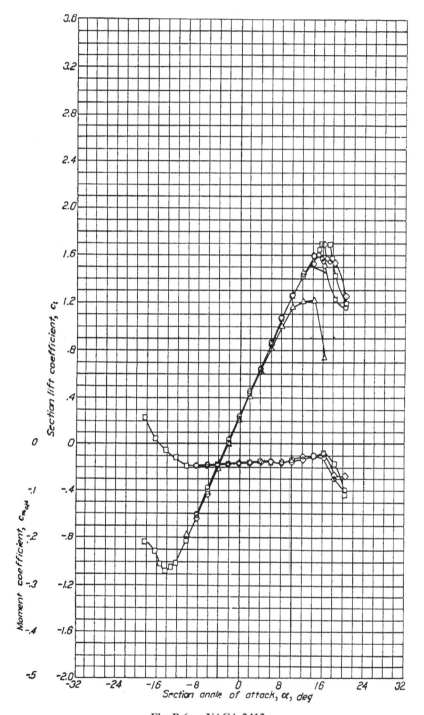

Fig. B.6a   NACA 2412.

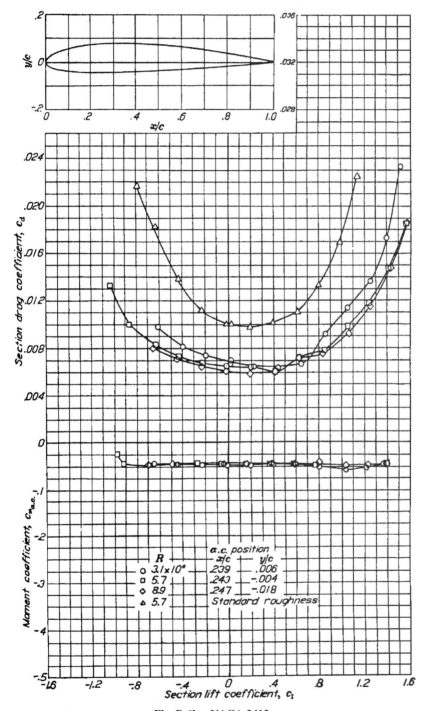

Fig. B.6b   NACA 2412.

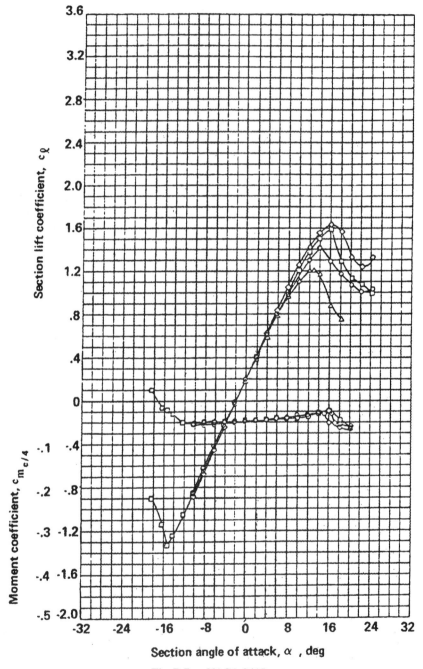

Fig. B.7a   NACA 2415.

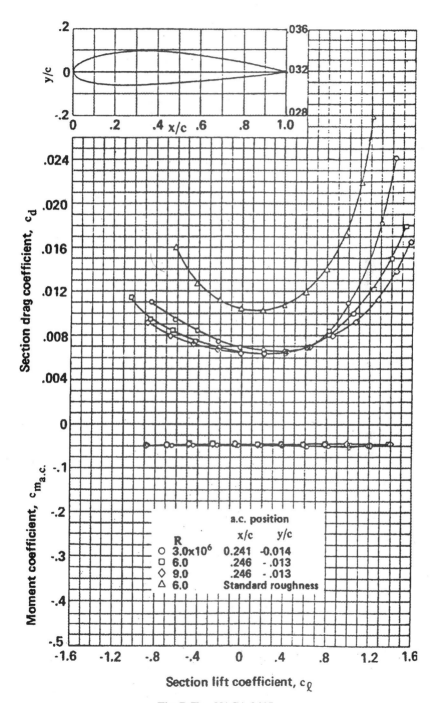

Fig. B.7b NACA 2415.

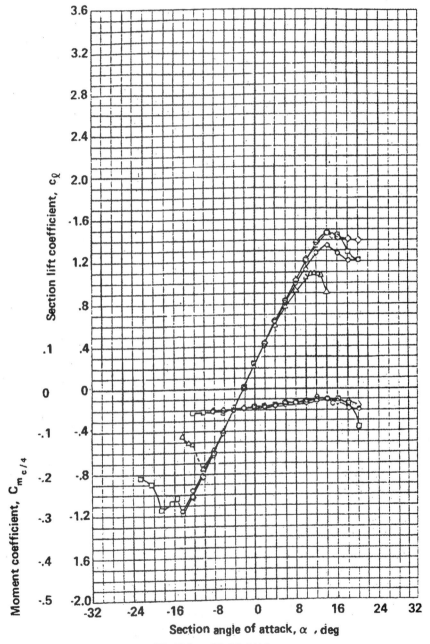

Fig. B.8a   NACA 2418.

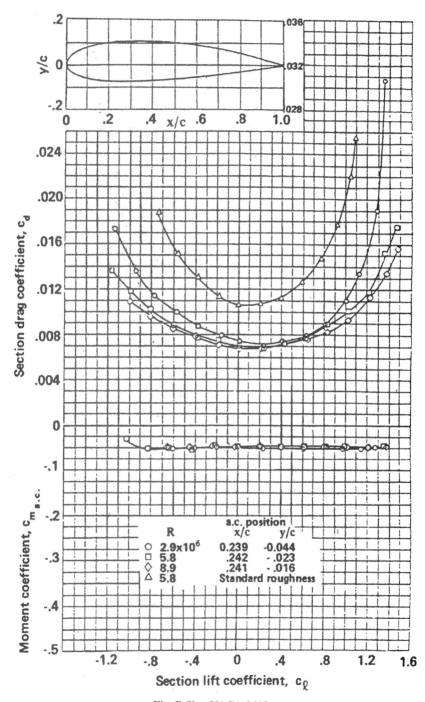

	R	a.c. position x/c	y/c
○	2.9x10⁶	0.239	-0.044
□	5.8	.242	- .023
◇	8.9	.241	- .016
△	5.8	Standard roughness	

Fig. B.8b   NACA 2418.

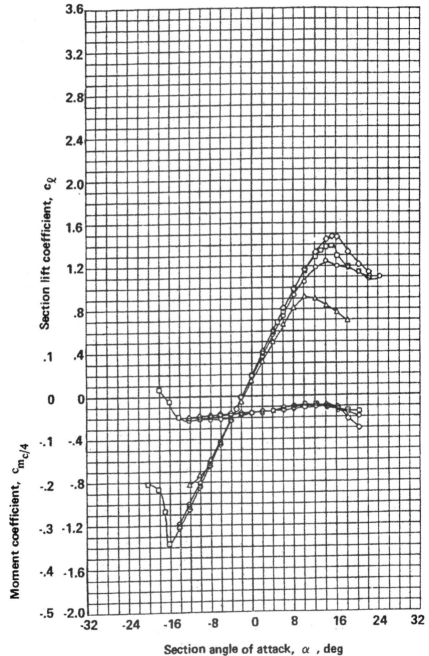

Fig. B.9a   NACA 2421.

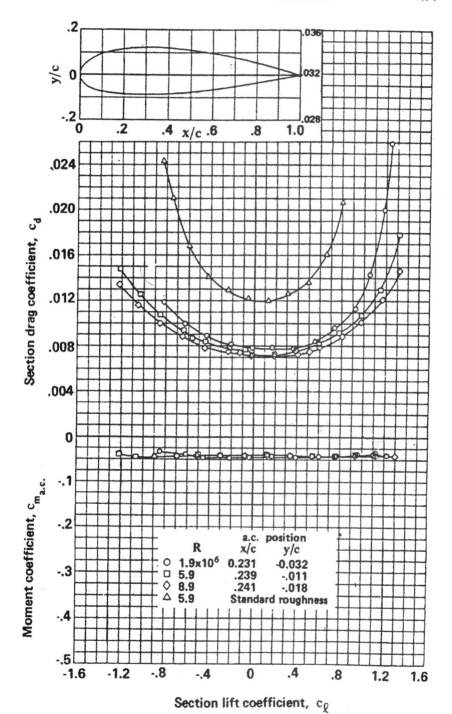

Fig. B.9b   NACA 2421.

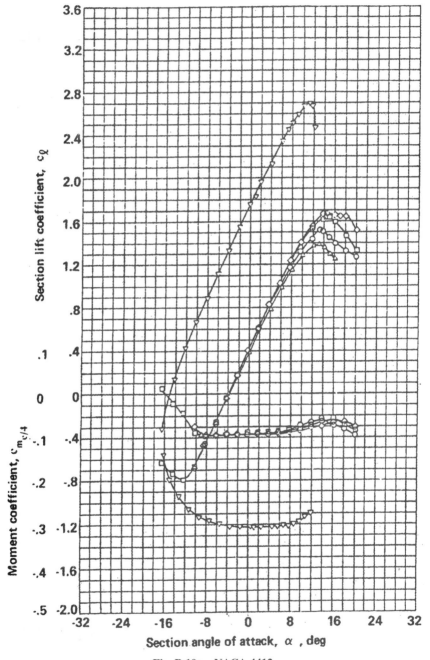

Fig. B.10a   NACA 4412.

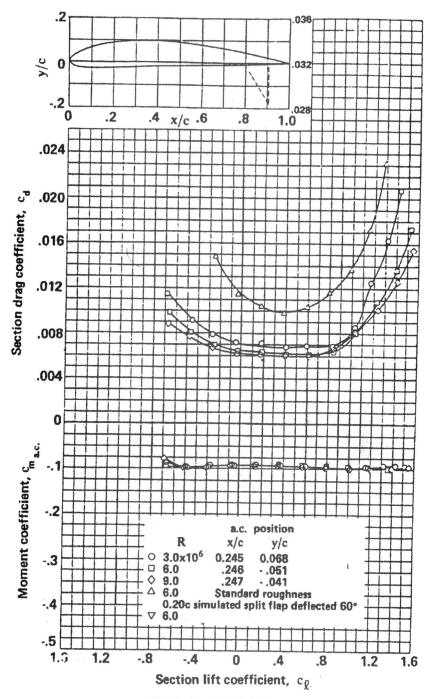

Fig. B.10b   NACA 4412.

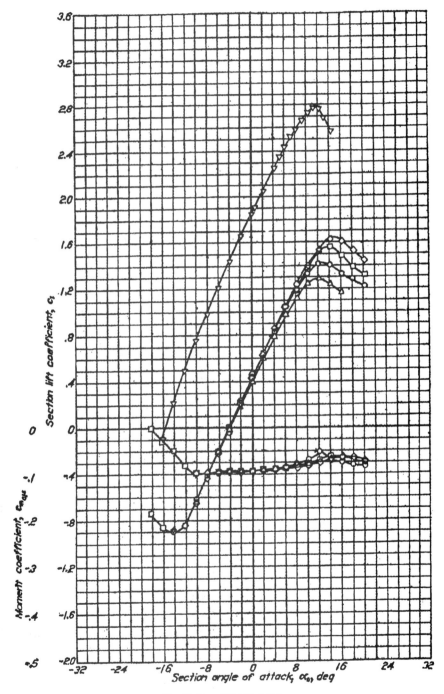

Fig. B.11a   NACA 4415.

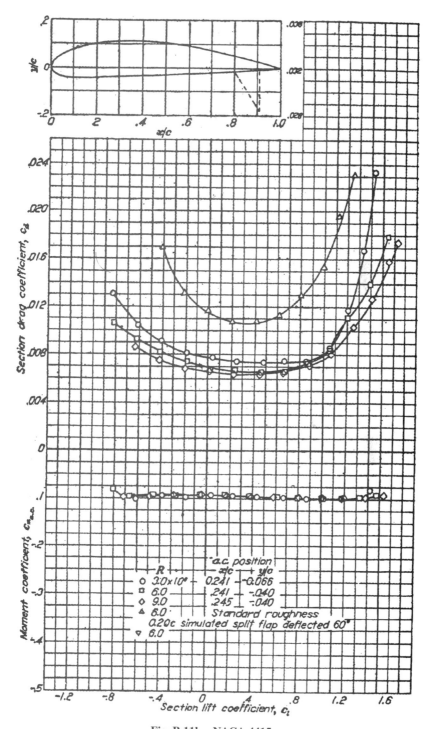

Fig. B.11b   NACA 4415.

T-38 Performance Data

T-38 Airframe Data

Table B.4   Airframe dimensions

Wing

Total area	170 ft^2
Span	25 ft 3 in.
Aspect ratio	3.75
Taper ratio	0.20
Sweepback (quarter chord)	24 deg
Airfoil section	NACA 65A004.8
Mean aerodynamic chord	92.76 in.
Dihedral	0 deg
Span/thickness ratio	51.1

Horizontal tail

Total area	59.0 ft^2
Exposed area	33.34 ft^2
Aspect ratio (exposed)	2.82
Taper ratio (exposed)	0.33
Sweepback (quarter chord)	25 deg
Airfoil section	NACA 65A004
Span/Thickness ratio (exposed)	58.6

Vertical tail

Total area	41.42 ft^2
Exposed Area	41.07 ft^2
Apect Ratio (exposed)	1.21
Taper Ratio (exposed)	0.25
Sweepback (quarter chord)	25 deg
Airfoil Section	NACA 65A004
Span/Thickness Ratio	42.2

Airplane

Height	12 ft 11 in.
Length	43 ft 1 in.
Tread	10 ft 9 in.

T-38 Powerplant Characteristics

Table B.5   J-85-GE-5A powerplant description

Number	2
Model	J85-GE-5
Manufacturer	General Electric
Type	Turbojet
Augmentation	Afterburning
Compressor	Axial Flow
Exhaust Nozzle	Variable Area
Length (overall)	107.4 in.
Maximum Diameter (afterburner tailpipe)	20.2 in.
Dry Weight	477 lb
Fuel Grade	JP-4
Fuel Specific Weight	6.2 to 6.9 lb/gal

Table B.6   J-85-GE-5A powerplant ratings[a]

Power setting	Normal power	Military power	Maximum power
Augmentation	None	None	Afterburner
Engine speed[b]	96.4	100	100
Thrust per engine, lb			
No losses	2140	2455	3660
Installed	1770	1935	2840
Specific fuel consumption[c]			
Installed	1.09	1.14	2.64

[a]Sea level static ICAO standard conditions with a fuel specific weight of 6.5 lb/gal.
[b]Units are % RPM where 100% = 16,500 RPM.
[c]Units are lb/hr per lb thrust.

Table B.7   J-85-GE-5A powerplant operating limitations

Power setting	Normal	Military	Maximum
Turbine discharge total temp., °F	1050	1220	1220

Lift and Drag Curves

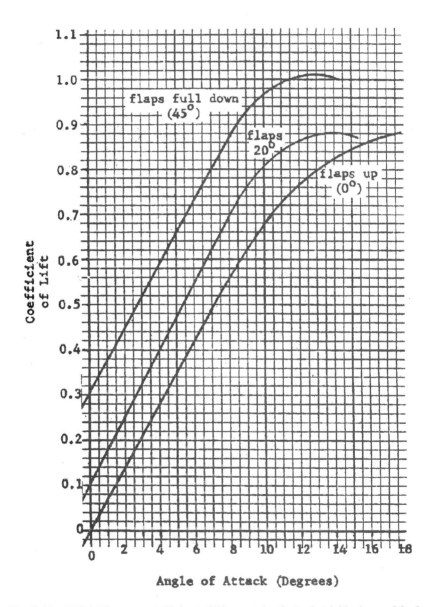

Fig. B.12 T-38A lift curve: coefficient of lift vs angle of attack (rigid wing-and-body model, Mach = 0.4 out of ground effect).

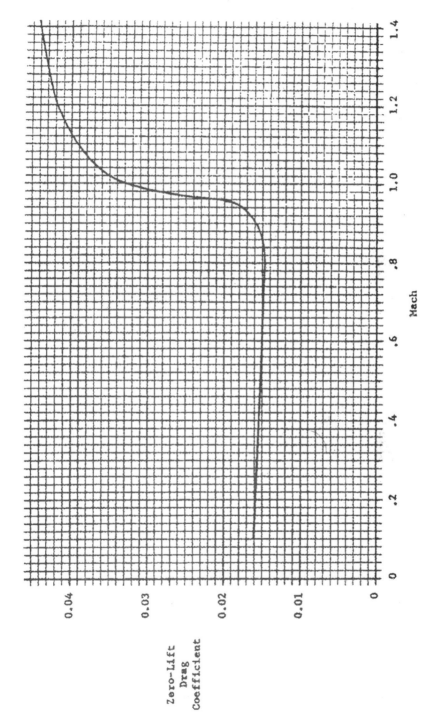

Fig. B.13   T-38A zero-lift drag coefficient variation with Mach number (full-scale model, clean configuration).

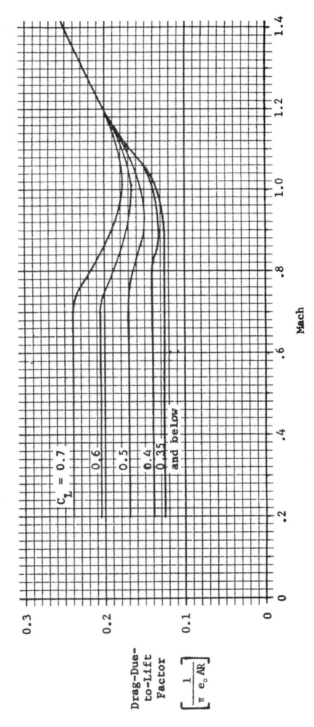

Fig. B.14   T-38A drag-due-to-lift factor variations with Mach number for five values of lift coefficient (full-scale model, clean configuration).

Thrust and Drag Curves

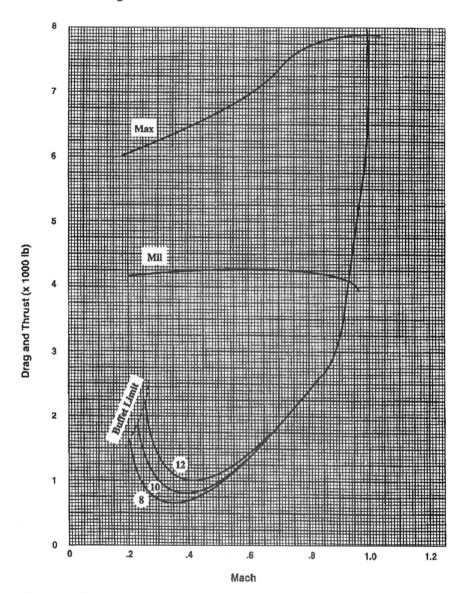

Fig. B.15   T-38A thrust required and thrust available for two J85-GE-5A engines at aircraft weights of 12,000, 10,000, and 8000 lb at sea level.

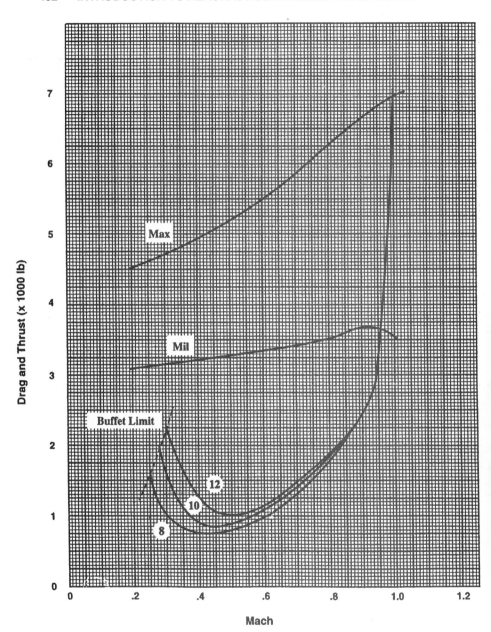

Fig. B.16   T-38A thrust required and thrust available from two J85-GE-5A engines at aircraft weights of 12,000, 10,000, and 8000 lb at an altitude of 10,000 ft.

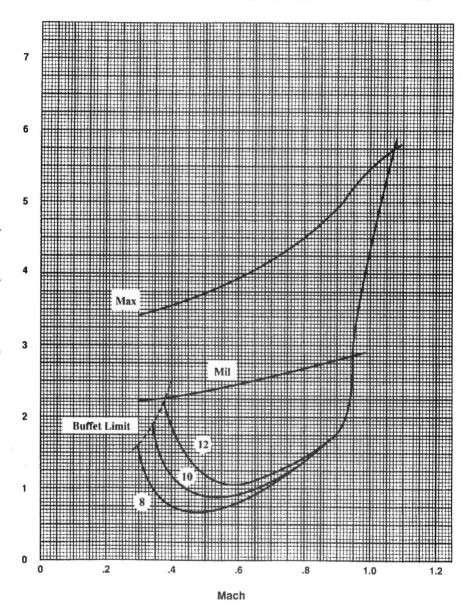

Fig. B.17 **T-38A thrust required and thrust available from two J85-GE-5A engines at aircraft weights of 12,000, 10,000, and 8,000 lb at an altitude of 20,000 ft.**

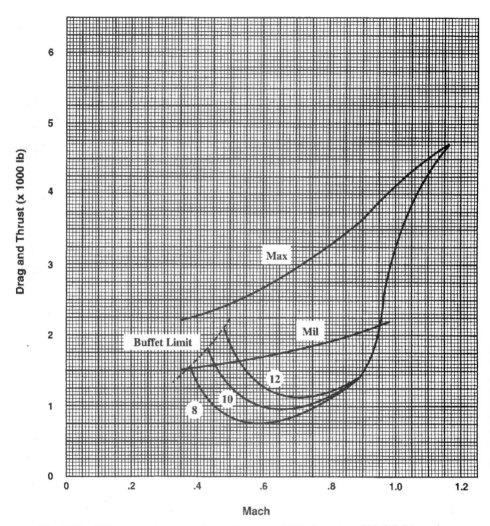

Fig. B.18   T-38A thrust required and thrust available from two J85-GE-5A engines at aircraft weights of 12,000, 10,000, and 8000 lb at an altitude of 30,000 ft.

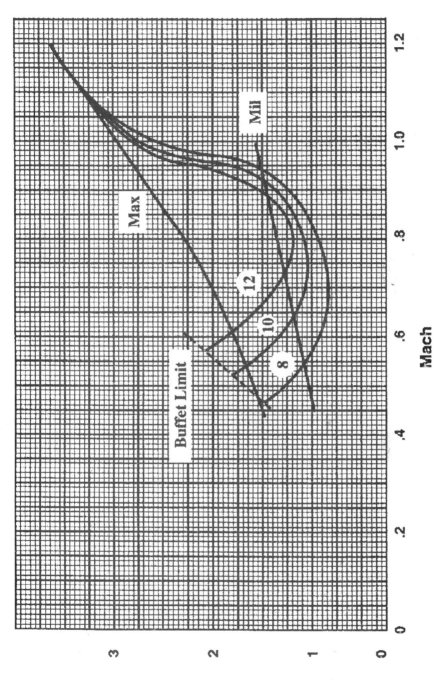

Fig. B.19   T-38A thrust required and thrust available from two J85-GE-5A engines at aircraft weights of 10,000 and 8000 lb at an altitude of 40,000 ft.

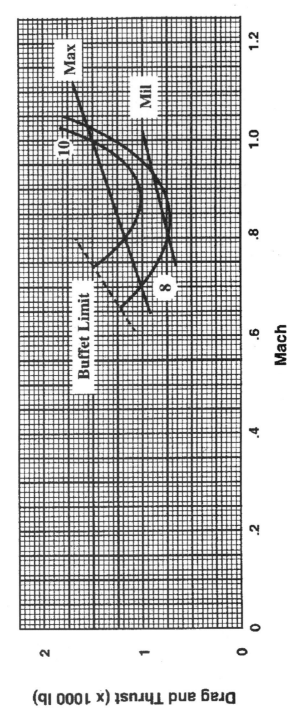

Fig. B.20   T-38A thrust required and thrust available from two J85-GE-5A engines at aircraft weights of 10,000 and 8000 lb at an altitude of 50,000 ft.

Power Curves

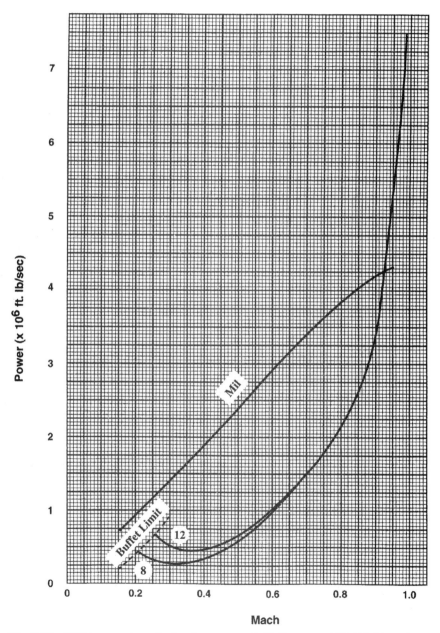

Fig. B.21   T-38A power required and power available from two J85-GE-5A engines at aircraft weights of 8000 and 12,000 lb at sea level.

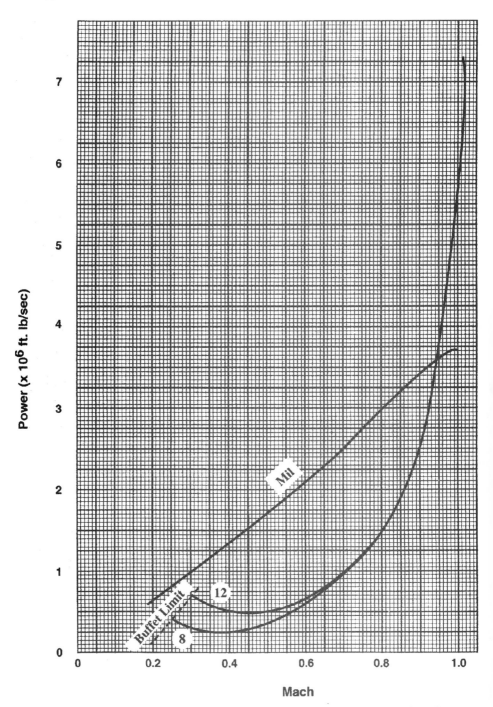

Fig. B.22   T-38A power required and power available from two J85-GE-5A engines at aircraft weights of 8000 and 12,000 lb at an altitude of 10,000 ft.

V–n Diagrams

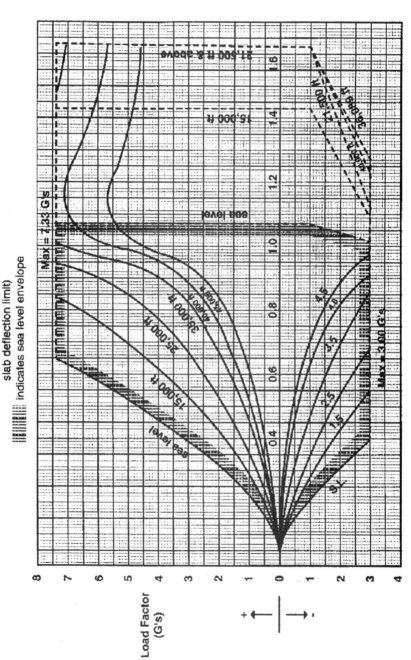

Fig. B.23   T-38A flight strength. Symmetrical loading for 96,000 lb gross weight flight test data.

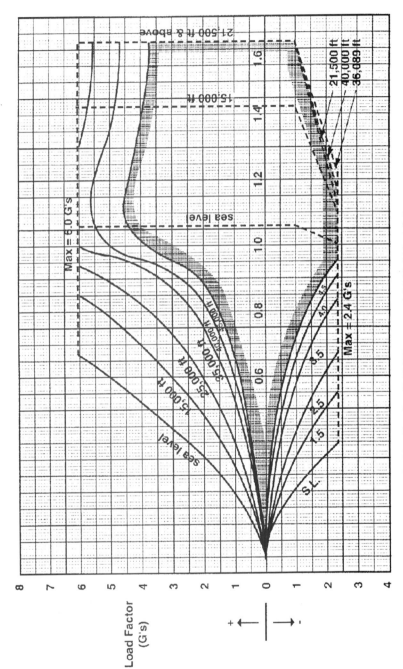

Fig. B.24 T-38A flight strength. Symmetrical loading for 12,000 lb gross weight flight test data.

Maneuverability Diagrams

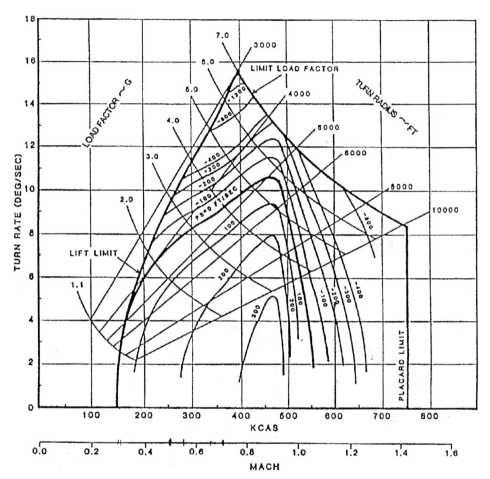

Fig. B.25   Maneuverability diagram at 15,000 ft altitude, fuel weight of 1400 lb, maximum thrust, and clean configuration.

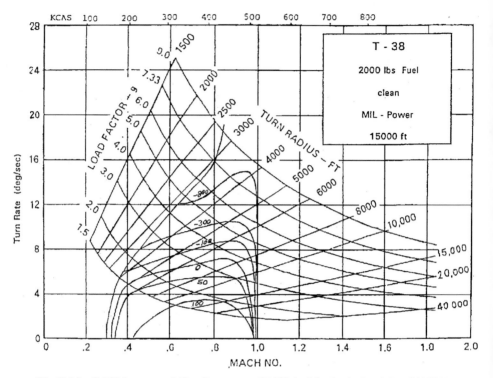

Fig. B.26   T-38 Maneuverability diagram at 15,000 ft altitude, fuel weight of 2000 lb, MIL power, and clean configuration.

Appendix C
Answers

Chapter 2 Answers

A-2.1. 226.9 Kg

A-2.2. $P = 1195$ lb/ft^2
Density $= 0.1496 \times 10^{-2}$ slug/ft^3
$T = 465.2°$R

A-2.3. 9000 ft

A-2.4. 0.001907 slug/ft^3

A-2.5. a) $h_P \approx 22{,}000$ ft
$h_T \approx 16{,}500$ ft
$h_\rho \approx 23{,}000$ ft
b) $h_p = h_T = h_\rho$

A-2.6. a) 2116.2 lb/ft^2
b) 2740.9 lb/ft^2
c) 2.33 $\times 10^{-3}$ slug/ft^3

A-2.7. a) 62.4 lb/ft^2
b) 249.6 lb

A-2.8. 5.157×10^{19} N
5.26×10^8 kg

A-2.9. 124,494 lb/ft^2 = 878 psi !

Chapter 3 Answers

A-3.1. Incompressible flow: density = constant.
For air, flow velocities of less than 100 m/s, 330 ft/s, or 225 mph (all $M = 0.3$) permit the incompressible flow assumption to be made.

A-3.2. Properties at a point are invariant with time (so is mass flow).
Examples:
Steady: inlet of an engine on aircraft flying at constant airspeed and altitude
Unsteady: air filling a balloon

A-3.3. a) Mass flow rate $= 244.4$ kg/s
b) Density $= 1.086$ kg/m^3
c) V_1 and $V_2 > 100$ m/s, so flow is compressible

A-3.4. Flow properties (P, T, ρ, and V) are constant across a cross section (plane perpendicular to flow) of flow.
Example: Wind tunnel ignoring regions close to wall

A-3.5. a) Mass flow rate $= 1.91$ slug/s
b) $A = 4.59$ ft^2

A-3.6. $A = 0.538$ ft^2
$V_1 = 293.3$ ft/s
$V_2 = 78.9$ ft/s
$P_1 = 973.3$ lb/ft^2
$P_2 = 1023.9$ lb/ft^2

A-3.7. $D_2 = 0.466$ in.
$V_2 = 112.8$ ft/s

A-3.8. $F = 387.5$ N (outward)

A-3.9. $V = 667.8$ kn
$V_G = 717.8$ kn
$t = 1.95$ h

A-3.10. $q = 33.8$ lb/ft^2

A-3.11. 79,048 Pa

A-3.12. $V_g = 430$ kn

A-3.13. $q = 136$ lb/ft^2

A-3.14. $V_i = 212$ kn

A-3.15. $V = 20.9$ m/s

A-3.16. 2141.5 lb/ft^2

A-3.17. a) $P = 79,501$ N/m^2
$T = 275.16$ K
$\rho = 1.0066$ kg/m^3
(from Standard Atmosphere Tables)
b) $P = 79,501$ N/m^2
c) $P_o = 83478$ N/m^2
d) Same. While the pressures at a point may differ, the pressure *differentials* between the upper and lower surfaces, which cause the aerodynamic forces, will be the same, because the dynamic pressure is the same.

A-3.18. a) $Re = 7.95 \times 10^6$
b) $X_{crit} = 0.403$ ft

A-3.19. $L = 570.48$ lb
$D = 6.37$ lb

A-3.20. $Re = 10.17 \times 10^6$
Turbulent

A-3.21. a) Laminar
b) Turbulent
c) A and B
d) D
e) C
f) 1) Downstream (aft, right)
   2) Upstream (forward, left)
   3) Downstream

A-3.22. a) A region of increasing pressure in the direction of flow ($dP/dx > 0$). On an airfoil it occurs generally aft of the max thickness (min pressure) point.
b) Low energy flow in the boundary layer has insufficient momentum to overcome the adverse pressure gradient.
c) 1) Decreased lift
   2) Increased drag
d) Dimples create surface roughness that "trips" the boundary layer causing an early transition to turbulent flow. The turbulent boundary layer has a "fuller" velocity profile near the wall that allows it to penetrate further into the adverse pressure gradient region on the back side of the ball, delaying separation. This, in turn, reduces pressure drag. There is an increase in skin friction drag, but the total drag is reduced and the ball travels farther.
e) Transition is the change from a laminar to a turbulent boundary layer (stays "attached" to wall). Separation is the "detaching" of a boundary layer (can be laminar or turbulent) from a surface.

A-3.23. a) Max camber $= 0.028$ m
   Location $= 0.56$ m aft of leading edge
   Max thickness $= 0.21$ m
b) Symmetrical airfoil with zero camber
   Max thickness $= 0.023$ m

A-3.25. a) -2 deg
b) Positive
c) 16 deg
d) 1.63
e) 0.1/deg

A-3.26. a) Point on an airfoil where the moment resulting from aerodynamic forces is zero.
b) Equal to zero.

A-3.27. a) *Fixed* point on an airfoil where the moment resulting from aerodynamic forces does not change with changes in angle of attack.
b) 0.245 m aft of leading edge
Yes, only slightly (0.002 m)

A-3.28. a) $Re = 8.99 \times 10^6$
b) $L = 98,740$ N
   $D = 802$ N
   $M = -24,050$ N · m
c) 1) 13 deg
   2) -4 deg
   3) 0.1/deg

A-3.29. a) $Re = 6 \times 10^6$
b) $c_l = 0.61$, $c_d = 0.0069$, $c_m = -0.05$
c) $l = 5006.2$ N, $d = 56.6$ N, $m = -549.9$ N · m
d) $c_{l\,max} = 1.6$, $\alpha_{l=0} = -2$ deg, $\alpha_{0\,min} = 0$

Chapter 4 Answers

A-4.1. a) $Re = 6 \times 10^6$
b) $C_L = 0.393$, $C_D = 0.0215$
c) $L = 8,062$ N, $D = 442$ N
d) $\alpha_{l=0} = -2$ deg, $C_{Do} = 0.0064$

A-4.2. a) $C_L = 0.698$, $C_D = 0.0247$
b) $L = 16,259$ lb, $D = 576$ lb

A-4.3. 1) Increased drag (must add "induced" drag)
2) Decreased lift ($C_{L\alpha} < c_{l\alpha}$)

A-4.4. a) Lift
b) Zero
c) C_L is larger at low speeds, and so induced drag is increased at low speeds.

A-4.6. a) 6.6
b) 1,833 kn

A-4.7. $M_{crit} = 0.689$

A-4.9. a) $\mathcal{R} = 3$
b) $L = 50,000$ lb
c) $C_D = 0.02326$
d) $D = 5,464$ lb

A-4.10. 50% but it can't sustain flight if it is not generating lift

A-4.11. $L/D = C_L/C_D = 10.3$

A-4.12. $C_{L\alpha} = 0.0609$ per degree, $C_{L_{max}} = 0.9135$, $k = 0.097$

A-4.13. $C_{D_0} = 0.0214$ at $M = 0.2$ and $C_{D_0} = 0.0377$ at $M = 2.0$

Chapter 5 Answers

A-5.1. $C_L = 0.419$
$C_D = 0.0159$

A-5.2. a) $T_R = 11{,}751$ N
b) $T_R = 10{,}523$ N

A-5.3. $(L/D)_{\max} = 27.32$ at $C_L = 0.546$

A-5.5. a) 800 lb
b) 3,275 lb
c) 5,200 lb
d) 2,475 lb
e) 750 lb
f) $M = 0.25$ caused by buffet (stall) limit
g) $M = 1.0$

A-5.6. a) 14.8
b) $Ve = 215.6$ ft/s

A-5.7. 209 ft/s

A-5.8. a) 256.8 ft/s buffet limited
b) 522.8 ft/s thrust limited

A-5.9. $T_A = 1593$ lb
From chart: 1800 lb

A-5.10. a) $P_R = 4.5 \times 10^5$ ft lb/s
b) $P_A = 1.65 \times 10^6$ ft lb/s
c) $P_X = 1.2 \times 10^6$ ft lb/s
d) $M_{PR_{\min}} = 0.36$
e) $P_{R_{\min}} = 4.5 \times 10^5$ ft lb/s

A-5.12. a) Slower
b) Faster
c) Slower

A-5.13. a) $(L/D)_{\max} = 27$, glide 27 km at 28.1 m/s
b) 18.25 min at 21.4 m/s
c) Altitude would increase at 2.087 m/s

A-5.14. a) 37.5 n mile
b) 37.5 n mile

A-5.15. Weight

A-5.16. a) $ROC_{\mathrm{mil}} = 110$ ft/s
$ROC_{AB} = 199$ ft/s
b) 22.65 deg
c) $M_{ROC_{\max}} = 0.7$
d) $ROC_{\max \, \mathrm{in \, mil}} = 193.75$ ft/s

A-5.17.   a) $M = 0.53$, $V = 549.6$ ft/s $= 325.2$ kn
b) 2.29 h
c) 744.7 n mile

A-5.18.   1.54 h

A-5.19.   a) 643.8 ft/s
b) 794 n mile

A-5.20.   298 n mile

A-5.21.   a) 10,000 ft
b) 5823 ft

A-5.22.   a) 223.3 ft/s
b) 0.79
c) 290 ft/s
d) 2383 ft
e) s_L will increase

A-5.23.   519 m

A-5.24.   a) Turn rate $= 8.7$ deg/s
     $r = 5544$ ft
b) Turn rate $= 13.1$ deg/s
     $r = 3696$ ft
c) Turn rate $= 10.7$ deg/s
     $r = 4526$ ft

A-5.25.   a) 118 n mile
b) 0.231 deg/s
c) 43 n mile

A-5.26.   a) 2588 ft
b) 11.1 deg/s
c) 1533 ft
d) 18.4 deg/s

A-5.27.   173.2 kn

A-5.28. $n = 1.14$
Bank angle $= 28.7$ deg
$r = 2330$ ft

A-5.29. $V = 899.9$ ft/s, $n = 6$

A-5.31.   a) 3.0
b) 7.33
c) 1.06
d) 1.43
e) 0.94
f) 0.65

A-5.32. a) B52
b) B52
c) 373 ft/s
d) T38
e) 14.2 ft/s^2

A-5.33. a) 22.93 ft/s
b) 57,564 ft

A-5.34. a) Yes, you can accelerate faster (19 ft/s^2)
b) 1) Pilot capability/training/   11) Mission objectives
      experience   12) Battle damage
    2) Cockpit visibility   13) Low observables (stealth)
    3) Radar/sensor capability   14) Weather/day vs night
    4) Weapons capability   15) Terrain
    5) Formation/mutual support   16) Paint scheme
    6) Tactics   17) SAMs/AAA
    7) Fuel available
    8) Ground/air radar control
       (AWACS and GCI)
    9) ECM/ECCM/flares/chaff/etc.
   10) Force ratio

A-5.35. a) B, H, E, I
c) 60,000 ft
d) 57,000 ft
e) 96,000 ft
f) 96,000 ft
g) Yes, but cannot sustain it.
h) No.

A-5.36. a) 0.2 (buffet limit)
b) 0.33 (L/D_{max} or D_{min})
c) 0.33 (maximum excess thrust or L/D_{max})
d) 0.33
e) 0.44 (tangent to T_R)
f) 0.93 ($T_R = T_A$)
g) <0.33 (left of L/D_{max})
h) Approximately 0.59 (maximum excess power)

Chapter 6 Answers

A-6.1.

Motion	Control surface	Axis
Roll	Aileron	x
Pitch	Elevator	y
Yaw	Rudder	z

A-6.2. Six degrees of freedom (3 degrees of freedom translational and 3 degrees of freedom rotational)

A-6.3. Static: *Initial* tendency of body following a disturbance from equilibrium (positive if it returns to equilibrium).
Dynamic: *Time history* of response to disturbance (positive if body eventually returns to equilibrium).

A-6.4. When not aligned with the wind, the tail is at some angle of attack to the wind and generates a lift force. This force creates a moment about the pivot that tends to align it with the wind. In an aircraft, this is called directional stability.

A-6.5. A very stable aircraft will resist motion including those induced by control inputs. Thus, it is not very maneuverable; as stability decreases, maneuverability increases.

A-6.6. $L_t = 2000$ lb
$W = 42{,}000$ lb

A-6.7. $L_c = 10{,}000$ lb
$W = 50{,}000$ lb

A-6.8. $L = 45{,}000$ lb
$W = 60{,}000$ lb

A-6.9. a) Increasing V_H makes C_{M_α} more negative implying increased stability.
b) Using the equation from A-6.8a, moving center of gravity forward makes C_{M_α} more negative, which means increased stability.
c) An increase in $\bar{x}_{ac}$ makes C_{M_α} more negative so stability is increased.

A-6.10. a) 0.495
b) 0.045
c) Yes, static margin is positive.
d) $C_{M_\alpha} = -0.0036$/deg, $C_{M_0} = 0.034$
e) $C_{L_{trim}} = 0.755$
f) $V_{trim} = 118$ ft/s

Chapter 7 Answers

A-7.1. 60 in.^{2}

A-7.2. Required spar cap/skin area decreases to 56.1 in.^{2}

A-7.3. The one in the vice took fewer cycles to break. This is because you could bend it more sharply in the vice, putting more stress in a smaller area of the clip. Higher stress means fewer cycles to failure.

A-7.4. The air is bumpier down low, and so the B-52 structure experienced many more loading cycles per flight or per hour of flying time than it was designed for. More cycles per sortie means more fatigue per sortie, and the cracks developed sooner.

Chapter 8 Answers

A-8.1. 227.5 ft^2

A-8.2. Yes

A-8.3. $16 million 1996 U.S. dollars

Index

absolute angle of attack, 151
acceleration of gravity, 44
accelerations, 355–356
aerodrome, 18
AeroDYNAMIC, 169, 373, 383
aerodynamic analysis, 380, 383–386
aerodynamic center, 90, 91, 150, 302–303
aerodynamic chords, 302, 303
aerodynamic forces, 32, 41, 61, 63, 70,
    71, 75, 174, 220, 225, 228, 275, 287,
    295
aerodynamic loads, 325, 326
aerodynamic properties, 95
aerodynamic twist, 115, 118
aerodynamics, 20, 61
    applications, 65–77
    basic, 61–66
    molecular collisions, 54–57
aeronautics, 1, 10, 11, 24
afterburners, 180–182
ailerators, 270–272
ailerons, 269, 271
aircraft
    design, 31, 427–433
    reference lines, 173
airfoil, 19, 26, 61–63, 72, 74–75, 85–95,
    113
    aerodynamic center, 90
    angle of attack, 87–88
    cambered, 89–90, 136, 139
    compressibility, 101–102
    data charts, 91–95, 454–475
    drag coefficients, 85–88
    equations, 98
    freestream, 98
    lift coefficients, 85–88, 123
    moment coefficient, 90
    NACA, 98–99, 103
    planform area, 86
    Reynolds-number effects, 91
    shapes, 19, 21, 85–86, 137–139
    span, 86
    symmetrical, 89
    terminology, 98
    thickness, 98
airmanship, 18
airplanes, 15–24, 26, 32
airships, 16
airspeed, 68–70

calibrated, 68–69, 100
    equivalent, 69–70
    ICeT, 68–70, 100–102
    indicated, 61, 68, 100–102
    indicators, 61, 66–68
    true, 68, 70
altimeters, 41, 52
altimetry, 51–52, 53
altitude, 41–42, 215–217
    indicated, 41
    measurement, 41
amplitude, 273–274
analysis, 1–3, 8-9, 12, 25, 35, 402–404
    computer, 14
    constraint, 247–253, 379–380, 387–389
    control, 267, 276, 305–311
    cost, 380, 392
    finite element, 340
    longitudinal control, 275
    mission, 354, 355
    performance, 173–184, 253–255
    sizing example, 367–373
    stability, 285
    stability and control example, 305–311
angle of attack, 75, 87–88, 90, 151, 279
angles, 283, 287, 288, 291
area rule, 138
aspect ratio equation, 145
atmosphere, 42–44
axial stress, 341

balance, 365–366, 377
balloons, 16
Balzer, Stephen M., 17, 26
barometers, 45–46
Bernoulli's equation, 65–66, 96
best cruise altitude/best cruise Mach, 215, 217
Blériot, Louis, 19
body axis system, 269
Boeing 247, 408–409
Boeing 247D, 413–415
boundary layer, 79–80
boundary-layer control, 126–128
bow shock, 135, 148
brainstorming, 9, 25
brake specific fuel consumption, 186
Breguet endurance equation, 357
buckling, 336–338, 341
bypass ratio, 182, 183, 188

Supporting Materials

A complete listing of titles in the AIAA Education Series and other AIAA publications is available at http://www.aiaa.org.